AF614718

Methods in Molecular Biology™

Series Editor
John M. Walker
School of Life Sciences
University of Hertfordshire
Hatfield, Hertfordshire, AL10 9AB, UK

For further volumes:
http://www.springer.com/series/7651

Lipoproteins and Cardiovascular Disease

Methods and Protocols

Edited by

Lita A. Freeman

Cardiovascular & Pulmonary Branch, National Heart, Lung, and Blood Institute
National Institutes of Health (NIH), Bethesda, MD, USA

Editor
Lita A. Freeman
Cardiovascular & Pulmonary Branch
National Heart, Lung, and Blood Institute
National Institutes of Health (NIH)
Bethesda, MD, USA

ISSN 1064-3745 ISSN 1940-6029 (electronic)
ISBN 978-1-60327-368-8 ISBN 978-1-60327-369-5 (eBook)
DOI 10.1007/978-1-60327-369-5
Springer New York Heidelberg Dordrecht London

Library of Congress Control Number: 2013938950

Printed on acid-free paper

Humana Press is a brand of Springer
Springer is part of Springer Science+Business Media (www.springer.com)

Preface

Methods in Molecular Biology: Lipoproteins and Cardiovascular Disease: Methods and Protocols is a compendium of advanced and classical molecular biology methods targeted towards lipoprotein, atherosclerosis, and vascular biology research.

Lipoprotein, atherosclerosis, and vascular biology studies present unique challenges to the molecular biologist. The lipid-rich and otherwise challenging nature of many key tissues complicate the isolation of high-quality RNA for gene expression analysis, for example, and the unique nature of lipoproteins and their biological effects has engendered unique methodologies. To date, no volume has yet encompassed these lipoprotein-centered cutting-edge methods in molecular biology.

This book brings together in a single volume an updated set of protocols and strategies for methods now driving advances in lipoprotein and atherosclerosis research, along with classical methods that are still widely used. The chapters are written for researchers at any level, from graduate students to established investigators with no prior experience in the described techniques, and may be of interest to molecular biologists outside the lipoprotein field using similar techniques.

Of particular interest to readers are methods chapters on quantitative real-time PCR, microarrays, RT-PCR laser capture microdissection, and tissue-specific gene overexpression, knockout, and knockdown methodologies, including AAV as a liver-directed gene delivery vehicle. Special topics include an overview of next-generation and third-generation sequencing, antisense technology, chromatin immunoprecipitation, streamlined LCAT activity assays, and native HDL subpopulation analysis. Updated methods for 5′ and 3′ RACE cloning of full-length cDNAs and Northern analysis have been added. Overviews, strategic considerations, and background information are included for particularly novel or complex methods.

This edition complements its classic predecessor, "Lipoprotein Protocols," edited by Jose Ordovas, by incorporating cutting-edge methodological advances developed over the past decade. The two volumes together provide a complete, up-to-date set of methods for any researcher with an interest in lipoproteins and their biological effects.

I would like to thank the following people for their contributions to this volume: John Walker, the Series Editor, for his invaluable guidance and support, Gregory Kato, Robert Shamburek, and my colleagues in the Pulmonary and Vascular Medicine Branch at NIH for their support, patience, and encouragement, and Silvia Santamarina-Fojo and H. Bryan Brewer for their guidance and contributions to lipoprotein metabolism over the years. My former mentors and colleagues who taught me molecular biology over the years will find their sage advice sprinkled throughout this volume—a small token of gratitude for their efforts and encouragement. Many, many thanks as well to the Wolffe lab members. Finally, this work would not have been accomplished without the bottomless support, encouragement, and help from my friends, neighbors, and family.

Bethesda, MD, USA *Lita A. Freeman*

Contents

Contributors

GRANT D. BARISH • *Gene Expression Laboratory, Howard Hughes Medical Institute, The Salk Institute for Biological Studies, La Jolla, CA, USA*
ROSANNE M. CROOKE • *Cardiovascular Disease Research, Antisense Drug Discovery, Isis Pharmaceuticals, Carlsbad, CA, USA*
LENA DIAW • *Pulmonary and Vascular Medicine Branch, National Heart, Lung, and Blood Institute, National Institutes of Health, Bethesda, MD, USA*
YUBIN DU • *iPSC and Genome Engineering Core, National Heart, Lung, and Blood Institute, National Institutes of Health, Bethesda, MD, USA*
JONATHAN E. FEIG • *The Marc and Ruti Bell Vascular Biology Disease Program, Department of Medicine (Cardiology), New York University School of Medicine, New York, NY, USA; Department of Cell Biology, New York University School of Medicine, New York, NY, USA*
EDWARD A. FISHER • *The Marc and Ruti Bell Vascular Biology Disease Program, Department of Medicine (Cardiology), New York University School of Medicine, New York, NY, USA; Department of Cell Biology, New York University School of Medicine, New York, NY, USA*
LITA A. FREEMAN • *Cardiovascular & Pulmonary Branch, National Heart, Lung, and Blood Institute, National Institutes of Health, Bethesda, MD, USA*
MARK J. GRAHAM • *Antisense Drug Discovery, Isis Pharmaceuticals, Carlsbad, CA, USA*
CHANGYUN GUI • *iPSC and Genome Engineering Core, National Heart, Lung, and Blood Institute, National Institutes of Health, Bethesda, MD, USA*
XIAN-CHENG JIANG • *Department of Anatomy and Cell Biology, SUNY Downstate Medical Center, Brooklyn, NY, USA*
JULIE C. JOHNSTON • *Penn Vector Core, Department of Pathology and Laboratory Medicine, University of Pennsylvania School of Medicine, Philadelphia, PA, USA*
WILLIAM R. LAGOR • *Institute for Translational Medicine and Therapeutics, University of Pennsylvania School of Medicine, Philadelphia, PA, USA; Cardiovascular Institute, University of Pennsylvania School of Medicine, Philadelphia, PA, USA*
CHENGYU LIU • *iPSC and Genome Engineering Core, National Heart, Lung, and Blood Institute, National Institutes of Health, Bethesda, MD, USA*
MARTIN LOCK • *Penn Vector Core, Department of Pathology and Laboratory Medicine, University of Pennsylvania School of Medicine, Philadelphia, PA, USA*
DANIEL J. RADER • *Institute for Translational Medicine and Therapeutics, University of Pennsylvania School of Medicine, Philadelphia, PA, USA; Cardiovascular Institute, University of Pennsylvania School of Medicine, Philadelphia, PA, USA*
NALINI RAGHAVACHARI • *Genetics and Developmental Biology, National Heart, Lung, and Blood Institute, National Institutes of Health, Bethesda, MD, USA*

ALAN T. REMALEY • *Lipoprotein Metabolism Section, Pulmonary and Vascular Medicine Branch, National Heart, Lung, and Blood Institute, National Institutes of Health, Bethesda, MD, USA*

RAJENDA K. TANGIRALA • *Division of Endocrinology, Diabetes, and Hypertension, David Geffen School of Medicine, University of California Los Angeles, Los Angeles, CA, USA*

JAMES G. TAYLOR VI • *Pulmonary and Vascular Medicine Branch, National Heart, Lung, and Blood Institute, National Institutes of Health, Bethesda, MD, USA*

BORIS L. VAISMAN • *Lipoprotein Metabolism Section, Cardiovascular-Pulmonary Branch, National Heart, Lung and Blood Institute, National Institutes of Health, Bethesda, MD, USA*

LUK H. VANDENBERGHE • *Penn Vector Core, Department of Pathology and Laboratory Medicine, University of Pennsylvania School of Medicine, Philadelphia, PA, USA*

ELKE M. WAGNER • *Plasma Analytics/Development and Optimization, Baxter AG, Wien, Austria*

WEN XIE • *iPSC and Genome Engineering Core, National Heart, Lung, and Blood Institute, National Institutes of Health, Bethesda, MD, USA*

VICTORIA YOUNGBLOOD • *Pulmonary and Vascular Medicine Branch, National Heart, Lung, and Blood Institute, National Institutes of Health, Bethesda, MD, USA*

Part I

RNA and Gene Expression

Chapter 1

Cloning Full-Length Transcripts and Transcript Variants Using 5′ and 3′ RACE

Lita A. Freeman

Abstract

Gene transcripts and transcript variants must be cloned to characterize gene function and regulation. However, obtaining full-length cDNAs with accurate sequences from the 5′ end through to the 3′ end can be challenging. Here we describe a reverse-transcriptase-based method for obtaining full-length cDNAs using the SMARTer ("Switching Mechanism At RNA Termini") RACE technology developed by Clontech. RNA is isolated from the tissue of interest and annealed to a primer (a modified oligo(dT) primer for polyA+ transcripts; random hexamers or a gene-specific primer for polyA− transcripts). A modified MMLV-reverse transcriptase uses the primer to initiate cDNA synthesis from RNA transcript(s) annealed to the primer and continues cDNA synthesis (reverse transcription) towards the 5′ end of the transcript(s). Importantly, this reverse transcriptase possesses terminal transferase activity, so when it reaches the 5′ end of a transcript it adds a 3–5 residue "tail" to the newly synthesized cDNA strand. Included in the reverse transcriptase reaction mix is an oligonucleotide containing a sequence tag as well as a terminal series of modified bases that anneal to the 3–5 residue tail on the newly synthesized cDNA. The reverse transcriptase proceeds from the end of the transcript onwards into the modified bases and the rest of the sequence-tagged oligo. The newly synthesized cDNA now has a sequence tag attached to it and can be used as a template for PCR, with one primer complementary to the sequence tag and the second primer specific to the gene of interest. The fragment can be cloned and sequenced or just sequenced directly. If high-quality, undegraded RNA is used, obtaining the true 5′ end of a transcript is greatly enhanced. In combination with 3′ RACE, full-length transcripts are easily cloned. This method provides sequence information on important regulatory regions, such as 5′ and 3′ UTRs and flanking regions, and is ideal for detecting transcript variants, including those with alternative transcriptional start sites, alternative splicing, and/or alternative polyadenylation.

Key words RT-PCR, RNA, cDNA, RACE, 5′ UTR, 3′ UTR, mRNA, Noncoding RNA, Cloning, Sequencing

1 Introduction

The central dogma of one gene, one transcript, one protein has been transformed by systems biology. We now know that one protein-encoding gene can produce multiple transcripts through a variety of transcriptional or post-transcriptional processes, including the use of alternative transcriptional start sites, alternative exon

Lita A. Freeman (ed.), *Lipoproteins and Cardiovascular Disease: Methods and Protocols*, Methods in Molecular Biology, vol. 1027, DOI 10.1007/978-1-60327-369-5_1, © Springer Science+Business Media, LLC 2013

splicing, alternative splicing at short-distance tandem sites just a few nucleotides apart, alternative polyadenylation, post-transcriptional processing, and, rarely, RNA editing [1–9]. By definition, a protein-encoding gene will have at least one of its transcripts translated into protein. Alternative transcripts from the same gene, which share some of the same exons as the major protein-coding transcript and have protein-coding potential, may or may not be translated. These alternative transcripts are often tissue-specific and their regulation is not well understood. Protein-encoding genes can also contain independent transcription units that do not share exons with the protein-coding transcripts. In fact, small RNAs (<200 nt) that do not encode protein are transcribed not only from protein-coding genes but also from intergenic regions [6]. Some are promoter-associated small RNAs with an apparent role in transcription initiation [6, 10]; some are microRNAs with regulatory functions, for example miR-122 [11], and the function of many small noncoding RNAs remains to be determined [1, 2, 6]. Noncoding long RNAs with potential regulatory roles in transcription have also been described [12]. Additional noncoding RNAs (ncRNAs) such as rRNAs, tRNAs, telomerase RNA, and small nucleolar RNAs (snoRNAs) are well characterized.

With the flood of information from systems biology, is cloning and sequencing gene transcripts still necessary? For genes involved in lipoprotein metabolism, the answer at this point is emphatically yes. Proteomics and microarray analysis as well as exons genome and RNA sequencing continue to reveal new potential therapeutic targets for atherosclerosis and dyslipidemia, and altered lipoproteins or lipoprotein metabolism gene variants have been associated with diseases as various as AIDS, pulmonary hypertension, kidney disorders, and schizophrenia, as well as atherosclerosis. Moreover, investigation of lipoprotein metabolism transcript variants has uncovered unique regulatory mechanisms such as apoB mRNA editing, and it is likely that additional co- or post-transcriptional RNA processing mechanisms relevant to lipoprotein metabolism remain to be discovered. To fully understand a protein-coding gene's regulation and function in lipoprotein metabolism and disease, all of its transcripts and control mechanisms must be characterized, in all relevant tissues. Only by cloning and sequencing of all of its transcripts can a gene be fully characterized.

Most protein-encoding transcripts have a 5′ cap, a 5′ UTR, a translation start site, a protein-coding sequence, a stop codon, a 3′ UTR, and finally a 3′ polyA tail (*see* **Note 1**). Cloning protein-coding transcripts involves reverse transcription of RNA, starting from the polyA tail, to form cDNA. Obtaining a full-length transcript and determining the true 5′ end of a transcript is not straightforward, however. Strong secondary structure and/or high GC content within the transcript can impede the progress of the reverse transcriptase, which may fall off the transcript before reaching the true 5′ end. Incomplete 5′ end sequence information can misidentify a downstream

AUG as the N-terminus of the protein, leading to cloning and sequencing of an incomplete protein lacking its correct N-terminus. Incomplete 5′ end sequence information will also lead to inaccurate identification of the promoter and 5′ UTR sequences. The 3′ end of the transcript must also be accurate to identify the C-terminus of the protein and the 3′ UTR sequence of the transcript, which can contain regulatory signals (for example, *see* ref. 13). The true 3′ end of a typical polyA+ mRNA transcript is not difficult to determine, since a long run of A's marks the end of the transcript. Obtaining the true 5′ end of protein-coding transcripts is generally the major hurdle.

RACE ("Rapid Amplification of cDNA Ends"), in which the cDNA ends are sequence-tagged and then PCR-amplified using primers complementary to the sequence tag or poly(dT) primer, was an important advance in obtaining full-length transcripts. However, even with improved reverse transcriptase technology, 5′ RACE was still subject to incomplete reverse transcription and inaccurate 5′ end formation. We describe in this chapter a novel technology, SMART ("Switching Mechanism At RNA Termini") RACE (now termed SMARTer RACE; *see* **Note 2**), that enriches cDNA pools for 5′ end sequences.

The 5′ SMARTer RACE technology developed by Clontech is specifically designed to amplify cDNA transcripts in which the reverse transcriptase has reached a 5′ end. The strategy of SMART cDNA synthesis is shown in Fig. 1. PolyA+ RNA is annealed with a modified oligo(dT) primer ("5′-CDS primer A" for 5′ RACE) that binds the polyA+ tail (*see* **Note 3**). A modified MMLV-reverse transcriptase ("SMARTScribe RT") uses this primer to synthesize cDNA from the transcript. Importantly, this particular reverse transcriptase possesses terminal transferase activity, so when it reaches the 5′ end of the transcript it adds a 3–5 residue "tail" to the newly synthesized cDNA strand (Fig. 1a). Included in the reverse transcriptase reaction mix is the "SMARTer IIA Oligonucleotide" which contains a sequence tag as well as a terminal series of modified bases. These modified bases anneal to the new 3–5 residue tail on the cDNA. The SMARTScribe RT is tricked into believing that the SMART IIA oligo is contiguous with the original RNA transcript and reverse-transcribes it. The newly synthesized cDNA now has a sequence tag attached to it (Fig. 1a) and can be used as a template for PCR, with one primer complementary to the sequence tag and the second primer specific to the gene of interest (Fig. 1b). High-temperature PCR increases specificity and enhances processivity through difficult templates. Suppression PCR combined with step-out PCR [14] provides additional specificity. The resulting DNA fragments (*see* **Note 4**) can be cloned and sequenced or just sequenced directly. If high-quality, undegraded RNA is used, the odds of obtaining the true 5′ end of a transcript are considerably enhanced. For 3′ RACE a modified oligo(dT) primer containing a 3′ sequence tag ("3′-CDS primer A") is used as a primer for reverse transcription with

SMARTScribe RT and a gene-specific primer can be used to PCR the 3′ end of the gene (Fig. 1b) (*see* **Note 3**). Conventional cloning or end-to-end PCR is used to generate full-length cDNA (Fig. 1c).

The procedure described in this chapter has limitations. Very strong secondary structure or GC-rich regions will interfere with reverse transcription and RACE, regardless of the system. It is always

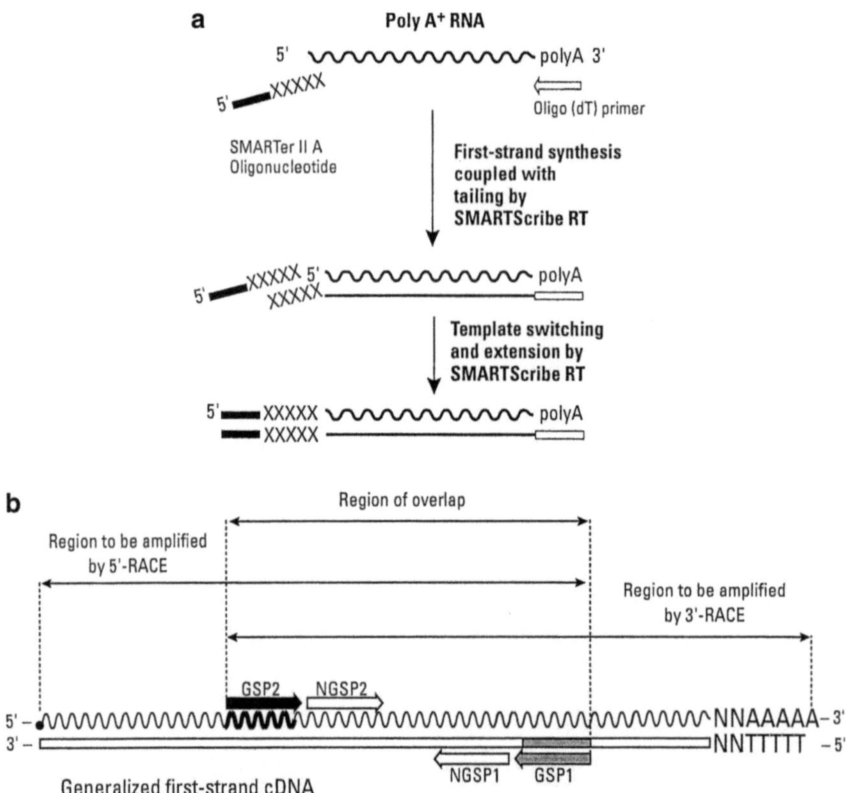

Fig. 1 Overview of SMART RACE. (**a**) Mechanism of SMARTer cDNA synthesis for 5′ RACE. First-strand synthesis is primed using a modified oligo(dT) primer. After SMARTScribe reverse transcriptase (RT) reaches the end of the mRNA template, it adds several nontemplated residues. The SMARTer II A Oligonucleotide anneals to the tail of the cDNA and serves as an extended template for SMARTScribe RT. The 5′ RACE-ready cDNA, containing extreme 5′ end sequences of all polyA+ transcripts present in the RNA, is ready for PCR amplification of specific genes. (**b**) Gene-specific primers used for 5′ and 3′ RACE amplification. Gene-specific primer 1 (GSP1) is used for 5′ RACE PCR and GSP2 is used for 3′ RACE PCR amplification for a specific gene. Nested PCR using nested gene-specific primers (NGSP) is necessary only when background or nonspecific amplification is high. Further details are available in the manufacturer's user manual. (**c**) Overview of steps involved in cloning full-length cDNA using SMARTer RACE. 5′ RACE generates 5′ RACE fragments containing extreme 5′ transcript ends; 3′ RACE generates 3′ RACE fragments containing extreme 3′ transcript ends. Options include cloning and sequencing the entire 5′ and 3′ RACE fragments (recommended) or alternatively sequencing just the 5′ end of the 5′ product and the 3′ end of the 3′ product to obtain sequences of the extreme ends of the transcript. This information is used to design 5′ and 3′ gene-specific primers to use in long-distance end-to-end PCR, with the 5′ RACE-ready cDNA as template, to generate the full-length cDNA. Overlap PCR is not recommended (*see* **Note 16**). Full-length cDNA can also be obtained by conventional cloning of overlapping RACE fragments if a suitable restriction site is present

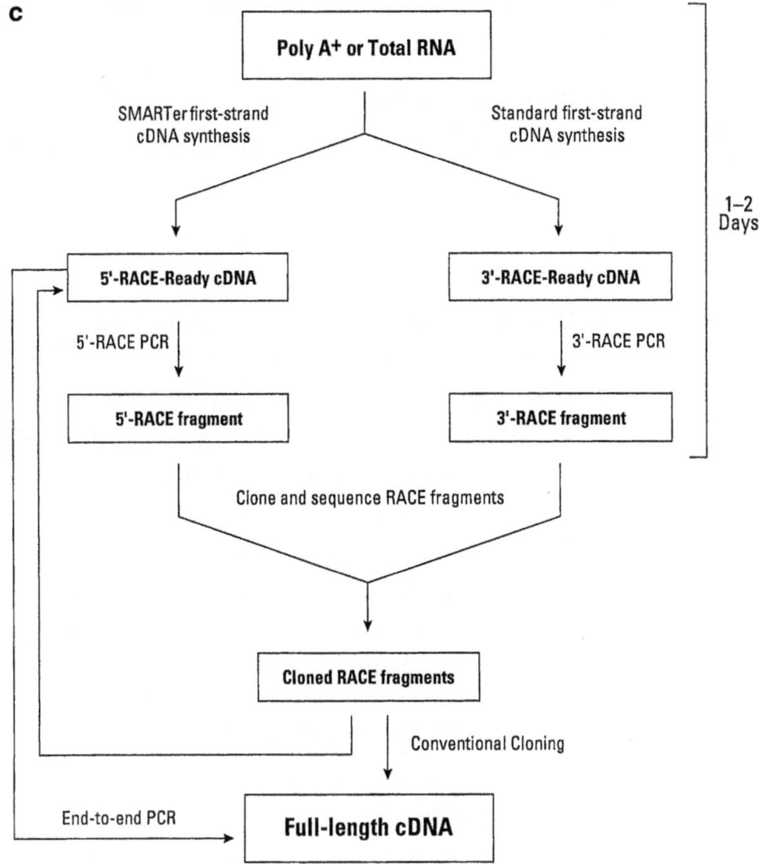

Fig. 1 (continued)

a good idea to align 5′ end sequences determined by 5′ RACE against the corresponding genomic sequence and inspect the upstream 5′ genomic sequence for GC-rich regions or predicted strong secondary structure. 5′ RACE can be repeated taking appropriate measures to overcome any GC-rich or strong secondary structure (*see* **Note 5**). Also, degraded RNA will present a 5′ end that will be sequence-tagged as if it were the true 5′ end, so RNA degradation must be avoided at all cost. Finally, as far as we are aware the SMARTer RACE procedure has not been used on small RNAs such as microRNAs, which fare well with specialized linker-ligation-based RACE methods [15, 16]. If the experimental goal is specifically to sequence cDNA ends, especially in a high-throughput manner, without cloning full-length cDNA, methods such as CAGE [17] or Deep-RACE [18] should be considered. Chapters 6 and 7 of this volume describe sequencing of PCR-amplified DNA and next-generation (high-throughput) DNA sequencing methods [19, 20].

In general, however, the Clontech SMARTer RACE procedure described below [14] is a simple and robust method that reliably provides the most 5′ sequence information, enabling cloning and sequencing of full-length gene transcripts and their variants.

2 Materials

2.1 First-Strand cDNA Synthesis Using the SMART RACE cDNA Amplification Kit

1. RNA isolation kit.
2. RNA quantitation and quality control reagents and equipment.

2.1.1 Before Beginning

2.1.2 First-Strand cDNA Synthesis

1. Hot-lid thermal cycler (e.g., Gene Amp PCR System 9700 thermal cycler, Applied Biosystems, Carlsbad, CA).
2. Thin-walled PCR tubes for thermal cycler.
3. SMARTer™ RACE cDNA Amplification Kit (Clontech Laboratories, Mountain View, CA).
4. RNA isolated from tissue or cell type of interest.
5. (Optional) For polyA– transcripts: Poly(A) polymerase (Catalog #2180A, Takara Bio, Madison, WI).
6. Molecular biology (MB)-grade, sterile water (certified RNase-free, but not DEPC-treated).

2.2 RACE PCR Reactions [14]

Make sure the PCR is programmed in advance.

1. Molecular biology grade, sterile water (certified RNase-free, but not DEPC-treated).
2. PCR polymerase, e.g., Advantage® 2 PCR Kit (Clontech), Advantage® GC 2 PCR Kit (Clontech), Advantage® HF 2 PCR Kit (Clontech), or other PCR system.
3. Thin-walled PCR tubes.
4. Hot-lid thermal cycler.
5. Gene-specific primers (*see* **Note 6**):

   23–28 nt.

   50–70 % GC.

   $T_m \geq 65$ °C; for best results $T_m > 70$ °C.

   Not complementary to the 3′ end of the Clontech Universal Primer.

   Specific to gene of interest (check by BLAST).
6. If not using a hot-lid cycler, PCR-grade mineral oil (Sigma).
7. (Optional) For extreme RNA secondary structure: a thermostable reverse transcriptase such as ThermoScript Reverse Transcriptase (Invitrogen, Carlsbad, CA) and the GeneRACER™ kit (Invitrogen) with additional materials according to manufacturers' instructions.

2.3 Cloning and Sequencing RACE Products

1. Agarose gel electrophoresis reagents and equipment.
2. Ethidium bromide.
3. Kit to extract DNA fragments from agarose gel slices for cloning. For automated DNA purification (up to 12 samples) from agarose gel slices in a spin-column format, the QIAcube from Qiagen is convenient. For nonautomated DNA purification from gel slices, the NucleoTrap Gel Extraction Kit Gel supplied with the SMARTer RACE kit is a good choice, as is the QIAquick or QIAEX Gel Extraction Kit (Qiagen, Valencia, CA).
4. TA-type cloning vector (for example the TOPO®-TA Cloning Kit from Invitrogen). If the PCR products are blunt-ended, use a ZeroBlunt® TOPO® cloning kit (Invitrogen) for cloning. Use plates, liquid media, and other materials as recommended by the supplier.

3 Methods

3.1 First-Strand cDNA Synthesis Using the SMART RACE cDNA Amplification Kit

3.1.1 Before Beginning

1. Isolate RNA from the tissue of interest. RNA used for 5′ RACE or 3′ RACE should be intact and free of DNA. The RNA isolation methods described in Chapters 2 and 3 of this volume [21, 22] are appropriate for 5′ and 3′ RACE as long as a DNase-treatment step has been included. Other methods that produce RNA of similar quality are completely acceptable. RNA integrity and concentration can be assessed as described in Chapters 2 and 3 of this volume [21, 22].
2. Program your PCR machine with all constant-temperature and cycling programs that will be used throughout the procedure.
3. Just before beginning the procedure, thaw needed reagents on ice and keep cold (on ice). Keep the RNA on dry ice until right before it is to be added to the reaction mix. Then thaw it quickly by hand, place immediately on ice after thawing, spin down briefly at 4 °C in a chilled, RNase-free microfuge, add the RNA to the reverse-transcriptase tube, and refreeze IMMEDIATELY in dry ice if it is to be reused (for something other than RACE) at a later date. (Try not to reuse RNA that has been thawed and re-frozen for RACE.)
4. Ensure that all procedures are performed in an RNase-free manner and take great care not to cross-contaminate samples, since PCR is used later in the procedure.

3.1.2 First-Strand cDNA Synthesis [14]

1. Preheat a hot-lid thermal cycler (e.g., Applied Biosystems Gene Amp PCR System 9700 thermal cycler) to 70 °C.
2. Prepare Buffer Mix.
   - For one 10 μl cDNA synthesis reaction, mix:

     2.0 μl 5× First-Strand Buffer.

1.0 μl DTT (20 mM).

1.0 μl dNTP mix (10 mM).

Prepare enough for all the cDNA synthesis reactions, plus one extra reaction (or 10 % extra), in a clean microtube. Spin briefly in a microcentrifuge and then leave at room temperature until used in Subheading 3.1.2 **step 8**, below.

3. Combine in a separate 0.5 ml thin-walled PCR tube:
   (a) *For 5′ RACE-ready cDNA*:
      - 1 – 2.75 μl total or polyA+ RNA (10 ng to 1 μg). For the negative control, substitute an equivalent volume of MB-grade water (*see* **Notes 7** and **8**).
      - 1.0 μl 5′-CDS primer A.
      - Sterile MB-grade water to a final volume of 3.75 μl for each reaction (*see* **Note 8**).

   (b) *For 3′ RACE-ready cDNA*:
      - 1 – 3.75 μl total or polyA+ RNA (10 ng to 1 μg). For the negative control, substitute an equivalent volume of MB-grade water (*see* **Notes 7** and **8**).
      - 1.0 μl 3′-CDS primer A.
      - Sterile MB-grade water to a final volume of 4.75 μl for each reaction (*see* **Note 8**).
4. Mix contents and spin tubes briefly in a microcentrifuge.
5. Incubate the tubes at 72 °C in a hot-lid thermal cycler for 3 min, and then cool the tubes to 42 °C for 2 min.
6. Spin tubes briefly for 10 s at 14,000 × *g* in a microcentrifuge.
7. (FOR 5′ RACE REACTION ONLY) Add 1.0 μl SMARTer IIA oligo per reaction.
8. Prepare a Master Mix (enough for all cDNA synthesis reactions, plus one extra reaction).
   - To prepare Master Mix for one reaction, combine:

     4.0 μl Buffer Mix (from **step 2**, above).

     0.25 μl RNase inhibitor (40 U/μl).

     1.0 μl SMARTScribe Reverse Transcriptase (100 U/μl).

     Mix gently but thoroughly and centrifuge briefly to collect contents at the bottom of the tube.
9. Add 5.25 μl Master Mix to each reaction tube containing denatured RNA (**step 6** for 3′ RACE cDNA; **step 7** for 5′ RACE cDNA).
10. Mix the contents of each tube by gently pipetting and close top. Spin briefly to collect the contents at the bottom of the tube.

11. Incubate the tubes at 42 °C for 1.5 h in a hot-lid thermal cycler (or air incubator). Do not use a water bath.
12. Remove tubes from thermal cycler, and increase temperature of thermal cycler to 70 °C.
13. Heat tubes at 70 °C for 10 min.
14. Dilute the first-strand reaction product with Tricine-EDTA buffer:

    Add 20 μl if you started with ≤200 ng of total RNA.

    Add 100 μl if you started with ≥200 ng of total RNA.

    Add 250 μl if you started with polyA+ RNA.
15. Samples can be used immediately for 5′ RACE and 3′ RACE or stored at −20 °C for up to 3 months (*see* **Notes 9** and **10**).

3.2 RACE PCR Reactions [14]

1. *Make sure the PCR is programmed in advance.*
2. For each 50-μl PCR reaction, prepare a Master Mix:

    34.5 μl PCR-grade water.

    5 μl Advantage 2 PCR buffer.

    1 μl dNTP mix (10 mM).

    1 μl 50× Advantage 2 Polymerase mix (*see* **Note 11**).

    Prepare enough for all the PCR reactions, plus one extra reaction (or 10 % extra), in a clean microtube. Mix well to vortex, avoiding bubble formation, and spin briefly.

    (a) *FOR 5′ RACE*:

    To each 0.5 ml PCR tube add (in the order shown):

    2.5 μl 5′ RACE-ready cDNA from Subheading 3.1.2 **step 14**.

    5.0 μl UPM.

    1.0 μl GSP1 (10 μM).

    41.5 μl Master Mix.

    (b) *FOR 3′ RACE*:

    To each 0.5 ml PCR tube add (in the order shown):

    2.5 μl 3′ RACE-ready cDNA from Subheading 3.1.2 step 14.

    5.0 μl UPM.

    1.0 μl GSP2 (10 mM).

    41.5 μl Master Mix.

    The final volume will be 50 μl per tube.
3. Mix gently.
4. Centrifuge briefly (in a PCR minifuge if available).
5. Begin PCR using the appropriate program and a hot-lid thermal cycler.

(*NOTE*: If not using a hot-lid cycler, overlay contents with two drops of mineral oil.)

For Program 1 (if GSP T_m > 70 °C) (RECOMMENDED)

5 Cycles:
94 °C 30 s.
72 °C 3 min (if the expected fragment will be >3 kb, add 1 min for each additional 1 kb).

5 Cycles:
94 °C 30 s.
70 °C 30 s.
72 °C 3 min (if the expected fragment will be >3 kb, add 1 min for each additional 1 kb).

20 Cycles (polyA+) or 25 cycles (total RNA).
94 °C 30 s.
68 °C 30 s.
72 °C 3 min (if the expected fragment will be >3 kb, add 1 min for each additional 1 kb).

NOTE: If the GSP T_m = 60–70 °C, use:

20 Cycles (polyA+ RNA) or 25 cycles (total RNA):
94 °C 30 s.
68 °C 30 s.
72 °C 3 min (if the expected fragment will be >3 kb, add 1 min for each additional 1 kb).

3.3 Cloning and Sequencing RACE Products

1. Analyze by gel electrophoresis, stain with ethidium bromide (*see* **Note 12**), and excise *all* bands (*see* **Notes 13** and **14**).
2. Isolate DNA fragments from the gel slices using your preferred gel extraction method. For automated DNA purification (up to 12 samples) from agarose gel slices in a spin-column format, the QIAcube from Qiagen is convenient. For non-automated DNA purification from gel slices, the NucleoTrap Gel Extraction Kit Gel supplied with the SMARTer RACE kit is a good choice, as is the QIAquick or QIAEX Gel Extraction Kit (Qiagen).
3. Clone the gel-purified fragments into a TA-type cloning vector (for example using a TOPO®-TA Cloning Kit from Invitrogen). If you used a PCR polymerase that gives blunt-ended PCR products, use a ZeroBlunt® TOPO® cloning kit for cloning the PCR product.
4. To obtain the maximum amount of 5′ end sequence, sequence 8–10 clones from EACH fragment excised from the gel. Some genes have multiple initiation sites over a ~100 bp region rather than one discrete transcription initiation site. If this appears to be the case, sequence more clones to ensure the 5′ end has really been reached (*see* **Note 15**).

5. To obtain a full-length clone, perform long-distance PCR with a high-fidelity polymerase using primers from the extreme 5′ and 3′ ends and the 5′ RACE-ready cDNA (Subheading 3.1.2 **step 14**) (*see* **Note 16**).
6. Several databases provide sequence information on gene transcripts and transcript variants, either predicted in silico or experimentally verified. (*See* ref. 23 for a recent list of transcript databases. New databases are likely to emerge, so check the literature.) It is always a good idea to check your sequence results against existing databases and/or previously published papers, whether you are seeking confirmation that you have the "right" sequence or whether you are looking for novel variants.

4 Notes

1. A notable exception is the class of replication-dependent histone mRNAs, which encode protein but lack a polyA tail. Also note that while protein-coding transcripts are generally capped, a transcript with a 5′ cap will not necessarily be translated into protein. That is, many noncoding RNAs are capped [6].
2. The original SMART RACE cDNA Amplification Kit was replaced in February 2009 by a new, improved version termed the SMARTer RACE cDNA Amplification Kit. The principles behind the procedure have not changed but some experimental details have. For example, the newer kit uses a new reverse transcriptase and modified oligonucleotide sequences, and less RNA can be used in the RT step. If you refer to an online procedure when using the kit, make sure it's the updated protocol for the SMARTer RACE kit.
3. Many noncoding RNAs lack a polyA tail or are "poorly polyadenylated" [9]. To amplify transcripts that lack a poly(A) tail, a poly(A) tail can be added to the 3′ end of the transcript using Poly(A) polymerase (Takara). SMARTer RACE can then be carried out as usual, using 5′-CDS or 3′-CDS primer A to prime the reverse transcriptase reaction for 5′ or 3′ RACE, respectively. Traditional linker-ligation RACE can also be used to determine the 5′ and 3′ ends of polyA– transcripts. If you don't need 3′ end information and just need the 5′ end of a polyA– transcript, use the SMARTer RACE kit with random primers or a primer of known sequence to prime the reverse transcriptase reaction.
4. Usually multiple bands representing different transcript variants are apparent on an agarose gel after RACE. *All* bands should be excised from the gel and cloned and sequenced. It is sometimes assumed that the longest 5′ RACE product size contains the

"real" 5′ end and that smaller 5′ RACE products are merely incomplete reverse-transcription products, but this is not a valid assumption. Transcript heterogeneity is commonplace.

5. For truly intractable RNA secondary structure, reverse transcription should be performed at higher temperatures using a thermostable reverse transcriptase such as ThermoScript Reverse Transcriptase (Invitrogen). ThermoScript RT can generate cDNA transcripts from 100 bp to >12 kb at temperatures ranging from 50 to 65 °C. Higher temperatures will promote denaturation of secondary structure. This enzyme can be used in Invitrogen's GeneRACER™ kit according to manufacturers' instructions. Somewhat to our surprise, we have found that kits designed to specifically amplify transcripts with a 5′ cap do not necessarily yield more 5′ sequences than the SMARTer RACE kit, perhaps because of the increased handling required to capture the capped transcripts. Ease of use, selection for full-length cDNAs, and high specificity have led to our preference for SMARTer RACE technology. Efficient PCR amplification of GC-rich regions or sequences with strong secondary structure is a separate issue that is easily remedied by adding a cosolvent or increasing the temperature during the PCR step (*see* **Note 11**, below).
6. The recommended length of 23–28 nt refers to the length of overlap with the target sequence. Additional nucleotides, for example restriction sites for cloning, can be placed on the 5′ ends of the oligonucleotide. If the PCR product is to be restriction-enzyme digested and cloned directly into the vector of interest without prior subcloning into a TA-type vector (*see* Subheading 3.3), an additional 6–8 nt of random sequence should be added 5′ of the restriction site to ensure complete digestion by the restriction enzyme. If possible, use the same nucleotides in opposite directions for 5′ and 3′ RACE to minimize formation of a chimeric DNA fragment after PCR.
7. Negative and positive controls: It is essential to include a negative control (no RNA added) every time a reverse-transcription reaction is performed to check for cross-contamination. The positive control included in the kit (control human placental DNA) is also highly recommended, especially for the beginner, to be used according to manufacturer's instructions. As an additional positive control to test cDNA made from your own RNA sample, use the cDNA created from in this step from your sample RNA to perform RACE for a gene whose 5′ end has already been determined.
8. Do not use DEPC-treated water. Residual DEPC may inhibit the reverse transcriptase.

9. Samples can be stored long-term at −70 °C. Avoid repeated freeze-thaw cycles.
10. The samples can be saved and used for other applications that use cDNA as a starting material.
11. For efficient amplification through GC regions, Clontech recommends the Advantage GC 2 PCR kit (#633119 & #639120). Alternatively, a cosolvent additive (e.g., PCRx Enhancer, Invitrogen, Carlsbad, CA) [24] that helps to "melt" DNA can be added to the PCR reaction according to manufacturer's instructions; optimization may be required. For highest fidelity Clontech recommends the Advantage HF 2 PCR kit (#639123 & #639124). We suggest using either the Advantage 2 PCR or Advantage GC 2 PCR kits for the initial 5′ RACE or 3′ RACE PCR amplifications (and for nested PCR if necessary), since the SMARTer RACE kit has been optimized using these reagents. Once the extreme 5′ and 3′ ends are known, the user can use the Advantage HF 2 PCR kit, or alternatively the high-fidelity PCR enzyme of their choice, along with primers corresponding to the 5′ and 3′ ends, to amplify full-length fragments from cDNA for cloning and/or sequencing.
12. If no bands are visible on an ethidium bromide-stained agarose gel, perform nested PCR using the Nested Universal Primer A (NUP; 10 μM), which is included in the SMARTer RACE kit, and a second gene-specific primer.
13. *Cut out all bands*, not just the band with the highest molecular weight, and clone and sequence them. Multiple bands result from alternative transcription initiation sites, alternative splicing, and alternate polyadenylation events, and it cannot be assumed that the highest-MW band, or even the most intense band, corresponds to the major transcript that is translated into the major protein isoform. Cloning and sequencing alternative transcript variants as well as the primary transcript can frequently yield new insights into gene function and regulation. Just as frequently, rather bizarre transcript variants, lacking any apparent function, turn up as well. Such transcripts are regarded as valid, albeit mysterious, and should not cause alarm. Importantly, if the number of bands seems excessive, ensure that the RNA is not degraded and troubleshoot the procedure as described in the SMARTer RACE cDNA Amplification Kit User Manual.
14. Shotgun cloning of 5′ RACE or 3′ RACE reactions is not recommended since shorter clones will be overrepresented.
15. As a check, alternative RACE methods (*see* **Note 5**) or 5′ CAGE [17, 25] using a gene-specific primer or other methodologies that provide RNA sequences adjacent to the 5′ cap can also be investigated. Note, however, that 5′ CAGE sequences

are not always promoter adjacent [1, 6]. Chromatin IP with an antibody specific for RNA Polymerase II phosphorylated at Ser 5 can be used to verify transcription initiation within a ~300–500 bp region [26], and promoter activity can be tested directly by cloning the promoter in front of a luciferase gene, for example using one of the pGL4 vectors from Promega, transfecting into the appropriate cell type and assaying luciferase activity.

16. Although some protocols suggest using overlap PCR between the 5′ and 3′ RACE products to obtain the full-length PCR product, this method introduces the possibility that two transcripts with closely related but nonidentical overlap sequences may co-amplify to form a chimeric transcript. It is better to use the 5′ and 3′ RACE sequence information to design primers and then perform 5′ end-to-3′ end PCR with a high-fidelity PCR enzyme from the cDNA prepared in Subheading 3.1.2 **step 14**. The full-length fragment is then cloned and sequenced.

References

1. Gustincich S, Sandelin A, Plessy C, Katayama S, Simone R, Lazarevic D, Hayashizaki Y, Carninci P (2006) The complexity of the mammalian transcriptome. J Physiol 575:321–332
2. Hume DA (2008) Our evolving knowledge of the transcriptional landscape. Mamm Genome 19:663–666
3. Singer GA, Wu J, Yan P, Plass C, Huang TH, Davuluri RV (2008) Genome-wide analysis of alternative promoters of human genes using a custom promoter tiling array. BMC Genomics 9:349
4. Kim E, Goren A, Ast G (2008) Alternative splicing: current perspectives. Bioessays 30:38–47
5. Hiller M, Platzer M (2008) Widespread and subtle: alternative splicing at short-distance tandem sites. Trends Genet 24:246–255
6. Affymetrix/CSHL ENCODE Project (2009) Post-transcriptional processing generates a diversity of 5′-modified long and short RNAs. Nature 457:1028–1032
7. Powell LM, Wallis SC, Pease RJ, Edwards YH, Knott TJ, Scott J (1987) A novel form of tissue-specific RNA processing produces apolipoprotein-B48 in intestine. Cell 50:831–840
8. Chen SH, Habib G, Yang CY, Gu ZW, Lee BR, Weng SA, Silberman SR, Cai SJ, Deslypere JP, Rosseneu M (1987) Apolipoprotein B-48 is the product of a messenger RNA with an organ-specific in-frame stop codon. Science 238:363–366
9. Carninci P (2006) Tagging mammalian transcription complexity. Trends Genet 22:501–510
10. Core LJ, Waterfall JJ, Lis JT (2008) Nascent RNA sequencing reveals widespread pausing and divergent initiation at human promoters. Science 322:1845–1848
11. Esau C, Davis S, Murray SF, Yu XX, Pandey SK, Pear M, Watts L, Booten SL, Graham M, McKay R, Subramaniam A, Propp S, Lollo BA, Freier S, Bennett CF, Bhanot S, Monia BP (2006) miR-122 regulation of lipid metabolism revealed by in vivo antisense targeting. Cell Metab 3:87–98
12. Carninci P (2009) Molecular biology: the long and short of RNAs. Nature 457:974–975
13. Hao S, Baltimore D (2009) The stability of mRNA influences the temporal order of the induction of genes encoding inflammatory molecules. Nat Immunol 10:281–288
14. Clontech (2009) SMARTer RACE cDNA amplification kit user manual. Clontech, Mountain View, CA, pp 1–33
15. Kong W, Zhao JJ, He L, Cheng JQ (2009) Strategies for profiling microRNA expression. J Cell Physiol 218:22–25
16. Sdassi N, Silveri L, Laubier J, Tilly G, Costa J, Layani S, Vilotte JL, Le PF (2009) Identification and characterization of new miRNAs cloned from normal mouse mammary gland. BMC Genomics 10:149
17. Kodzius R, Kojima M, Nishiyori H, Nakamura M, Fukuda S, Tagami M, Sasaki D, Imamura K, Kai C, Harbers M, Hayashizaki Y, Carninci P (2006) CAGE: cap analysis of gene expression. Nat Methods 3:211–222

18. Olivarius S, Plessy C, Carninci P (2009) High-throughput verification of transcriptional starting sites by Deep-RACE. Biotechniques 46:130–132
19. Youngblood V, Taylor JV (2010) Sequencing PCR-amplified DNA in lipoprotein and cardiovascular disease research. In: Freeman LA (ed.) Methods in molecular biology: lipoproteins, 2nd edn, Walker JM (series ed.). Humana, Totowa, NJ
20. Diaw L, Youngblood V, Taylor JV (2010) Introduction to next-generation nucleic acid sequencing in cardiovascular disease research. In: Freeman LA (ed.) Methods in molecular biology: lipoproteins, 2nd edn, Walker JM (series ed.). Humana, Totowa, NJ
21. Wagner EM (2010) Monitoring gene expression: quantitative real-time RT-PCR. In: Freeman LA (ed.) Methods in molecular biology: lipoproteins, 2nd edn, Walker JM (series ed.). Humana, Totowa, NJ
22. Raghavachari N (2010) Microarray technology: basic methodology and application in clinical research for biomarker discovery in vascular diseases. In: Freeman LA (ed.) Methods in molecular biology: lipoproteins, 2nd edn, Walker JM (series ed.). Humana, Totowa, NJ
23. Koscielny G, Le TV, Gopalakrishnan C, Kumanduri V, Riethoven JJ, Nardone F, Stanley E, Fallsehr C, Hofmann O, Kull M, Harrington E, Boue S, Eyras E, Plass M, Lopez F, Ritchie W, Moucadel V, Ara T, Pospisil H, Herrmann A, Reich G, Guigo R, Bork P, Doeberitz MK, Vilo J, Hide W, Apweiler R, Thanaraj TA, Gautheret D (2009) ASTD: the alternative splicing and transcript diversity database. Genomics 93:213–220
24. Sabol SL, Brewer HB Jr, Santamarina-Fojo S (2005) The human ABCG1 gene: identification of LXR response elements that modulate expression in macrophages and liver. J Lipid Res 46:2151–2167
25. Shiraki T, Kondo S, Katayama S, Waki K, Kasukawa T, Kawaji H, Kodzius R, Watahiki A, Nakamura M, Arakawa T, Fukuda S, Sasaki D, Podhajska A, Harbers M, Kawai J, Carninci P, Hayashizaki Y (2003) Cap analysis gene expression for high-throughput analysis of transcriptional starting point and identification of promoter usage. Proc Natl Acad Sci U S A 100:15776–15781
26. Xu YX, Manley JL (2007) Pin1 modulates RNA polymerase II activity during the transcription cycle. Genes Dev 21:2950–2962

Chapter 2

Monitoring Gene Expression: Quantitative Real-Time RT-PCR

Elke M. Wagner

Abstract

Two-step quantitative real-time RT-PCR (RT-qPCR), also known as real-time RT-PCR, kinetic RT-PCR, or quantitative fluorescent RT-PCR, has become the method of choice for gene expression analysis during the last few years. It is a fast and convenient PCR method that combines traditional RT-PCR with the phenomenon of fluorescence resonance energy transfer (FRET) using fluorogenic primers. The detection of changes in fluorescence intensity during the reaction enables the user to follow the PCR reaction in real time.

RT-qPCR comprises several steps: (1) RNA is isolated from target tissue/cells; (2) mRNA is reverse-transcribed to cDNA; (3) modified gene-specific PCR primers are used to amplify a segment of the cDNA of interest, following the reaction in real time; and (4) the initial concentration of the selected transcript in a specific tissue or cell type is calculated from the exponential phase of the reaction. Relative quantification or absolute quantification compared to standards that are run in parallel can be performed.

This chapter describes the entire procedure from isolation of total RNA from liver and fatty tissues/cells to the use of RT-qPCR to study gene expression in these tissues. We perform relative quantification of transcripts to calculate the fold-difference of a certain mRNA level between different samples. In addition, tips for choosing primers and performing analyses are provided to help the beginner in understanding the technique.

Key words RT-PCR, qPCR, RNA extraction, TaqMan® probe, Endogenous control, Liver, Macrophages, Relative quantification, Single-well reaction

1 Introduction

Lipoproteins have a vital role in transporting hydrophobic lipids through the aqueous milieu of the circulation. Triglycerides, cholesterol (free cholesterol and cholesteryl esters), and phospholipids are the major lipid constituents of lipoproteins, while other small hydrophobic molecules such as fat-soluble vitamins or lipophilic drugs may be carried in lipoproteins as well [1–4].

Given their critical roles in supplying hydrophobic molecules to cells and tissues as well as in preventing accumulation of excess lipid throughout the body, it is not surprising that genes involved

Lita A. Freeman (ed.), *Lipoproteins and Cardiovascular Disease: Methods and Protocols*, Methods in Molecular Biology, vol. 1027, DOI 10.1007/978-1-60327-369-5_2, © Springer Science+Business Media, LLC 2013

in lipoprotein metabolism are highly regulated. Gene variants, gene under- or over-expression (for example in transgenic or knockout mice), dietary influences, and therapeutic agents that alter lipoprotein metabolism can change gene expression patterns in many cells and tissues: liver and intestine, which produce apoB-containing lipoproteins and HDLs; macrophages, endothelial cells, inflammatory cells, and smooth muscle cells in the artery wall; and other cell types such as adipose tissue, lungs, pancreatic β-cells, and cytokine/chemokine-secreting cells. In addition, the brain has its own separate lipoprotein metabolism due to the inability of lipids/lipoproteins to cross the blood–brain barrier [5]. As different tissues may regulate the same gene quite differently, it can be necessary to characterize expression of the same gene(s) in multiple tissues. Clearly, a fast, high-throughput, cost-effective, quantitative, and reproducible method for measuring gene expression, such as quantitative real-time PCR (qPCR)—specifically, qPCR in combination with reverse transcription (RT-qPCR) [6–8]—will be advantageous to any state-of-the-art laboratory investigating lipoprotein metabolism.

RT-qPCR, also known as real-time RT-PCR, kinetic RT-PCR, or quantitative fluorescent RT-PCR, combines traditional RT-PCR [9] with the phenomenon of fluorescence resonance energy transfer (FRET) using fluorogenic primers [10]. While RT-qPCR is not without limitations, it can provide unmatched sensitivity and speed for determining levels of specific transcripts when using proper controls, careful initial reaction characterization, and quality control.

In order to measure gene expression by RT-qPCR, RNA is isolated and reverse-transcribed using random hexamers as primers to form cDNA (*see* **Note 1**). Gene-specific PCR primers are then used to amplify a segment of the cDNA of interest. The reaction is followed in real time by detecting fluorogenic tags that are covalently linked to modified primer(s); alterations in fluorescence intensity during each PCR reaction cycle enables the user to follow the PCR reaction in real time. Several fluorogenic PCR assays with a variety of innovative primer designs are available to date (TaqMan® hydrolysis probes, Molecular Beacons, LUX primers, hybridization probes, Scorpion primers, Sunrise primers, etc.); a good overview for common qPCR assays can be found in Wong and Medrano [10]. For the purpose of this basic protocol, the TaqMan® hydrolysis probe (ABI) is described (*see* Fig. 1a).

The TaqMan® probe is a linear, dual-labeled oligonucleotide of 20–30 nucleotides (nts) that anneals to the sequence between the two traditional PCR primers. It contains a fluorogenic reporter dye on its 5′ end and a quencher molecule on its 3′ end (Fig. 1a). The reporter dye is hydrolyzed by the 5′ nuclease activity of the PCR enzyme during each new strand synthesis and emits fluorescence once it is liberated and out of range of the quencher dye on the

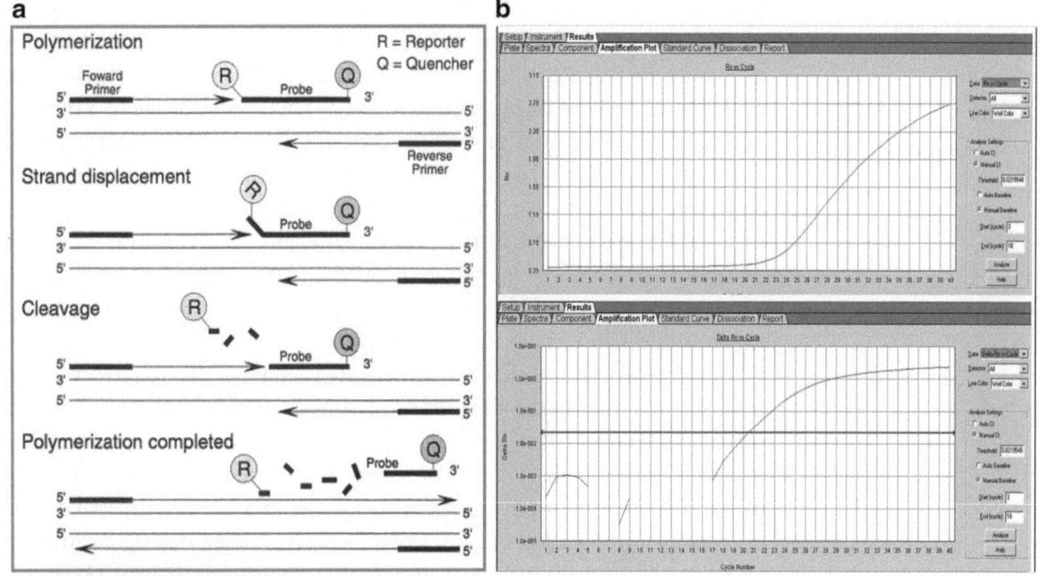

Image kindly provided as a courtesy of Applied Biosystems.

Fig. 1 Principle of qPCR using a TaqMan® probe. (**a**) Schematic of a qPCR reaction. In addition to the traditional two PCR primers, a third primer with two covalently linked tags on its 5′ and 3′ end anneals to the DNA. When the 5′ fluorogenic reporter dye (e.g., "FAM," "VIC," or other dyes) is excited by light, its emission will be quenched while bound to the primer and in close proximity to the nonfluorescent quencher molecule on the 3′ end. As the 5′ exonuclease activity of the Taq DNA polymerase hydrolyzes the 5′ reporter dye during strand extension, the dye is liberated and emits fluorescent light. The amplification can thus be followed using a fluorescent light detector. The most basic outfit of a real-time instrument would consist of a PCR running in optical reaction tubes on a thermo cycler with a lamp (excitation) and a detector (fluorescence). (**b**) Data analysis of TaqMan® assay using the 7300 SDS software. *Upper panel*: Linear view of a real-time PCR amplification curve. *Lower panel*: Logarithmic view of the same curve. The exponential phase seems now "linearized." During this exponential increase, we analyze the qPCR, reading the number of cycles at a chosen "threshold" reporter signal intensity (Delta Rn). The cycle at threshold (C_t) for the curve is 20.5. It can be compared to other samples at the same threshold; relative gene expression levels between, for example, a wild-type and a transgenic organism, can easily be derived from the number of PCR cycles needed to achieve the same quantity of fluorescent light ("threshold"). Rn, measure of fluorescent reporter signal. Delta Rn, measure of fluorescent reporter signal corrected for baseline/background

probe. The rise in fluorescence intensity is detected and presented graphically as light intensity (Delta Rn) vs. PCR cycles (cycle number) (*see* Fig. 1b). The use of three gene-specific oligonucleotides in the TaqMan® assay assures highly specific amplification of even short cDNA sequences (typically 60–100 nts).

When studying gene expression, "relative" quantification is used to compare the gene expression levels between different experimental conditions or tissues. In this case, differences are expressed as "fold-changes." Absolute quantification of DNA copy numbers may be of interest for the determination of viral loads or transgene integration (*see* Chapter 11 of this volume [11]). For the analysis of a qPCR experiment, a "threshold" level of fluorescence is chosen in the exponential phase of the PCR, and the number of cycles required for fluorescence to reach this threshold level [termed cycle at threshold (C_t) or cycle at crossing point (C_p)] is determined (*see* Fig. 1b). Relative amounts of copy numbers of a transcript in different samples can be assessed by comparing their C_ts, with smaller C_ts corresponding to higher initial copy numbers [12]. In order to assure that the same amount of total cDNA is compared, a loading control (or "housekeeping gene") is detected in addition to the gene of interest in each sample.

In qPCR, the ability to follow and analyze the reaction progress during the early exponential phase minimizes confounding factors that affect late cycles (e.g., in the end-point quantification of traditional PCR), such as reagent consumption, slower reactions, and reaction product degradation [13]. This improves the reproducibility and accuracy of the initial transcript copy number, and reduces hands-on steps and hence the possibility of contamination. However, it is important to note that RNA species are not directly visualized by this method and that unspecific reactions, isoforms, and splice variants will go undetected.

The following chapter describes the isolation of total RNA of high quality from liver and fatty tissues/cells and the use of TaqMan® Gene Expression Assays to study gene expression in these tissues (*see* **Note 2**). Tips for choosing primers are provided to help the first-time user in understanding the technique. We have included a detailed description of manual data analysis for performing relative quantification with a standard curve. While this may appear laborious compared to automated software modules, we feel this leads to a basic understanding of the methodology and will help to estimate the results and to keep the user informed of hidden biases in automated analysis.

We encourage the reader to inform himself/herself about PCR and qPCR as well as the instrument/software to be used. Very concise and useful brochures are accessible for all users on the ABI home page [12–14]. Finally, we discuss potential pitfalls for the reader to keep in mind when setting up a specific assay for a certain tissue/cell line for the first time.

2 Materials

If materials change, make sure to get comparable material or contact provider and ask for substitute products. We assume that pipettes, RNase- and DNase-free pipette tips (filtered), centrifuges, agarose gel apparatus, and gloves are available in the lab.

2.1 Total RNA Extraction from Tissue Using TRIzol

IMPORTANT: Wear clean gloves while preparing all reagents and equipment for RNA extraction and during the extraction. Change gloves as needed. The workspace and all tools listed below should be wiped with RNase*Zap* and/or 70 % ethanol. Purchase several liters of RNase-free water (Ambion Inc., Austin, TX), absolute ethanol (Sigma-Aldrich, St. Louis, MO) and prepare 70 % wash alcohol with RNase-free water.

2.1.1 Isolation and Stabilization of Tissue

1. Tissue/paper towels.
2. RNase*Zap* (Ambion Inc., Austin, TX).
3. 70 % Ethanol, prepared with RNase-free water in a wash bottle.
4. RNA*later* (Ambion Inc.) or equivalent (e.g., Allprotect Reagent, Qiagen, Valencia, CA).
5. Optional: a Styrofoam lid and pins to hold the mouse corpse in place.
6. Two pairs of scissors, wiped clean.
7. Two blunt forceps and two bent, blunt forceps, wiped clean.
8. Small glass plate (about 10 cm × 10 cm), wiped clean.
9. RNase-free tubes, 2 ml (e.g., Safe-Lock Eppendorf; Fisher Scientific, Pittsburgh, PA). Use with clean gloves. Dedicate a box or bag of tubes for work with RNA only. Keep bag closed.
10. Liquid nitrogen or dry ice/ethanol bath for snap-freezing the tissues after excision.

2.1.2 RNA Extraction from Tissue (TRIzol)

1. 50 ml BD/Falcon tubes (Fisher Healthcare, Pittsburgh, PA).
2. Polytron/Ultra-Turrax [Kinematica Inc., Bohemia, NY (formerly Brinkmann Instruments)] or equivalent homogenizer, rinsed with RNase*Zap* or 70 % alcohol (in 50-ml tubes). Wipe dry.
3. Aliquot 30 ml of RNase-free water into 50 ml tubes (one per sample, used to rinse homogenizer).
4. TRIzol reagent (Invitrogen, Carlsbad, CA): avoid skin contact, wear coat and gloves, keep in refrigerator, and always clean the outside of the bottle well after use. Use appropriate thick-walled waste containers, as organic compounds dissolve some plastic containers (and plastic bags) readily!
5. Centrifuge tubes appropriate to the centrifuge available: e.g., for the SA-600 rotor on a Sorvall floor-centrifuge, use sterile 13-ml Sarstedt round-bottom polypropylene tubes with push caps (Sarstedt Inc., Newton, NC).
6. Chloroform (Sigma-Aldrich, St. Louis, MO). A separate, dedicated 100- or 500-ml glass bottle with chloroform labeled

"RNA only" is strongly recommended. Avoid skin contact, wear gloves, and, if possible, work in a hood. Use appropriate thick-walled waste containers, as organic compounds like chloroform dissolve some plastic containers (and plastic bags) readily!

7. Any 5-ml glass pipettes, graduated (attention: do not dip too far into the liquid, graduation marks dissolve in chloroform!).
8. Isopropanol (Sigma-Aldrich). A separate, dedicated 100- or 500-ml bottle of isopropanol, labeled "RNA only" is highly recommended. Avoid skin contact with isopropanol, wear gloves, and if possible, work in a hood.
9. Absolute ethanol (Sigma-Aldrich). A separate, dedicated 100- or 500-ml bottle of absolute ethanol labeled with "RNA only" is strongly recommended. Prepare 50 ml 75 % alcohol for washing RNA, made up with RNase-free water, and store in a 50-ml Falcon tube.
10. Refrigerated centrifuge for centrifugation up to 12,000 × *g*, e.g., Eppendorf 5417R (Fisher Scientific).
11. RNase-free tubes, 1.5 and 2 ml (e.g., Safe-Lock Eppendorf; Fisher Scientific). Use with clean gloves. Dedicate a box or bag of tubes for work with RNA only. Keep bag closed.
12. Thermal block for microcentrifuge tubes set to 55–60 °C.

2.2 Isolation of RNA from Cultured Cells and RNA Cleanup Using the Qiagen RNeasy Mini Kit

1. RNase*Zap* (Ambion Inc.).
2. A clean ("RNase*Zap*"-wiped) tabletop centrifuge for microcentrifuge tubes.
3. Individually wrapped cell scrapers (e.g., BD Falcon, 18-cm handles), one per cell culture plate.
4. QIAshredder (Qiagen, Valencia, CA) for homogenization (one per cell culture plate).
5. A few milliliter of β-mercaptoethanol (Sigma-Aldrich). This reagent should be stored in the refrigerator. Use only in a hood, or keep on ice if no hood is available. Work fast, close containers quickly, and get rid of waste in a closed container/bag.
6. RNeasy Mini Kit (Qiagen). Restore buffers with absolute ethanol as indicated in the kit.
7. Qiagen RLT buffer (included in the kit). Prepare the necessary amount in a 15-ml tube: 350 μl per sample for tissue RNA cleanup; 600 μl per sample for RNA extraction from cells. Add 10 μl of β-mercaptoethanol per 1 ml of RLT buffer under a hood.
8. RNase-free DNase (Qiagen No. 79254): Add 550 μl of RNase-free water to the lyophilized DNase I. *Mix gently; do not vortex!* Aliquot in 5 × 110 μl portions (you will need 10 μl per sample)

and freeze at −20 °C. Thawed aliquots can be stored at 4 °C for up to 6 weeks. Frozen aliquots can be stored at −20 °C for 9 months. Buffer RDD is stable in the refrigerator for at least 9 months.

2.3 Quantification and Storage of Total RNA

1. UV spectrophotometer (or equal instrumentation, for example a NanoDrop ND-1000 spectrophotometer, Thermo Scientific, Chicago, IL).
2. Clean quartz microcuvette(s) with 100 μl volume for UV spectrophotometer.
3. Aliquots of RNase-free water (e.g., 100 ml, Ambion Inc.).
4. RNase-free 1.5-ml tubes for quantification.
5. Pressure air to dry cuvette between samples, if available.
6. Ultra Pure 1 M Tris–HCl, pH 7.5 (Invitrogen).
7. RNase-free 10 mM Tris–HCl, pH 7.5: to prepare, dilute 1 ml Ultra Pure 1 M Tris–HCl, pH 7.5 with 99 ml of RNase-free water.
8. Aerosol-tight pipette tips for pipetting RNA, cDNA, or other RT-PCR components.
9. Optional: Prepare 1 or 1.5 % agarose gel(s) with RNase-free buffer to check quality of RNA samples. Any 1–1.5 % agarose gel made up with RNase-free gel reagents (agarose, water, buffer) will do. Use 0.5× TAE as gel/running buffer, and ethidium bromide or any dye used in your lab to visualize nucleic acids.

2.4 Reverse Transcription (RT)

1. TaqMan Reverse Transcription Reagents (Part no. N808-0234, protocol no. 402876, ABI). Just before use, prepare a bucket with wet ice. Thaw the components of the mix, spin down, and keep on wet ice. Keep the enzyme in the freezer until ready to use. Use random primers, not oligo(dT) (*see* **Note 1**). Read kit instructions before use; versions/concentrations may change.
2. Aerosol-tight pipette tips for pipetting RNA, cDNA, or other RT-PCR components.
3. Molecular biology grade water (Ambion Inc.).
4. Thermo cycler programmed as follows: 10 min at 25 °C, 60 min at 37 °C, 5 min at 95 °C.
5. 200-μl Thin-walled RNase-free tubes that fit the thermo cycler (if necessary, use an adaptor plate).

2.5 Real-Time PCR (qPCR)

1. MicroAmp 96-well optical reaction plate (ABI) and MicroAmp optical adhesive film (ABI; part number 4314320, MicroAmp adhesive film applicator, part number 4333183). *Touch only the sides of the cover films! Impurities such as smudges absorb light and will lead to incorrect readings and quantitations of the reactions in the sample wells.*

2. Dark plate support (ABI, 96-well plate support base, part number 4379590).
3. TaqMan® Gene Expression Assays (for the choice of primers *see* **Notes 3–6**). Keep frozen until use, thaw aliquots, and keep on wet ice, and keep dark (e.g., cover tubes in aluminum foil). A traditional assay for many lipoprotein researchers will be to detect the murine low-density lipoprotein receptor (LDLr, e.g., Mm00440169_m1, ABI) and β-actin as the endogenous control (4352341E, ABI) in liver or macrophages.
4. TaqMan® Universal PCR Mastermix w/o UNG (*see* **Note 7**) (Part no. 4324018, protocol no. 4304449, ABI) on wet ice. Store in refrigerator after thawing.
5. 2 % Agarose gel to check specificity of assays/tissues used for the first time. Use 0.5× TAE or 1× TBE as gel/running buffer, use ethidium bromide or any dye used in your lab to visualize nucleic acids, and use a 100-bp ladder as size standard (e.g., Gene Ruler™ Ultra Low Range, Fermentas, Glen Burnie, MD).

3 Methods[1]

RNA is readily degraded by RNase, an enzyme present ubiquitously. Therefore, it is critical to work clean and fast, from the beginning to the end of the procedure. When working with mice (*see* **Note 8**), avoid bleeding of tissues as far as possible; use only clean tools and wear a clean lab coat, gloves, and a mask. RNA isolation always requires fast treatment, but RNA from tissues such as lung, intestine, and adipose tissue are especially prone to degradation. Tissue pieces more than 5 mm × 5 mm in size will not be totally immersed with RNA*later*, as the passive diffusion of the liquid will not proceed that far. Therefore, prepare several small pieces of tissue, immerse in RNA*later*, and snap-freeze samples as well. For adipose tissue, consider using Allprotect (Qiagen) in place of RNA*later*.

For RNA isolation from cells: A confluent 100-mm Petri dish will hold a few million cells, depending on the size and confluency. Lipid-loaded fat cells are quiescent and very large, and more than one confluent 100-mm plate might be needed per experimental condition. For liver cells or macrophages, one 60- or 100-mm dish will provide enough RNA for qPCR. For example, pooled peritoneal macrophages from five unboosted mice, plated onto one

[1]The processes described in this chapter were developed during the author's fellowships at the University of Graz, Austria, and the NHLBI, NIH, Bethesda, MD, USA. The processes described here do not represent procedures specified by Baxter.

100-mm Petri dish in fresh medium and rinsed twice after 3 hours, provide ~10 μg total RNA. Read the Qiagen RNeasy Handbook carefully. It provides a valuable overview of the considerations involved in RNA work, as well as the use of common cells/tissues for RNA extraction (e.g., table of RNA quantities extracted from common cell lines).

We assume that the user is familiar with basic methods in molecular biology, such as running agarose gels and performing enzymatic assays (RT-PCR).

3.1 Total RNA Extraction from Tissue Using TRIzol

Isolation and stabilization of mouse tissue is described, but the methodology is applicable to any mammalian species.

3.1.1 Isolation and Stabilization of Tissue

1. Wipe work space, glass plate, scissors, and forceps with RNase*Zap* and 70 % ethanol before handling each mouse. Pre-label RNase-free tubes containing RNA*later*—for example, 1.5 ml RNA*later* per 2-ml tissue-sample tubes. Tubes are kept closed on clean tube racks. Optionally, larger tubes can be used, and up to 10 ml RNA*later* may be added to fully immerse precious samples (a 10- to 15-fold volume of RNA*later* is recommended by the manufacturer—still, pieces that are too large will not be immersed by diffusion).
2. (Optional: Pin down dead mouse body on Styrofoam lid). Wipe mouse abdomen with alcohol. Cut open outer belly skin/fur by making an X-like incision on the belly towards the legs. Flip skin/fur to the side (optional: and pin down). Abdomen is now free.
3. Use a new pair of scissors and forceps to open the abdominal wall. Excise a piece of tissue. Work from fat and intestine to lung and liver (liver bleeds most).
4. Liver must be cut into small pieces up to 5 mm × 5 mm (one piece per extraction is enough), intestine into 15-mm-long pieces, and lung, fat, and brain into 3–4 pieces of 5 mm × 5 mm each per sample. Intestine needs to be cleaned by carefully striking out the diet with forceps onto a clean glass plate. Excision of brain tissue is tricky and needs practice: have two bent forceps to get the whole brain out immediately after the skull is opened crosswise with sharp scissors. Work fast.
5. Always immerse tissue immediately in RNA*later* solution and throw cups into liquid nitrogen or shock-freeze in a dry ice/ethanol bath (the latter dissolves ink labels on tubes—make sure tubes are labeled on lid as well). Store at −70 °C till further use.
6. Work through **steps 1–5** above for each mouse, and then proceed to the next mouse.
7. Frozen tissue in RNA*later* is stable for several months. Thus, RNA extraction from these tissues can be done any time after excision and storage of tissues.

3.1.2 RNA Extraction from Tissue (TRIzol)

1. For RNA extraction from tissue, prepare a clean workspace where homogenization takes place.
2. Set up the Polytron/Ultra-Turrax, rinse the homogenizer with RNase*Zap* or 70 % cleaning alcohol, and wipe dry before each new sample.
3. Wearing clean gloves, pre-label the 12-ml tubes for homogenization (*see* **Note 9**), add 1 ml TRIzol, and keep tubes on ice. I recommend not isolating more than 8–12 samples at a time. If samples are still degraded, try isolating fewer samples at a time, working faster and make sure samples are kept clean and cold.
4. Wearing clean gloves, pre-label RNase-free 1.5-ml tubes for storage of the isolated RNA samples. For number of tubes/aliquots *see* Subheading 3.3, **step 9**.
5. Preheat a thermal block for microcentrifuge tubes to 55–60 °C.
6. Thaw tissues once everything is prepared for RNA extraction (*see* Subheading 2). Add 50–100 mg tissue per 1 ml TRIzol (not more than 10 % w/v per sample). Pick the pieces out of the tube and weigh them. If necessary, chop off excess pieces of tissue until the desired weight is reached.
7. Homogenize tissues in TRIzol with Polytron/Ultra-Turrax for 5–10 s. Place back on ice immediately. After another 20 s of chilling, homogenize for a second time. Rinse homogenizer in RNase-free water (dip in 50-ml Falcon tube with water, turn on briefly), and then rinse with 70 % alcohol. Wipe dry.
8. After all samples are homogenized in TRIzol, let samples sit at room temperature for 5 min.
9. Add 0.2 ml chloroform, mix by vigorous shaking, and let stand at room temperature for 2–3 min.
10. Centrifuge for 15 min at 4 °C at 12,000 × *g* (*see* **Note 10**).
11. Transfer colorless upper phase containing the RNA into a new, labeled centrifugation tube.
12. Precipitate RNA with 0.5 ml isopropanol. Let stand for 10 min at room temperature.
13. Centrifuge for 10 min at 4 °C at 12,000 × *g*.
14. The RNA pellet is a colorless, often invisible pellet on the bottom of tube, against the outer wall of the tube. Carefully discard supernatant, holding the tube with the pellet up so that the risk of discarding the pellet is minimal. Carefully wash pellet with 1 ml of 75 % EtOH, vortex, and centrifuge at max 7,500 × *g* for 5 min at 4 °C.
15. Briefly air-dry RNA pellet (5–10 min; place tube upside down on tissue, making sure pellet remains in the tube). Do not let the pellet completely dry out.

16. Dissolve pellet in RNase-free water, resuspend by pipetting, and transfer into 1.5 ml tube. For adipose tissue, intestine, and lung use 100 μl RNase-free water (up to 300 μl may be used). For liver, preferably 500 μl (but up to 1,000 μl) RNase-free water may be used.
17. Incubate at 55–60 °C for 10 min; all of the pellet should have dissolved. Put back on ice. Adding more water, reheating and vortexing may help to dissolve the pellet. However, if the sample doesn't dissolve (e.g., because it was too dry) it must be discarded.
18. Measure concentration at 260 and 280 nm according to Subheading 3.3 and proceed to RNA cleanup.

3.2 Total RNA Extraction from Cells and RNA Cleanup Using the Qiagen RNeasy Mini Kit

3.2.1 For Cleanup of Total RNA from Tissue After TRIzol Extraction

1. Adjust the volume of about 5–20 μg RNA (depending on the availability and volume of the isolated RNA) to 100 μl with RNase-free water (or use 100 μl of the original RNA extract, if total RNA content <100 μg/100 μl).
2. Add 350 μl of buffer RLT (containing β-mercaptoethanol), and mix well.
3. Add 250 μl of absolute ethanol and mix well by pipetting. Do not vortex. Proceed with **step 6** below (application to spin column).

3.2.2 Cell Culture, Adherent Cells

1. Label RNase-free 1.5-ml microcentrifuge tubes, have necessary aliquots of RLT buffer with β-mercaptoethanol prepared, and proceed to cell culture hood.
2. To harvest adherent cells, e.g., 3T3-L1 fat cells, HepG2 liver cells, or murine peripheral/abdominal macrophages, remove medium by aspiration and rinse cells to eliminate dead cells or red blood cells (hemoglobin may inhibit enzymatic reactions).
3. For one confluent Petri dish (100 mm) add 600 μl of RLT buffer containing β-mercaptoethanol, collect lysate with cell scraper, and transfer into a pre-labeled RNase-free tube. Genomic DNA makes the lysate viscous at this stage, so use a wide-bore pipette tip for transfer. Proceed back to the lab bench.
4. Homogenize the lysate by transferring 600–700 μl onto a Qiashredder and spin for 2 min at full speed at room temperature. Discard shredder column. If more homogenate is available, use a new Qiashredder and repeat step.
5. Add 600 μl of 70 % EtOH to the homogenate and mix by pipetting. *Do not vortex.*
6. Apply 700-μl aliquots of the homogenate (or tissue RNA after Trizol extraction from above) onto the pre-labeled spin column (RNA binding capacity is 100 μg), and spin at ≥8,000 × *g* for 15 s to bind RNA to membrane (*see* **Note 10**). Discard flow-through (or use new collection tube—check with your

current Qiagen RNeasy Mini Kit Handbook Manual for the amount of collection tubes provided). If more than 700 μl of homogenate are present, repeat loading of the same spin column.

7. Make sure all buffers are restored with the proper amount of ethanol before first use. Mix by inverting. Add 350 μl of buffer RW1, and spin for 15 s at ≥8,000 × *g*.
8. DNase treatment on column: Prepare a master mix containing 70 μl of RDD buffer and 10 μl of DNase stock solution per sample and mix gently. *Do not vortex.* Carefully add 80 μl of the master mix onto the middle of the membrane without touching membrane, and incubate at RT for 10–15 min. Do not incubate more than 15 min.
9. Add 350 μl of buffer RW1, spin for 15 s at ≥8,000 × *g*, and discard collection tube.
10. Wash with 500 μl of buffer RPE, spin for 15 s at ≥8,000 × *g*, discard flow-through, or use new collection tube. Add again 500 μl of buffer RPE, spin for 2 min at ≥8,000 × *g*, and discard flow-through again. Spin another minute at full speed to air-dry membrane.
11. Pre-label elution tubes (provided with kit). Discard collection tubes and place the spin column on a new, pre-labeled, 1.5-ml tube RNA elution tube.
12. Elute: Carefully add 30 μl of RNase-free water onto the middle of the membrane without touching it, incubate for 1 min, and spin for 1 min at ≥8,000 × *g* (*see* **Note 11**). Repeat this step. Discard the spin column and keep the tubes with the eluted RNA on wet ice.

3.3 Quantification and Storage of Total RNA

Eluted total RNA in RNase-free water can be quantified as such on a spectrophotometer or any instrument capable of UV spectroscopy. It is important to dilute the RNA in a neutral solution, preferably a buffer, or water. Wear gloves. The purity of the RNA is measured through the ratio of the absorption at 260–280 nm (A_{260}/A_{280}). This value should come close to 1.8 (optimally 2) to assure clean RNA. However, traces of protein impurity (e.g., DNase) will reduce this value (*see* **Note 12**). If necessary, dilute samples several times, such that the absorption at 260 nm falls between 0.3 and 1. Note that dilution factors may vary between tissues.

1. Label RNase-free tubes and add 98.0 μl of 10 mM Tris–HCl, pH 7.5 (or RNase-free water).
2. Add 2.0 μl of RNA, and mix. Put RNA tube back on ice.
3. Vortex and spin down.

4. Choose an RNA quantification program on your UV spectrophotometer, or measure each sample first at 260 nm, and then at 280 nm.
5. Blank the UV spectrophotometer with 100 μl of the Tris buffer (or the RNase-free water) in a 1 cc (1 ml) quartz cuvette.
6. Measure the absorbance of the samples, rinse in between with RNase-free water, and blow-dry with pressured air, if available. If necessary, repeat samples in different dilutions if absorbance at 260 nm (A_{260}) is out of range (less than 0.3 or greater than 1).
7. RNA concentration (μg/ml) = 44 μg/ml × dilution factor × absorption at 260 nm (A_{260}), where 44 μg/ml is the concentration for RNA with an $A_{260}=1$ at neutral pH in a 1-cm pathway cuvette. When diluting 2.0 μl into 100 μl of total volume, the dilution factor is 50.
8. Make sure the purity is acceptable ($A_{260}/A_{280}=\sim1.6\text{–}2.0$).
9. Store the purified RNA at −80 °C in aliquots of 2–6 μg for precious RNA, or in larger portions for abundant RNA samples (*see* **Note 13** for the amounts of RNA that will be needed).
10. RNA integrity is best checked by agarose gel electrophoresis. Use ~2–3 μg of RNA. One strong band for the 28S rRNA and one band for the 18S rRNA that is about half as intense as the 28S rRNA band, as shown in Fig. 2, shall be present (*see* **Note 14**). The 5S rRNA band is usually weak.

3.4 Reverse Transcription (RT)

Total RNA is reverse-transcribed according to the user manual (TaqMan® Reverse Transcription Reagents, ABI). Two micrograms of RNA are reverse-transcribed in a total volume of 100 μl [62.5 μl master mix, 37.5 μl sample (2 μg RNA adjusted to 37.5 μl with RNase-free water)], using random hexamers as primers. Wear gloves.

1. Program thermo cycler.
2. Label tubes. Include one tube for the RT$^-$ control (no reverse transcriptase included, *see* **Note 14**) and one extra tube for a second reverse transcription of one of the normal/wild-type samples used for the standard curve (*see* **Note 15**).
3. Thaw reagents and place on ice, except for the reverse transcriptase enzyme, which should be stored in the freezer (−20 °C) until use.
4. Calculate the volume of 2.0 μg RNA for each of your isolated RNA samples to be reverse-transcribed. Subtract this amount from 37.5 μl; this is the amount of RNase-free water you have to pipette into the corresponding tube for reverse transcription. (RNA will be added last.)

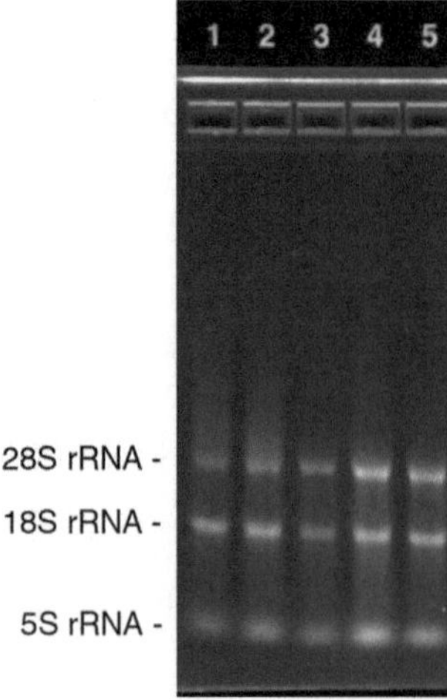

Fig. 2 Adipose tissue RNA after DNase cleanup. Approximately 3 μg of freshly isolated RNA samples were adjusted to 20 μl with RNase-free water and were subjected to agarose gel electrophoresis on a 1.2 % E-gel (Invitrogen)

5. Thaw RNA samples on ice.
6. Mix a master mix for all samples, plus one extra volume:

   Per 1× reaction volume (100 μl), combine:

   10 μl 10× buffer.

   22 μl $MgCl_2$.

   20 μl dNTPs.

   5 μl random hexamers.

   2 μl RNase inhibitor.

   Mix by vortexing.

   Per 1× reaction volume, add 3.5 μl reverse transcriptase.

   Mix carefully by pipetting. Avoid air bubbles or stressing the enzyme by mechanical forces (such as vortexing).
7. Pipette 62.5 μl of Master Mix to each sample tube already containing water. Add the 2.0 μg of the corresponding sample RNA into the mix. This should result in 100 μl/tube.
8. In the RT^- tube prepare one reaction volume without RT:

   10 μl 10× buffer.

   22 μl $MgCl_2$.

   20 μl dNTPs.

   5 μl random hexamers.

   2 μl RNase inhibitor.

   3.5 μl RNase-free water.

   Add 2 μg of the desired RNA for the RT^- control in a volume of 37.5 μl. Vortex.

9. Spin down tubes or flick tubes to ensure that no air bubbles accumulate at the bottom of the tube (they inhibit heat transfer). Enter the reaction volume and then start the reaction on the thermo cycler.
10. Freeze RNA aliquots at −70 °C.
11. When the RT is finished, freeze cDNA at −20 °C, or use immediately for real-time PCR (keep on wet ice).

3.5 Real-Time PCR (qPCR)

The use of the absolute quantification software module in order to do a relative quantification with standard curve is described (*see* **Note 16**). The term "standard curve" may be misleading, if one is thinking of an external standard in absolute numbers (e.g., exact copy numbers). Here, one of the samples, typically a wild-type/control sample, is used in a serial dilution to relate the RNA concentration (μg/ml) measured by UV spectroscopy with the C_t. However, we will not know the absolute copy number of a transcript (now cDNA) present in our sample.

This protocol describes a single PCR setup (*see* **Note 17**) with the detection of a housekeeping gene/endogenous control (*see* **Note 5**) for normalization to the "RNA load" in separate reactions. We also assume that the SDS7300 (ABI) with the corresponding 7300 SDS version 1.3 software is being used and that the user has access to Microsoft Office Excel. Any other qPCR machine and software can be used. The sample description and assay setup for a 96-well plate remain the same. An "assay setup run" to define the dilution range for the new assay/tissue combination is performed before running all samples (*see* **Note 18**). Fewer standards are necessary for known assay/tissue combinations with preestablished dilution ranges and thus the assay setup differs slightly (*see* **Note 19**). When handling fluorescent assays, keep in mind that energy excites fluorescence. Keep tubes dark (*see* **Note 6**) and cool. For the choice of your primers and TaqMan assay, *see* **Notes 3–5**.

3.5.1 Program Setup

1. Start the SDS7300 and under file/new choose "absolute quantification." If software was already open, restart it.
2. Choose the appropriate assay (called "detector" when using the ABI software). Make sure to choose the according reporter dye (FAM, VIC, etc.) used in your assay. You may want to define your assays first (new detectors).
3. Choose ROX as passive reference (included in buffer to normalize for pipetting errors).
4. Put in sample setup for the "assay setup run" (*see* **Note 18**): The SDS software analyzes data row-wise so it is easier to design the sample application row-wise, from A1–A12, B1–B12, etc. Each sample must be measured at least in duplicate, preferably in triplicate. For example, when testing the

setup for murine LDL-receptor primers (ldlr) and β-actin as housekeeping primers in a new tissue (*see* **Note 18**), a typical plate layout might be:

Wells that will contain the ldlr Master Mix

- A1–B3 standard concentrations 1–5, each in triplicates.
- B4 RT^- control.
- B5 NTC (non-template control = water).

Wells that will contain the β-actin Master Mix

- C1–D3 standard concentrations 1–5, each in triplicates.
- D4 RT^- control.
- D5 NTC.
- (For an assay that has already been established, use the layout described in **step 10**, below.)

5. Choose assay (detector) per well/rows: select wells, go to view/well inspector, and check according assay (e.g., ldlr). Repeat step for each assay.
6. Define sample status for each sample/row/selected wells in the well inspector (unknown, standard, NTC). It is OK to leave wells empty and still measure them (define as unknown or NTC and assign an assay). Defining all wells of the plate assures that samples are measured even if plate is mounted in wrong position, e.g., well H12 in position A1; in this case, detectors and sample status can be corrected later.
7. Enter the quantities for the standards in the well inspector (e.g., for A1–A3 enter "100" for the theoretical 100 ng mRNA per well, for A4–A6 enter "10" (*see* **Note 18**)).
8. Thermal program setting: On the Instrument Tab, check program data and correct if necessary [1 × 10 min 95 °C; 40× (15 s 95 °C, 1 min 60 °C)]; add a 1 × 10 °C forever step (sample will be used later for gel analysis when doing an assay setup for a new assay/tissue combination).
9. Save setup; Choose a self-explaining name, e.g., date, user, samples, and assay.
10. For setup of a preestablished assay/tissue combination, follow **steps 1–3**. The plate layout differs from that in **step 4** as fewer standards are necessary. A typical plate layout for a preestablished assay is outlined below, assuming that you test standards and samples each in triplicates (*see* **Note 19**).

    Wells that will contain the ldlr Master Mix

    - A1–A9 standard concentrations 1–3, each in triplicates.
    - A10–C3 samples 1–6, each in triplicates.

- C4 RT^-.
- C5 NTC.

Wells that will contain the β-actin Master Mix

- D1–D9 standard concentrations 1–3, each in triplicates.
- D10–F3 add samples 1–6, each in triplicates.
- F4 RT^-.
- F5 NTC.
- Follow **steps 5–9**, above, and then proceed to **step 11**, below.

3.5.2 cDNA Dilution

11. Dilute cDNA as needed for the standards and the samples (*see* **Notes 18** and **19**). Use siliconized 0.5 ml tubes. If dilutions are to be stored, thaw them only once more (*see* **Note 20**).
12. Use a total PCR reaction volume of at least 25 μl per well on the 96-well plate (*see* **Note 21**). A Master Mix with the TaqMan® Universal PCR Master Mix w/o UNG is prepared per assay ("detector," e.g., ldlr primers).

    For each 1× 25 μl reaction volume mix:

    12.5 μl 2× buffer.

    6.25 μl molecular biology grade water.

    1.25 μl 20× ABI TaqMan® Gene Expression Assays (e.g., ldlr or β-actin).

    Include one extra volume per every pipetted row (12 wells), plus extra volumes for the NTC and the RT^- controls.

    For example:

    - For an "assay setup run" (five standard concentrations in triplicate = 15 samples, plus RT^- and NTC controls) it is necessary to mix 20× volumes of each master mix for each assay (detector) master mix.
    - For an established assay of 12 experimental samples (six wild-type liver cDNA, six transgenic liver cDNA) and 3 standard curve dilutions, each in triplicates, calculate as follows: 12 × 3 samples = 36, 3 × 3 standard curve = 9, plus 1 RT^- and 1 NTC = 47. Add 4 extra volumes as you will pipette ~4 rows on the plate = 51× volumes for each assay (detector) Master Mix.
13. Place a labeled (barcoded) 96-well reaction plate onto the dark support. Add 20 μl of the appropriate Master Mix into the appropriate wells (on the side and close to but not touching the bottom of each well).
14. Add 5.0 μl of the according sample into the mix. Cover with optical film according to instructions, spin plate down briefly, or flick samples down.
15. Disconnect computer from network when using ABI.

16. On the Instrument Tab change volume to 25 μl and hit ENTER.
17. Save settings again.
18. Start PCR.
19. Freeze original cDNA samples at −20 °C after use.
20. After the qPCR is finished, you may immediately analyze your data or you may want to transfer the data to another PC where the software is installed (reestablish network if needed). If you want to run another qPCR, immediately restart the SDS software before beginning another qPCR run (only for ABI software). Cool/freeze the finished plate in the dark for analysis of bands on an agarose gel if not done immediately (*see* **Note 22**). You should check the gel *before* running precious samples with a new assay.

3.5.3 Analysis

The SDS software analyzes data according to the settings defined by the user. This is the advantage when doing the analysis "manually." You will define the settings, meaning the range of "baseline" (background fluorescence), and the wells to be analyzed together (usually you analyze per detector, as the PCR efficiency and linearity may differ between assays). Once the software has calculated all C_ts, we will export the raw data to, e.g., Excel, and proceed there.

21. In the data file, go to Results, amplification plot. Select the samples per assay/detector (e.g., choose all wells with the ldlr). Omit wells of assays that are not presently analyzed (select wells, in View/well inspector, and check omit).
22. Choose manual baseline. Set the baseline using the linear view of the results/amplification plot—change from "Delta Rn vs. Cycle" to "Rn vs. Cycle" on the upper right side of the window (*see* Fig. 1b). One can see the end of the baseline/background, where the fluorescence starts rising. Common settings for the baseline are between C_ts 3 and 15, e.g., 3–12, leaving enough space before fluorescence comes up. Leave at least three C_ts between end of baseline and start of exponential phase, to be seen in the "linear" view Rn vs. Cycle; *see* Fig. 1b. Otherwise, an undesirable "drop" in the baseline/background can occur; *see* Fig. 3. The minimum baseline setting is C_ts 3–6. Hit enter.
23. Go back to the "Delta Rn vs. Cycle" view. Set the threshold: move the threshold line to the middle of the "linearized" PCR range where all reactions of this assay seem to run parallel.
24. Hitting the analysis button, threshold line becomes green.
25. The intersection of the threshold line and the sample curve is called cycle at threshold (C_t); *see* Fig. 1b. It is an inverse measure of the mRNA concentration, with less cycles (lower C_t)

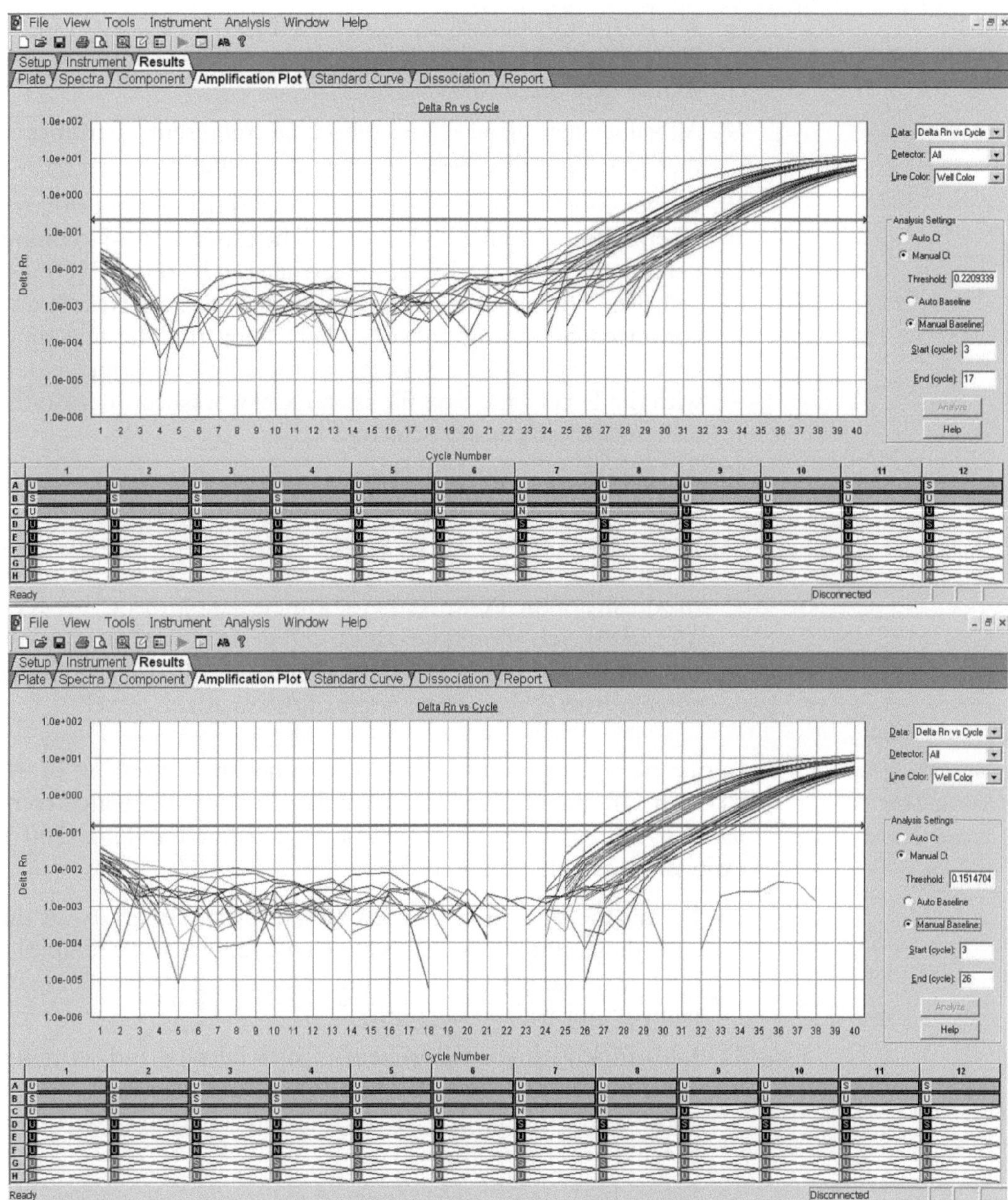

Fig. 3 *Upper panel*: Correct manual baseline setting. *Lower panel*: Incorrect baseline setting. Background was defined too close to exponential phase

corresponding to more mRNA transcripts present in the original sample.

26. Check the standard curve tab: a value of –3.33 for the slope indicates that between one log of sample concentration (ten-fold difference) are 3.33 cycles (or C_ts), which corresponds to a doubling of DNA per cycle (optimal reaction efficiency,

$2^{3.33} = 10$). Efficiencies are important for the $\Delta\Delta C_t$ method; *see* **Note 16**. The R^2 value is a measurement of the linearity, and should be close to 0.99.

27. All values appearing in the report tab represent quantities at that according threshold (reporter dye intensity).
28. In the report tab, export data. All samples that are not omitted when hitting the analysis button will be exported into a comma delimited .csv file. In Excel, the .csv file can be opened (choose "all files" in the OPEN window under "files of type:") and converted by hitting "next." When doing this the first time, check for appropriate settings, comparing the original data with the Excel data. Save file as an Excel worksheet. Repeat the analysis for each assay/detector.
29. At the end, copy all relevant sample data such as run specifications, sample names, well positions, quantities, and C_ts in one Excel master spreadsheet (*see* **Note 23**). Although C_ts won't be "used" in this protocol, they represent a control for the quality of the PCR. If you use sample data with a C_t between 35 and 40, you should be critical and decide if you want to use this value—are the curves representative, do the triplicates agree, is the C_t within the range of the standard curve?

3.5.4 Calculation of Relative Gene Expression Levels

In our example we were interested whether the expression of the LDLr in a transgenic liver was upregulated or downregulated compared to the wild-type liver (wild-type is defined as 1, or 100 % expression), *see* also **Note 24:**

30. Divide the quantities of each sample by the quantity of the according sample's housekeeping gene (e.g., qty ldlr sample1/qty β-actin sample1). The column containing these data can be named "normalized quantities."
31. Let's stick with the experiment of six wild-type and six transgenic mice: calculate the mean + standard deviation of the "normalized quantities" of each group. (Exclude assays that haven't worked properly.)
32. Divide both numbers by the *mean of the wild-type samples.* The mean of the wild-type samples will now become 1, as it was divided by itself. The mean of the transgenic livers will now have an x-fold higher or lower expression level compared to the control group (e.g., 0.2-fold means that 20 % of the wild-type level are present, meaning 80 % downregulation).
33. The standard deviations of the mean quantities must be divided by the mean of the control group as well, in order to give the appropriate standard deviations (**Note 25**).

MIQE guidelines [15] should be followed when publishing results of quantitative real-time PCR experiments.

4 Notes

1. Random hexamers are used as they guarantee equal reverse transcription of all mRNAs, while polyA can be used as primer only if the sequence to be amplified later is close to the 3′ end. However, since one is never sure which gene sequences have to be measured using the same cDNA further down the line, we recommend random hexamers.
2. ABI, as well as other companies, has designed primer and probe sets for virtually all human and mouse genes, and custom primers can be produced as well. The predesigned and specific "TaqMan® Gene Expression assays" are popular and are an excellent choice for the analysis of gene expression by RT-qPCR. *See* also **Note 4** for more details.
3. The choice of the primers/probe (also called qPCR primers, primer set) for a TaqMan® assay must be carefully chosen. In general, a region of 60–100 nts on the gene of interest (mRNA) is chosen, where two traditional PCR primers and one probe that binds in between are designed. If designing your own assay, get acquainted with the primer-design software (e.g., Primer Express, ABI). Exon–exon spanning probes are preferred for assays (*see* **Note 4**), as they exclude amplification of introns, e.g., from genomic DNA impurities. A gene with many isoforms or transcript sizes needs thorough examination of the sequences regulated and expressed in the relating tissue. The number of transcripts which a specific assay detects can be looked up for pre-developed assays from ABI—it is recommended to use an assay that detects the ref-sequence or the most transcripts known at present. In some cases, for example, ABCG1 which has many tissue-specific transcripts, it may be wise to use more than one assay, if real-time PCR is the technique of choice for mRNA quantification.
4. If you go to the ABI webpage and browse to the product guide, you can choose to search for gene expression assays (pre-developed) [16]. Enter only systematic names (low-density lipoprotein receptor), not common short names (LDLr). The extension "_m1" in ABI pre-developed gene expression assays refers to an exon–exon border-spanning assay, while "_g1" refers to detection of an exon–exon border which includes a short intron. The extension "_s1" refers to detection of genomic DNA. In the last two cases, DNase treatment of RNA is obligatory in order to not also detect genomic DNA. If you run your qPCR samples on a gel after using a new primer set/assay for the first time, and more than one PCR product is visible on the agarose gel, you need to increase the DNase treatment or use another assay, or make sure that there are no

other transcript sizes (do a Northern blot analysis). An assay will most reliably exclude amplification of introns when chosen towards the 5′ end of a sequence (long introns). If ordering assays from ABI, become acquainted with the ABI home page, call the helpline and have the search options explained well.

5. The actual RNA (or cDNA) "loaded" per well is determined by measuring an endogenous control or "housekeeping gene," which is constitutively expressed in a tissue. However, endogenous controls can have different properties in different tissues, so the endogenous control chosen for a given experiment must be appropriate. Housekeeping genes/endogenous controls for liver, macrophages, intestine, lung, adipose, and brain are as follows: (1) β-actin—note, however, that β-actin is neither recommended in adipose tissue nor in macrophages which have not been equally differentiated in vitro, due to different cell sizes/actin levels; (2) HPRT1—it is more highly expressed in brain, but is constitutively expressed in other tissues; and (3) 18S rRNA for the detection of very abundant genes of interest. The difference in cycles between the endogenous control and the gene of interest should not exceed 12 cycles in a sample; keep this in mind when measuring low-abundance genes, e.g., in liver, where 18S rRNA is very abundant. GAPDH is not a good choice, as it is involved in carbohydrate metabolism, which is linked to fat metabolism. Neither is cyclophilin, which may interact with the immune system. For testing of various endogenous controls, check available pre-developed assay sets (e.g., TaqMan® Express Endogenous Control Plate, for human samples, ABI [17]).

6. The TaqMan® Gene Expression Assays come as a 20× primer/probe mix that will give a final PCR concentration of 900 nM primer and 250 nM FAM-labeled probe. Aliquot into six portions after the first thawing. Use dark-colored tubes, optionally wrap in aluminum foil while used. Aliquots can be used for several (at least 3) years when properly stored at –20 °C, and can be thawed up to ten times.

7. UNG is an enzyme that digests specifically amplicons containing dUTP instead of dTTP. If you use dUTP for your qPCR, you can control for contamination of end products when using UNG in your PCR buffer. Keep in mind that negative controls (NTC, water) are still a must even when using UNG.

8. Mice are active at night; therefore, a metabolically active liver might be obtained at night. Consider whether it is necessary to fast mice before collecting tissue.

9. Brain tends to foam extremely. It is wise to use at least 10- to 12-ml tubes.

10. Conversion of $\times g$ (times relative centrifugal force; gravity) to rpm (revolutions per minute): rpm = sqrt $[(1.118 \times 10^5 \times g)/r]$, where r is the radius of the rotor used (in cm).
11. For centrifugation of the tubes, place elution tube with spin column in centrifuge, leave two positions spare, place next elution tube into position, and so forth. Turn tube lids to the side over the free positions, in order to support them during centrifugation. Clean rotor lid inside with 70 % ethanol; let dry. Close rotor lid, and spin.
12. For qPCR it will be acceptable if the ratio of absorption at 260–280 nm is lower, or if RNA is partially degraded (slight smear between 28S and 18S). The parallel measurement of the gene of interest and an endogenous control/housekeeping gene will control for RNA load during the qPCR. In general, after a few attempts at isolating RNA, handling becomes better and RNA cleaner (*see* Fig. 2). Fatty tissues like lung, fat, and intestine will easily be degraded when more than one tissue is taken from one animal at a time.
13. Two micrograms of RNA are used per reverse transcription reaction. Thawing more than three times is not recommended; discard RNA thereafter. RNA will be stable for at least 12 months at −80 °C under these conditions. For dilutions or small amounts (e.g., <1 μg/ml) of RNA or cDNA use 0.5 ml siliconized/low retention microcentrifuge tubes (Fisher Scientific) for storage. Do not thaw diluted samples more than once.
14. Contamination of RNA with genomic DNA can sometimes be seen on the agarose gel as a large band close to the origin. For a qPCR assay using probes that do not span exon–exon borders over large introns, DNase treatment is an essential step and must be carried out carefully. Furthermore, the use of a "minus reverse transcriptase" (RT^-) control reveals amplification of genomic DNA. (Note also that an unusually low efficiency of amplification in a sample containing all components, including RT and cDNA, might indicate unspecific competition by large quantities of genomic DNA.) Each set of RNA isolations should have an RT^- control included when doing the RT step.
15. For the standard curve, one defined cDNA sample will be used as a standard that will require larger quantities. It is recommended to add one extra RT reaction in order to reverse-transcribe this "normal" sample (e.g., a wild-type sample) twice and make sure one has enough material to work with.
16. Although we do a "relative" quantification of two or more conditions, we use the ABI "absolute" quantification module. This is preferred to the relative quantification module, as it is a less

automated procedure, helping to understand the calculation. Working with "absolute" quantities simplifies the calculations (relating concentrations to each other simply means dividing one by the other, and we lose the units, getting x-fold, relative expression levels). For high-throughput qPCR, the so-called $\Delta\Delta C_t$ method can be used for relative quantification [18] (appendix A). In this case, no quantities or standard curves are needed, only C_ts are used. However, it is a prerequisite of the $\Delta\Delta C_t$ method that the slopes of the standard curves of the assays that are to be compared are similar (e.g., –3.49 and –3.34): the slope represents the efficiency of the PCR, and similar slopes assure that target sequences are amplified equally. Therefore, assays have to be optimized carefully before using this method. In addition, an identical "calibrator" sample must be present on each plate. It serves to correct for fluctuations in fluorescent signal. Do not hesitate to call the ABI hotline in case of further questions; the according number can be found at http://www.appliedbiosystems.com/support/contact/index.cfm.

17. Single PCR refers to the common PCR approach of detecting one gene at a time in one well. For high-throughput PCRs, multiplex PCR is widely used, detecting two or more genes at a time (e.g., the gene of interest and the endogenous control in one PCR reaction, using different fluorescent markers, e.g., the reporter dye FAM for the gene of interest, and VIC for the endogenous control). However, multiplex PCR needs more optimization to overcome PCR efficiency competition and reach the same PCR efficiency. For routine high-throughput assays, multiplex optimization will save time and money in the long run.

18. First-time standard curve for a new assay/tissue combination: Usually, testing five logs of a sample dilution is recommended. This can be performed beforehand in a separate experiment (assay setup). Use the wild-type cDNA that was reverse-transcribed twice for this purpose (*see* **Note 15**), and dilute 1:10 four times in a serial dilution; this would comprise a theoretical RNA (cDNA) load of 100 ng (undiluted cDNA), 10, 1, 0.1, and 0.01 ng of cDNA per 25 μl reaction, when 5 μl sample ("template") is used. Enter these quantities according to sample position; do not add the units. We relate quantities and provide "fold changes" compared to a control level, which is arbitrarily set one by the user (e.g., mean of wild-type samples = 1). Importantly, when analyzing the data of a new assay/tissue combination, keep in mind that the gene of interest and the endogenous control must have their linear range within a maximum of 12 C_ts of range. If necessary, a greater difference in C_t range can be overcome by

using two different dilutions of the samples (one dilution for the gene of interest, one, e.g., for 18S rRNA), however, bearing a certain risk of inaccuracy.

19. Standard curve for a known assay/tissue combination: triplicates of three concentrations of one cDNA (*see* **Note 15**). The concentrations are chosen within the linear range of the standard curve tested above (**Note 18**). For example, a 1:3, 1:30, and 1:300 dilution of the cDNA (theoretically 33.5, 3.35, and 0.335 ng cDNA/reaction) can work for this purpose. Samples are measured in the middle concentration, e.g., 1:30 diluted.
20. Serial dilution of five standard concentrations with excess volume to freeze: wild-type sample x was reverse-transcribed twice, to be used for the standard curve (*see* **Note 15**). Label four 0.5 ml siliconized tubes from 1 to 4. Tube 1: Dilute 10 μl of cDNA sample x with 90 μl of RNase-free water, vortex, spin, and place on ice. Tube 2: Take 10 μl of this first dilution, and dilute further with 90 μl of RNase-free water; vortex, spin, and place on ice. Tube 3: Take 10 μl of the previous dilution and dilute further with 90 μl of RNase-free water, vortex, spin, and place on ice. Tube 4: Take 10 μl of the previous dilution and dilute further with 90 μl of RNase-free water, vortex, spin, and place on ice. When using the original cDNA sample x and the four 1:10 dilutions you have now generated a set of five serial dilutions, each will have 90 μl available, except for the last (100 μl volume). When using 3 × 5 μl per concentration for the ldlr standard curve, and 3 × 5 μl for the actin standard curve, you can still freeze at least ~60 μl for another testing.
21. Scaling up PCR volumes is usually not a problem except for the question of price. Downscaling to a volume below 25 μl starts generating errors, and standard deviations between wells with the same samples increase, according to my experience. Issues such as evaporation come into play, for example when using a 96-well plate with 340-μl well volume. When using a cycler with, for example a 100-μl well plate, small volumes can be used and accuracy may still be OK. Some studies claim that accuracy is lost as the price for the time won on the newer fast PCR blocks. For gene expression studies we usually talk about a few qPCRs to be run, so winning several minutes per qPCR run is not crucial, and I would opt for accuracy.
22. Run a 2 % gel when using a new assay/tissue combination. Pick some representative sample of each group as well as the RT^- control and run alongside a 100 bp ladder. Make sure only one band appears (typically around 60–100 nts; more information comes with each assay). PCR products should not be in open containers in the same room as the PCR setup

and run take place (good laboratory practice). Carefully cut open the position of the according wells using a scalpel (cut the adhesive cover crosswise), aspirate 15 µl, and subject to gel electrophoresis. Wrap plate in plastic bag and keep cool till no longer used. If more than one band appears, genomic DNA was present (extend DNase digestion to 15 min), or another transcript size is present under the experimental conditions (confirm by Northern blot as in Chapter 4 of this volume [19]). If none of these cases are true, and you only used one set of primers per master mix, the assay was not specific.

23. Making a master spreadsheet with relative quantities enables you to add data from other plates that use a standard curve, and the same endogenous control/housekeeping gene, without the necessity for a calibrator [a sample aliquot that has to be run on each plate (in triplicates) to correct for changes in fluorescence]. A sample C_t from one plate cannot be directly compared with the C_t of the same sample on another plate. The quantity of a sample, however, when normalized to the same housekeeping gene/endogenous control, can be compared to the quantity of this sample on another plate.

24. The ultimate goal is to quantitate changes in protein levels and even better, protein activity. Westerns require optimization, depend on the antibody, and are not high throughput. Sometimes a good antibody is just not available. Shotgun proteomics is high throughput but not necessarily quantitative. 2D DIGE is laborious. It is fair to state that most proteins do not have activity assays available, much less high-throughput activity assays. While it can't be assumed that the fold change found by RT-qPCR is going to translate to that exact fold-change in protein level or activity—it is a start and is usually an informative and productive line of research to follow. The researcher will inform himself/herself about the tissue-specific regulation of the genes of interest.

    Several proteins are regulated post-transcriptionally through de/phosphorylation or binding of factors, while the RNA moiety remains constant [e.g., the enzyme acetyl-coenzyme A carboxylase (ACC)].

25. For repeated use of certain assays/PCRs, one may want to get acquainted with the relative quantification software package (ABI) for a more automated analysis tool.

References

1. Gotto A, Pownall H (2003) Manual of lipid disorders: reducing the risk for coronary heart disease. Lippincott, Williams and Wilkins, Philadelphia, PA
2. Alonzi T, Mancone C, Amicone L, Tripodi M (2008) Elucidation of lipoprotein particles structure by proteomic analysis. Expert Rev Proteomics 5:91–104
3. Trevaskis NL, Charman WN, Porter CJ (2008) Lipid-based delivery systems and intestinal lymphatic drug transport: a mechanistic update. Adv Drug Deliv Rev 60:702–716
4. Wasan KM, Brocks DR, Lee SD, Sachs-Barrable K, Thornton SJ (2008) Impact of lipoproteins on the biological activity and disposition of hydrophobic drugs: implications for drug discovery. Nat Rev Drug Discov 7:84–99
5. Danik M, Champagne D, Petit-Turcotte C, Beffert U, Poirier J (1999) Brain lipoprotein metabolism and its relation to neurodegenerative disease. Crit Rev Neurobiol 13:357–407
6. Valasek MA, Repa JJ (2005) The power of real-time PCR. Adv Physiol Educ 29:151–159
7. Bustin SA (2000) Absolute quantification of mRNA using real-time reverse transcription polymerase chain reaction assays. J Mol Endocrinol 25:169–193
8. Nolan T, Hands RE, Bustin SA (2006) Quantification of mRNA using real-time RT-PCR. Nat Protoc 1:1559–1582
9. Veres G, Gibbs RA, Scherer SE, Caskey CT (1987) The molecular basis of the sparse fur mouse mutation. Science 237:415–417
10. Wong ML, Medrano JF (2005) Real-time PCR for mRNA quantitation. Biotechniques 39:75–85
11. Vaisman BL (2010) Genotyping of transgenic animals by real-time quantitative PCR with TaqMan probes. In: Freeman LA (ed.) Methods in molecular biology: lipoproteins, 2nd edn (Walker JM, series ed.). Humana, Totowa, NJ
12. Real-time PCR: Understanding Ct (Application Note, 2011), Applied Biosystems™ by *life* technologies (2009) http://www3.appliedbiosystems.com/cms/groups/mcb_marketing/documents/generaldocuments/cms_053906.pdf
13. Real-time PCR vs. traditional PCR (Tutorial), Applied Biosystems (2009) http://www3.appliedbiosystems.com/cms/groups/mcb_marketing/documents/generaldocuments/cms_042484.pdf
14. Essentials of real time PCR (General Document), Applied Biosystems (2009) http://www3.appliedbiosystems.com/cms/groups/mcb_marketing/documents/generaldocuments/cms_039996.pdf
15. Bustin SA, Benes V, Garson JA, Hellemans J, Huggett J, Kubista M, Mueller R, Nolan T, Pfaffl MW, Shipley GL, Vandesompele J, Wittwer CT (2009) The MIQE guidelines: minimum information for publication of quantitative real-time PCR experiments. Clin Chem 55:611–622
16. TaqMan® gene expression assay selection guide (2009) http://www.appliedbiosystems.com/absite/us/en/home/applications-technologies/real-time-pcr/taqman-probe-based-gene-expression-analysis/taqman-gene-expression-assay-selection-guide.html
17. Using TaqMan endogenous control assays to select an endogenous control for experimental studies (Application Note), *life* technologies (2009) http://www3.appliedbiosystems.com/cms/groups/mcb_marketing/documents/generaldocuments/cms_042279.pdf
18. Applied biosystems real-time PCR systems (Reagent Guide, 2010), Applied Biosystems (2008) http://www3.appliedbiosystems.com/cms/groups/mcb_support/documents/generaldocuments/cms_052263.pdf
19. Freeman LA (2010) Northern analysis. In: Freeman LA (ed.) Methods in molecular biology: lipoproteins, 2nd edn (Walker JM, series ed.). Humana, Totowa, NJ

Chapter 3

Microarray Technology: Basic Methodology and Application in Clinical Research for Biomarker Discovery in Vascular Diseases

Nalini Raghavachari

Abstract

Microarray technology is a novel tool in molecular biology, capable of quantitating hundreds or thousands of gene transcripts from a given cell or tissue sample simultaneously. A microarray has thousands of DNA fragments or oligonucleotides of known sequence arrayed in a known sequence of rows and columns on a chip. Hybridization of sample RNA that has been reverse-transcribed and labeled enables the detection and quantitation of specific transcripts. The ability to quantitate systemic gene changes in normal vs. diseased states has led to significant progress in many biomedical disciplines, including lipoprotein and atherosclerosis research, and can be used for discovery of diagnostic/prognostic and predictive biomarkers and to test the effectiveness of potential therapeutic agents. The design and analysis of microarray experiments present some unique problems to clinical medicine due to inherent issues related to biological sample procurement and processing, sensitivity and specificity of the assay, reliability and reproducibility of data, and applicability of the technology in multicenter-based clinical studies. This chapter will provide details on the methodologies that address these problems for successful microarray-based transcriptome analysis of tissues, whole blood, cell subpopulations, and cultured cells.

Key words Microarrays, Amplification, Globins, Whole blood, PBMC, Biomarkers, Gene chips, Affymetrix, Hybridization, Gene expression

1 Introduction

Discovery of transcriptional biomarkers represents a promising strategy in the field of translational medicine for early disease detection, for the development of personalized therapy for complex diseases, and for the definition of disease-specific signaling pathways [1]. Microarray-based transcriptome analysis quantitates thousands of gene transcripts simultaneously from a biological sample and is thus a frontier technology for the identification of potential biomarkers specific for the phenotypes under investigation.

Microarrays contain thousands of microscopic spots of DNA fragments (usually oligonucleotides) of known sequence individually

Lita A. Freeman (ed.), *Lipoproteins and Cardiovascular Disease: Methods and Protocols*, Methods in Molecular Biology, vol. 1027, DOI 10.1007/978-1-60327-369-5_3, © Springer Science+Business Media, LLC 2013

spotted onto a chip. RNA isolated from the tissue or cell type of interest is converted to labeled cDNA or cRNA and then hybridized to the chip. If a particular transcript present in the original RNA preparation contains a sequence that matches the sequence of a fragment/oligo on the chip, the labeled cDNA or cRNA will hybridize to that fragment or oligo. The position of the label on the chip identifies the transcript sequence and the intensity is proportional to transcript abundance. Thus, microarray analysis determines whether a particular transcript is present and also provides a quantitative readout for specific transcripts. Most importantly, since thousands of DNA fragments have been spotted onto the chip, thousands of transcripts can be analyzed in parallel. This technology is useful for finding differences in transcript levels under different experimental conditions or disease states.

Microarray analysis of RNA from animal models for dyslipidemia, atherosclerosis, and vascular disease, as well as cell culture microarray studies, has been particularly fruitful in identifying key genes and pathways involved in lipoprotein metabolism, atherosclerosis, and vascular disease. The ability to study gene expression in tissues such as liver, intestine, and aorta in normal vs. disease states or after nutritional or pharmaceutical intervention is a key advantage to using model organisms in the search for transcriptional biomarkers. Thus, microarray studies in animal models will continue to be attractive in basic lipoprotein/vascular disease research. Despite many similarities, however, transcriptional responses of mice, rats, and even monkeys cannot be assumed to perfectly mimic those of humans, and ultimately microarray technology must be applied to human samples in clinical trials.

Microarray analysis of gene expression from human samples presents special challenges. Although target disease tissues/cells such as biopsy materials from fine needle aspirates (FNA), cell subpopulations, or enriched isolates from laser capture microdissection (LCM) are ideal for such analyses, inherent problems are associated with the procurement of such tissue biopsies/cells and the low amount of RNA that can be isolated from these specimens for use in standard microarray assays. These challenges are exacerbated in large clinical trials.

Whole blood, cell subpopulations, and primary cell lines are practical and attractive surrogate tissues in clinical research [2–5]. Sample collection procedures that stabilize specimens and cell lysates for extended time periods prior to RNA isolation, as well as improved RNA isolation methodologies, have been developed, markedly improving RNA integrity and subsequent microarray analysis from clinical samples. Global transcriptome analysis of whole blood using microarrays has traditionally been difficult due to the high abundance of globin transcripts in the blood. Recent advances have made it possible to overcome these problems and perform microarray-based analysis of these biological specimens [6], which offer several advantages of blood to tissue biopsies/cells in measuring disease state and drug response.

This chapter will describe in detail methods that have been successfully applied to clinical samples for microarray analysis on the Affymetrix platform. An overview of the procedure is shown in Fig. 1. The procedures described are robust and can be applied to cells and tissues of model organisms as well.

2 Materials

All of the reagents and equipment for hybridization are from Affymetrix unless otherwise noted.

1. RNA*later* (Ambion, Austin, TX) or Allprotect Tissue Reagent (Qiagen, Valencia, CA) (*see* **Note 1**).
2. Liquid nitrogen.
3. PAXgene™ tubes (Qiagen, Valencia, CA).
4. CPT tubes (BD, Franklin Lakes, NJ).
5. Tabletop centrifuge for collecting cells (Beckman, Allegra X-15 and Eppendorf 1541-R).
6. Erythrocyte lysis buffer (Qiagen).
7. 1× PBS, chilled (4 °C or ice-cold).
8. Cell counter, for example Cell Dyn (Abbott Laboratories, Abbott Park, IL).
9. Qiashredder (purple shredder column) (Qiagen).
10. RNeasy Mini Kit (Qiagen) (*see* **Notes 2** and **3**).
11. RNase-free water (supplied with RNeasy Mini Kit).
12. Buffer RLT (lysis buffer)* (supplied with RNeasy Mini Kit). Contains guanidine isothiocyanate—use appropriate safety measures when handling.
13. Buffer RW1 (wash buffer)* (supplied with RNeasy Mini Kit). Contains guanidine isothiocyanate and alcohol—use appropriate safety measures when handling.
14. Buffer RPE (wash buffer) (supplied with RNeasy Mini Kit as a concentrate). Before using for the first time, add ethanol (90–100 %) as indicated on the bottle to obtain a working solution.
15. RNeasy mini columns (supplied with RNeasy Mini Kit).
16. Collection tubes for elution (supplied with RNeasy Mini Kit).
17. 14.3 M β-mercaptoethanol (β-ME).
18. Ethanol (96–100 %).
19. Ethanol (70 % in water).

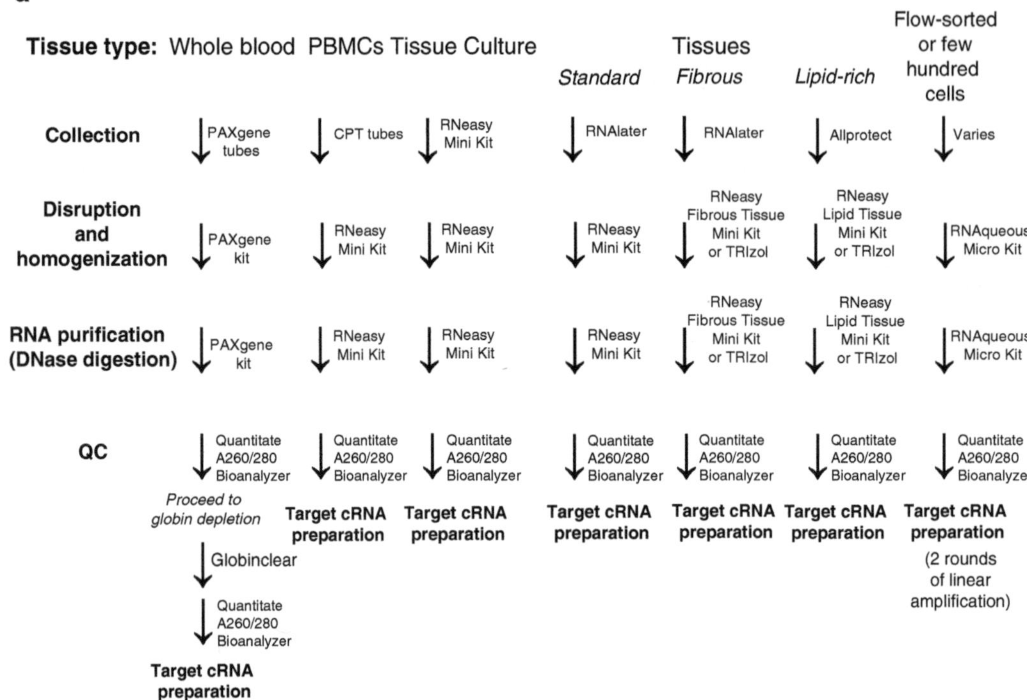

Fig. 1 Overview. (**a**) RNA isolation. Recommended RNA isolation procedures for whole blood, peripheral blood mononuclear cells (PBMCs), tissue culture cells, tissues, and flow-sorted cells, or other samples with only a few hundred cells. For methodological details, *see* Subheading 3. DNase digestion during RNA purification is mandatory in all cases. The RNeasy Fibrous Tissue kit can be used only for fibrous tissues that do not contain high levels of RNase. *Importantly, the RNeasy Fibrous Tissue kit must not be used to isolate RNA from intestine or pancreas*, or any other tissue rich in RNases. Also note that globin transcripts must be depleted from RNA purified from whole blood before proceeding to first-strand RNA synthesis and that RNA purified from flow-sorted or a few hundred cells is not abundant and must undergo two rounds of linear amplification prior to second-strand cDNA synthesis. (**b**) Biotin-labeled cRNA preparation, hybridization to probe array, staining, and scanning. Purified, quality-controlled RNA (*see* **c**) is converted first to cDNA and then to double-stranded cDNA (ds cDNA). A poly-A RNA is added to the cDNA synthesis reaction as a labeling control (not shown) (*see* Subheading 3.3.1). After cleanup, the ds cDNA is used as a template to synthesize biotin-labeled cRNA, which is then fragmented. Fragmented, biotinylated cRNAs [along with biotinylated hybridization controls (not shown); *see* **Note 4**] are placed into a hybridization cocktail, which is then added to a prehybridized probe array. Any biotin-labeled RNA fragment complementary to an oligonucleotide on the probe array will bind to that oligonucleotide. After washing away excess or unbound biotin-labeled RNA fragments, biotin-labeled RNA fragments that bound to oligonucleotides on the probe array are visualized by staining with a biotin-binding reagent (streptavidin) labeled with R-phycoerythrin and then scanning the microarray. Since the sequence of each oligonucleotide at every position on the microarray is already known, the presence of a signal at a given position identifies the sequence of the RNA fragment bound at that position. This information, along with the strength of the signal at each spot, is used to identify and quantitate the transcripts present in the RNA sample used for hybridization. (**c**) Quality control. *Step 1.* Assessment of RNA integrity and purity is an important first step in many gene expression analysis applications. It is preferable to use high-quality, intact RNA as starting material for microarray analyses. Initial QC involves determining the concentration and quality of the RNA samples by measuring the optical density (OD) at 260 and 280 nm (OD 260/280, or A_{260}/A_{280}) in a spectrophotometer. For RNA concentration, an OD_{260} of 1 corresponds to 0.040 μg/μl. For RNA purity, an A_{260}/A_{280} ratio between 1.8 and 2.1 is an acceptable ratio for downstream processing of samples. In addition, RNA integrity number (RIN) should also be determined on the Bioanalyzer. The RIN is a software tool designed to help scientists

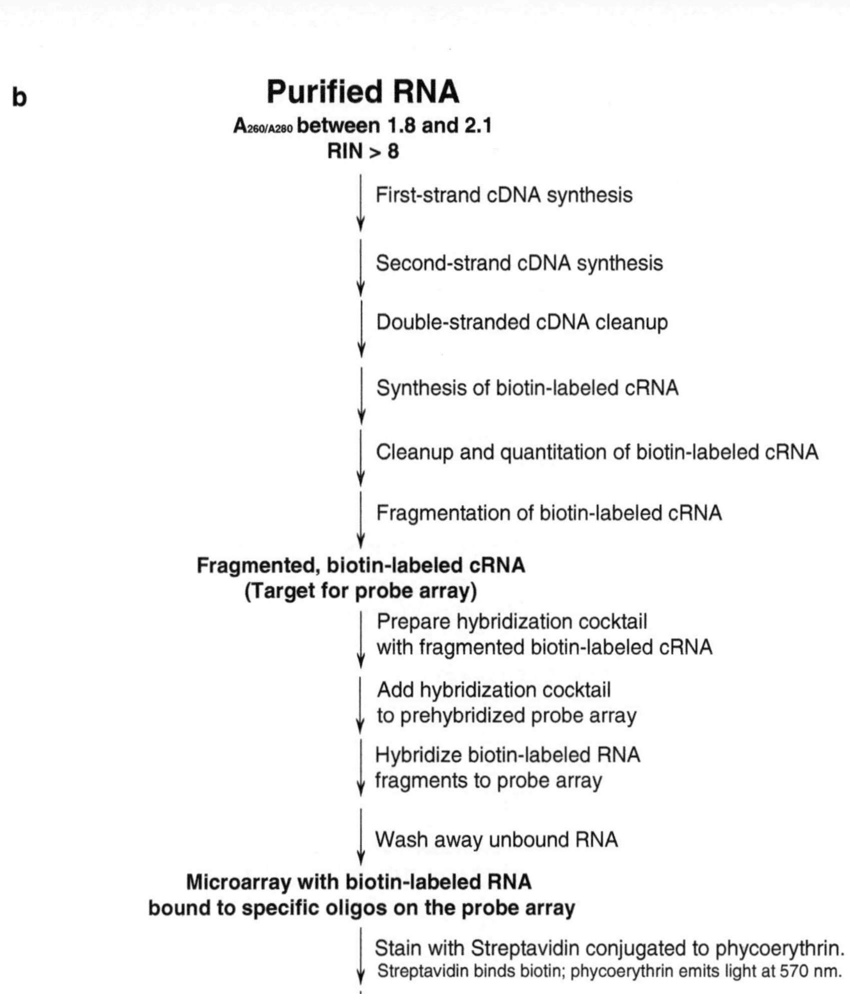

Fig. 1 (continued) estimate the integrity of total RNA samples. Using this tool, sample integrity is no longer determined by the ratio of the ribosomal bands, but by the entire electrophoretic trace of the RNA sample. This includes the presence or absence of degradation products. RIN above 8 denotes high-quality RNA. *Step 2*. In this step, the QC metrics of cRNA yield, OD 260/280 ratios are applied to the target cRNA prepared from total RNA. The Bioanalyzer analysis determines the size distribution of the cRNA rather than the RIN in this step. The median length of the cRNA before fragmentation should be 400–600 nt. The fragmented cRNA should show a size distribution less than 200 bp. *Step 3*. In this step, images obtained from the scanned chips will be inspected for chip uniformity and proper grid alignment. QC data from the report file with values on scale factor, background, present calls, 3′/5′ actin, and GAPDH ratios will be assessed to see if they fall within the acceptable range of Scale Factor (which is a measure of how the mean signal intensities vary between chips) SF < 5, BKG < 100, present calls (depending on the tissue type but >20 %), and 3′5′ ratios for actin, GAPDH < 3.0. Each chip—all samples and all controls—must meet stringent QC requirements before analyzing data from each chip, averaging data from replicate chips, and comparing experimental data to control data

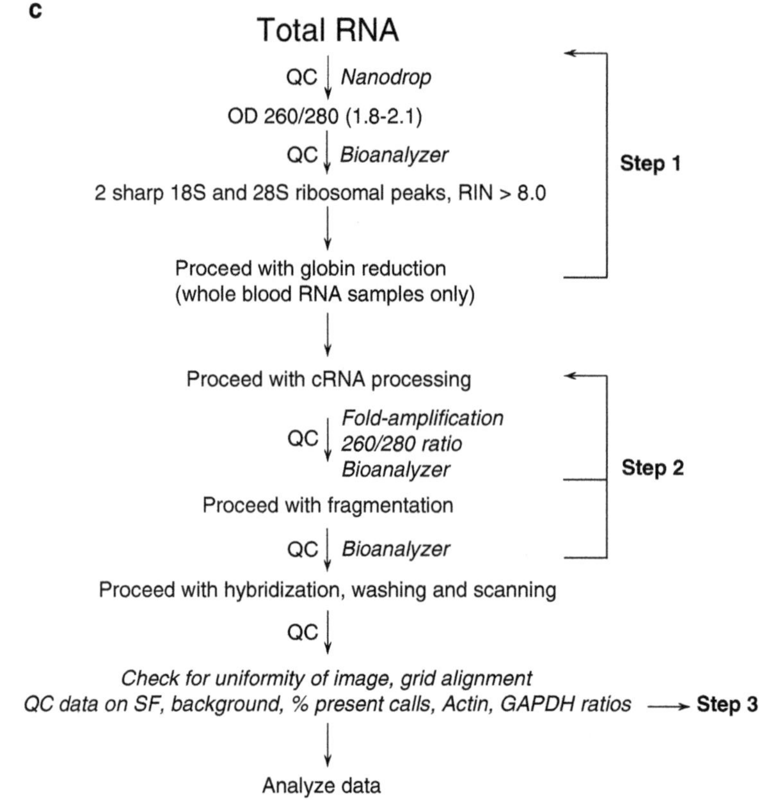

Fig. 1 (continued)

20. RNase-Free DNase (Qiagen).

    Dissolve solid DNase (1,500 Kunitz units) in 550 μl RNase-free water. Mix by inverting; do not vortex.

    Aliquot into 10-μl amounts and store at −20 °C for up to 9 months.

    Thawed aliquots can be stored at 2–8 °C for 6 weeks.

21. PAXgene™ Blood RNA kit (Qiagen).
22. GLOBINclear kit (Ambion). Make sure the correct species is chosen—mouse vs. human.
23. Agilent 2100 Bioanalyzer (Agilent Technologies, Palo Alto, CA).
24. Nanodrop ND-1000 Spectrophotometer (Thermo Scientific, Chicago, IL).
25. IVT Express Kit (Affymetrix, Santa Clara, CA).
26. RNAqueous-Micro Kit (Ambion) for RNA isolation from microscale samples. Kit includes the following:
    - One Micro Filter Cartridge Assembly.
    - 15 ml Wash Solution 1™ Concentrate.

      Add 10.5 ml 100 % ethanol before use!

- Wash Solution 2/3™ Concentrate.

  Add 22.4 ml 100 % ethanol before use!
- Lysis solution, containing guanidine thiocyanate. Store at 4 °C.
- DNase (2 U/μl). Store at –20 °C.
- 10× DNase I Buffer. Store at –20 °C.
- DNase Inactivation Reagent. Store at –20 °C.
- Elution Solution™.

27. NuGEN Ovation™ Biotin RNA Amplification and Labeling System (NuGEN, CA).
28. 5× Fragmentation Buffer (Affymetrix).
29. GeneChip Eukaryotic Hybridization Control (EHC) Kit (*see* **Note 4**). Completely resuspend the cRNA frozen stock by heating to 65 °C for 5 min before aliquotting.
30. Herring Sperm DNA at a concentration of 10 mg/ml (Promega Corporation, Madison, WI).
31. BSA solution (50 mg/ml) (Invitrogen, Carlsbad, CA).
32. 12× MES stock buffer: 1.22 M MES, 0.89 M [Na+], pH 6.5–6.7.

For 250 ml
MES free acid monohydrate 17.6 g
MES sodium salt 48.3 g
Add water to a final volume of 250 ml
pH MUST be between 6.5 and 6.7
Filter through a 0.2 μm filter
Keep at 4 °C. Shield from light

33. 2× Hybridization buffer: 200 mM MES, 2 M [Na+], 40 mM EDTA, 0.02 % Tween-20.

For 50 ml	
12× MES stock buffer	8.3 ml
5 M NaCl	17.7 ml
0.5 M EDTA	4 ml
10 % Tween-20	100 μl
Water	19.9 ml
Keep at 4 °C. Shield from light	

34. 1× Hybridization buffer: Combine equal volumes of RNase-free water and 2× hybridization buffer.
35. Oligonucleotide microarrays.
36. Hybridization Oven 640.
37. Affymetrix® Microarray Suite 5.0 (MAS 5.0) or GeneChip Operating Software (GCOS) on PC-compatible workstation.
38. Fluidics Station 400 or 450/250.
39. Wash Buffer A (non-stringent wash buffer) (1 L): 6× SSPE, 0.01 % Tween-20.
40. Wash Buffer B (stringent wash buffer) (1 L): 100 mM MES, 0.1 M [Na+], 0.01 % Tween-20.

For 1.0 l	
12× MES stock buffer	83.3 ml
5 M NaCl	5.2 ml
10 % Tween-20	1 ml
Water	910.5 ml
Filter through a 0.2 μm filter	
Store at 2–8 °C. Shield from light	

41. 2× Stain buffer: (final 1× concentration: 100 mM MES, 1 M [Na+], 0.05 % Tween-20).

For 250 ml
41.7 ml of 12× MES stock buffer
92.5 ml of 5 M NaCl
2.5 ml of 10 % Tween-20
113.3 ml of water
Filter through a 0.2 μm filter
Store at 2–8 °C and shield from light

42. Goat IgG Stock: resuspend 50 mg of Goat IgG (Sigma) in 5 ml of 150 mM NaCl. Store at 4 °C.
43. Streptavidin R-Phycoerythrin, 1 mg/ml (Invitrogen).
44. Anti-streptavidin antibody (goat), biotinylated (Vector Laboratories).
45. Affymetrix GeneChip® Scanner 3000 or Agilent GeneArray® Scanner.
46. Cell Dyn or equivalent cell counter.

3 Methods

3.1 Study Design

The design of microarray experiments depends on multiple factors, but should be mainly dictated by the biological question in mind. The five main factors to consider before setting up an array experiment are:

1. What biological questions will this study answer?
2. What specimens should be collected for this study?
3. How to get enough RNA from these biological specimens for cRNA target preparation?
4. What kind of arrays should be used for the study?
5. How many biological replicates are needed to obtain statistically significant data?
6. What quality control measures will be included?

For example, if you are interested in inflammation in humans, it is important to define the question carefully and design the study appropriately. For a clinical study that aims to determine transcriptional biomarkers for inflammation starting with blood, which is easily obtained, circulating mononuclear cells or T cells or white blood cells would be an ideal specimen for analysis. Whole blood might be another option, if abundant transcripts that interfere with microarray analysis (for example, globin) are removed. If instead the biological question is to investigate the effects of inflammation on transcriptional biomarkers in a particular cell type, for example endothelial cells, then isolation of endothelial cells would be advantageous, despite the greater difficulty in sample isolation. If such purified cell preparations are used for transcriptome analysis, only a limited amount of RNA would be available, and amplification of RNA would be required for microarray analysis. Some biological questions can be readily addressed using RNA isolated from an abundant source of human cells or tissues, e.g., adipose tissue, in which case RNA amplification would not be necessary.

Often a certain question, such as those requiring removal of liver or aortic tissue in healthy vs. diseased states, or before or after knockdown or knockout of a novel gene, can only be addressed in animal models, and experiments must be carefully designed to maintain as much physiological relevance as possible to humans.

What kind of arrays should be used for the studies? Several different types of arrays are available, including 3′ IVT expression arrays, whole-transcript gene expression arrays such as Gene 1.0 ST and Exon ST arrays, and microRNA expression arrays. Recently Affymetrix has many different whole genome and focus arrays available. In particular, a number of whole genome expression arrays are available for multiple organisms, including human, mouse, rat, canine, bovine, and chicken. In addition, ChIP-on-chip arrays and transcript mapping arrays that do not exactly measure the transcript levels but certainly do shed light on transcription can also be applied to study samples. Since the probe sets on 3′ IVT arrays are biased towards the 3′ end, it is recommended to use the whole-transcript gene or exon arrays. The major advantage of using exon arrays is in the analysis of alternatively spliced exons during disease processes.

With the recent development in the regulatory roles of micro RNAs and long noncoding RNAs, newer arrays such as Gene 2.0 ST arrays and miRNA version 3.0 arrays can be applied to studies where regulatory mechanisms need to be explored with respect to intergenic noncoding RNAs and microRNAs.

Since microarray experiments are biological in nature, it is important to have real—or biological—replicates in the designed study and to include well-matched controls. A biological replicate is an independent sample that varies all the variables that a colleague in another lab would not be able to control. In an imaginary "ideal" experiment, each replicate would be performed in a different lab—and so it may be advisable to approximate that situation as much as possible. While certain variables will not have an effect, one should not underestimate the sensitivity of microarrays. Variables such as batch of cells or time of day can very well have an observable effect that may become evident after performing replicates. Only observations that are general and reproducible should be reported. It is also critically important to prepare a perfectly matched control for every sample. For an exploratory analysis, three biological replicates of each sample and its perfectly matched control (for example mouse siblings/littermates with identical genotypes except for the presence of a transgene or knockout of a gene, or identical tissue culture samples split from the same flask of cells, treated identically except for the one controlled variable) are usually sufficient. More replicates are needed if the data are particularly noisy (e.g., samples from very small numbers of cells or samples from human subjects) or the expected effect is particularly small (e.g., changes occur only in very few, specialized cells in the sample). The greater the number of both samples and controls, and the better the match between the control and experimental groups, the more likely that the results will be true.

The study design should also include quality control measures applied to each of the steps involved in microarray analysis. These steps are sample procurement and handling RNA isolation, microarray sample processing, sample hybridization, microarray scanning, and data analysis (Fig. 1). The performance of some of these steps should be monitored with either specific control reagents or metrics (Fig. 1). *Only RNA that meets the highest standards of purity and integrity should be used for microarray analysis.*

3.2 Sample Collection, Cell Lysis, and RNA Isolation, with Maintenance of RNA Integrity

Many RNA purification methods have been developed but not all yield RNA that is suitable for microarray analysis. Moreover, specific types of biological specimens may require specific treatments and purification protocols to yield top-quality RNA. In this section we describe several tried-and-true methods for specimen stabilization and RNA purification from different types of samples.

RNA is very susceptible to degradation but must remain intact for microarray analysis. Thus, any samples that are not immediately processed for RNA extraction must be treated immediately with a reagent that permeates the tissue or cells and protects the RNA from degradation. Stabilization of specimens is essential in large multicenter clinical trials, when samples are collected but can't be processed immediately for RNA isolation, and is equally important for any sample that cannot instantly be used to purify RNA. Tissues and cells must also be disrupted in a manner that prevents RNA degradation. In most cases this is achieved by lysing cells and tissues in the presence of guanidinium/guanidine isothiocyanate, a chaotrope that denatures and inactivates any RNases, thus preventing RNA degradation.

Importantly, the specimen stabilization and RNA extraction method must be appropriate to the type of sample. The methodologies we recommend, described below in this section and summarized in Fig. 1a, are:

Whole blood: Collect in PAXgene tubes, purify RNA with PAXgene kit, deplete globin transcripts with GLOBIN*clear* (Subheading 3.2.1). Whole blood collected in PAXgene tubes is stabilized and can be stored for prolonged periods of time without RNA degradation. The stabilized whole blood can be shipped elsewhere for RNA isolation using the PAXgene kit. This method is thus very suitable for large multicenter clinical trials. Globin transcripts from erythrocytes which may interfere with microarray analysis are easily removed using the Ambion GLOBIN*clear* kit [6] (*see* **Note 5**).

Peripheral blood mononuclear cells (PBMCs): Collect PBMCs in CPT tubes, prepare a homogenized lysate and purify RNA with the RNeasy kit (Subheadings 3.2.2.1 and 3.2.2.2). PBMCs can be conveniently separated from erythrocytes and neutrophils in a single centrifugation step using CPT tubes from BD Biosciences. Whole blood is first drawn into a CPT tube and then centrifuged. PBMCs are easily collected as they form a cell layer just under the plasma and are physically separated from other blood cells by a waxy barrier. Any erythrocytes remaining in the PBMC fraction are lysed using a specific erythrocyte lysis buffer. The purified PBMCs are then disrupted in RLT, a buffer in the RNeasy kit containing guanidine isothiocyanate that disrupts cells and stabilizes RNA. The RLT lysate is homogenized using a Qiashredder, and PBMC RNA is isolated from the homogenized RLT lysate using the

RNeasy kit. Isolation of RNA using the RNeasy kit is described in Subheading 3.2.2.2.

Tissue culture cells: Prepare a homogenized lysate (Subheading 3.2.2.3) and purify RNA with the RNeasy Mini Kit (Subheading 3.2.2.2). To prepare a homogenized lysate and purify RNA from lipid-rich tissue culture cells, such as 3T3-L1 adipocytes, use the RNeasy Lipid Mini Kit (*see* **Note 2**) according to manufacturer's instructions.

Tissues: Stabilize in RNA*later* (or, for adipose tissue, Allprotect—*see* **Note 1**) and prepare a homogenized RLT lysate (Subheading 3.2.2.4) and purify RNA with the RNeasy Mini Kit (Subheading 3.2.2.2). (For RNA purification from lipid-rich or fibrous tissues, *see* **Note 2**.) Tissue samples must be stabilized immediately with RNA*later* (Ambion) or Allprotect (Qiagen) before homogenization to prevent RNA degradation within the tissue during homogenization. The stabilized tissue is disrupted in RLT and homogenized; mechanical lysis may be required for complete homogenization. RNA is then purified from the homogenized RLT lysate using the RNeasy Mini Kit procedure (Subheading 3.2.2.2). *See* also Chapter 2 of this volume [7].

Specialized cell populations: Isolate RNA with the RNAqueous-Micro Kit from Ambion (Subheading 3.2.3) and perform two rounds of amplification (*see* Subheading 3.3.6). Our experience is that Ambion's RNAqueous-Micro Kit is best suited for isolating RNA from microscale samples such as purified cell populations and flow-sorted cells (Subheading 3.2.3). For microarray gene profiling of laser-captured cells (laser capture microdissection; LCM), *see* **Note 6**. The low amount of RNA content in these samples requires special procedures to be applied for efficient recovery of RNA.

Failure to process samples correctly will lead to RNA degradation, invalid quantitation, and greater sample-to-sample variability in microarray analysis. The RNA isolation methods described below have been found to enhance RNA integrity and improve performance in subsequent analysis. Strict adherence to these sample collection and cell lysis procedures, if followed by appropriate quality control assurance, will yield RNA suitable for use in microarray experiments.

3.2.1 Whole Blood: Stabilization and Subsequent RNA Isolation Using the PAXgene™ Kit

Stabilization of Whole Blood in Qiagen-PAXgene™ Tubes

1. Collect blood specimens (2.5 ml) in PAXgene™ tubes.
2. Gently invert 8–10× to mix.
3. Incubate upright at room temperature for at least 2 h for RNA stabilization and then store at −80 °C.
4. Freeze upright in a wire rack; do not use Styrofoam.
5. Thaw the blood samples at room temperature before processing for RNA isolation.

Isolation of RNA from Human Blood

The following procedure is a modification of the "Manual Purification of Total RNA from Human Whole Blood Collected into PAXgene Blood RNA Tubes" section of the PAXgene Blood RNA Kit Handbook (Qiagen).

Important

- Once this isolation protocol is begun, continue through all the steps without any interruptions or hesitations.
- All centrifugation steps should be carried out at room temperature (15–25 °C).

1. Centrifuge the thawed PAXgene blood RNA tubes for 10 min at 3,000–5,000 × *g* using a swinging bucket rotor.

   Important: Ensure that the blood sample has been incubated for a minimum of 2 h at RT, in order to achieve complete lysis.

2. Remove the supernatant by decanting and blotting excess liquid on bench paper; discard. Add 5 ml of RNase-free water (supplied) to the pellet, and close the tube using a fresh secondary Hemogard closure.
3. Thoroughly resuspend the pellet by vortexing, and centrifuge for 10 min at 3,000–5,000 × *g* as in **step 1**.
4. Remove and discard the entire supernatant by decanting. Remove drops from the rim of the tube by dabbing the rim onto bench paper followed by wiping the rim with a clean Kimwipe. Residual supernatant will inhibit further processing steps.
5. Add 360 µl Buffer BR1 to the pellet and thoroughly resuspend by vortexing.
6. Transfer the sample, by pipetting, to a 1.5 ml centrifuge tube (not supplied). Add 300 µl Buffer BR2 and 40 µl of Proteinase K. Mix by vortexing and incubate for 10 min at 55 °C in a shaker-incubator, or shaking water bath.

   Use of a shaker-incubator with the speed set to maximum is recommended. Do not allow the temperature of the samples to decrease during vortexing.

   Important: DO NOT mix buffer BR2 and proteinase K together before adding to the resuspended pellet.

7. Centrifuge for 20 min at maximum speed (at least 10,000 × *g*) in a microcentrifuge. Increase temperature of shaker-incubator to 65 °C.
8. Transfer the supernatant to a fresh 1.5 ml tube (not supplied).

   NOTE: Transfer of small amounts of debris remaining in the supernatant will not affect further steps.

9. Add 350 µl of 100 % ethanol. Mix by vortexing. Remove drops from the inside of the tube lid with a pipet tip.

10. Apply 700 μl of sample to the PAXgene column sitting in a 2 ml processing tube, and centrifuge for 1 min at >8,000 × *g* (≥10,000 rpm). Place the column in a new 2-ml processing tube, and discard the flow-through and old tube.
11. Apply the remaining sample to the column and repeat the centrifugation. Place the column in a new processing tube, and discard the flow-through and old tube.
12. Apply 350 μl Buffer BR3 to the column and centrifuge for 1 min at >8,000 × *g*. Place the column in a new 2-ml processing tube, and discard the flow-through and old processing tube.
13. Add 10 μl DNase I stock solution to 70 μl Buffer RDD for each digestion. Include extra if preparing a Master Mix for several samples, to account for losses during handling. For example, for nine samples, make enough Master Mix for ten samples: Add 100 μl DNase I stock solution to 700 μl Buffer RDD to prepare 800 μl (10 × 80 μl) of DNase incubation mix. Mix by gently flicking the tube, and centrifuge briefly to collect residual liquid from the sides of the tube. Do not vortex.
14. Pipet 80 μl of the DNase I incubation mix prepared in the previous step directly onto the spin-column membrane, and place on the benchtop (20–30 °C) for 15 min.
15. Pipet 350 μl Buffer BR3 into the PAXgene spin column, and centrifuge for 1 min at >8,000 × *g*. Place the column in a new 2-ml processing tube, and discard the flow-through and old tube.
16. Make sure ethanol has been added to buffer BR4. Apply 500 μl Buffer BR4 to the column, and repeat centrifugation. Again, place in a new processing tube, discarding the flow-through and old processing tube.
17. Apply another 500 μl Buffer BR4 and centrifuge for 3 min at >8,000 × *g*.
18. Place the column in a new 2-ml processing tube (not supplied). Centrifuge again for 1 min to ensure removal of residual supernatant.
19. To elute the RNA from the column, transfer the PAXgene column to a 1.5 ml elution tube, and pipet 40 μl Buffer BR5 directly onto the column membrane. Centrifuge for 1 min at >8,000 × *g*.
20. Repeat the elution step with another 40 μl of Buffer BR5 and centrifugation.
21. Incubate the eluate for 5 min at 65 °C in a heating block or water bath to denature the RNA. Get a bucket of ice. Following incubation, chill immediately on ice.

22. Quantitate RNA on NanoDrop. Record μg/μl and 260/280 ratio. Typical yields: ~1 μg RNA/1 ml human blood. The A_{260}/A_{280} ratio should be between 1.8 and 2.1 (*see* **Note 7** and Fig. 1).
23. Aliquot 1.5 μl of 100 ng/μl concentration for BioAnalyzer. The RIN should be >8 (*see* **Note 7** and Fig. 1).
24. Store the eluted RNA at −80 °C.
25. Deplete globin transcripts (Subheading 3.2.1.3) before proceeding with target cRNA preparation and microarray hybridization (*see* **Note 5**).

Depletion of Globin Transcripts

Although the PAXgene blood RNA system employs an easy way to collect, store, transport, and stabilize RNA from whole blood, many studies have demonstrated that RNA prepared from the PAXgene blood tubes result in a significant increase in overall variability and decrease in the transcript detection sensitivity using microarrays [5, 8]. The observed anomalies have largely been attributed to the presence of predominant amounts of globin transcripts that constitute ~70 % of mRNA in whole blood samples [8, 9]. We also found that this represents a major problem for the study of vascular diseases such as hemolytic anemias, for example sickle cell disease, owing to the high abundance of globin transcripts in nucleated erythrocytes and reticulocytes. Upon evaluating methods to deplete globins from whole blood RNA, we found out that Ambion's GLOBINclear method increased the sensitivity and specificity of microarray data [6] and it is highly recommended in multicenter clinical trials. The technique basically employs incubation of total RNA from human whole blood mixed with a biotinylated Capture Oligo Mix in hybridization buffer for 15 min to allow the biotinylated oligonucleotides to hybridize with the globin mRNA species. Streptavidin magnetic beads are then added to capture the globin mRNA. The magnetic beads are pulled to the side of the tube with a magnet and the RNA, depleted of the globin mRNA, remains in solution. The RNA is further purified using a rapid magnetic bead-based purification method according to manufacturer's directions and is then used for target cRNA preparation (Subheading 3.3) followed by microarray hybridization (Subheading 3.4).

3.2.2 Purification of RNA with RNeasy Mini Kit: PBMCs, Cultured Cells, and Tissue Specimens

Peripheral Blood Mononuclear Cells (PBMCs): Isolation of PBMCs and Preparation of PBMC Stable Cell Lysate

1. Obtain blood into CPT tubes (blue top). Draw 8 ml blood into the CPT tube and invert the tube ten times immediately after draw. Do not shake. Keep at room temperature.
2. Spin at 1,960×*g* for 30 min at 5 °C. Plasma is at the top; PBMCs are under the plasma and above the wax.
3. Collect the layer under the plasma and above the wax into a 15-ml tube containing 2 ml cold PBS. Spin the tube at 640×*g* for 10 min at 5 °C. Aspirate PBS and then resuspend the washed pellet in 5 ml erythrocyte lysis buffer. Mix well gently to aid the lysis of any erythrocytes that may be present.

4. Spin at 640 × *g* for 10 min at 5 °C. Aspirate supernatant and resuspend the pellet containing PBMCs in 15 ml cold PBS. Centrifuge at 640 × *g* for 5 min at 5 °C.
5. Resuspend pellet (PBMCs) in 2 ml cold PBS.
6. Take an aliquot of 200,000–500,000 cells and place in 300 μl PBS to count in Cell Dyn (or equivalent) according to manufacturer's instructions for differential cell counts. Spin cells for 5–10 min at 1,200 rpm (150 × *g*) at 5 °C.
7. Aspirate the PBS and add RLT buffer: 1,000 μl for every 1×10^7 cells. Use RNase-free pipet.
8. The RLT buffer disrupts the cells, but it is also necessary to mechanically lyse the cells. Pass 700 μl at a time of RLT buffer and cells through a purple Qiashredder column twice at 15,300 × *g* for 1–2 min. This is the homogenized RLT lysate.
9. Store at −80 °C until ready for RNA purification using the RNeasy Mini Kit (Subheading 3.2.2.2). The homogenized RLT lysate can be stored for several months at −80 °C.

Isolation of Total RNA (>200 bp) from Homogenized RLT Lysate Using the RNeasy Mini Kit, with DNase Digestion

Important: The maximum binding capacity of the RNeasy Mini Spin Column is 100 μg RNA; do not overload the column. In addition, only RNA >200 nt are retained on the RNeasy Mini Spin Column; RNA smaller than 200 nt may be lost. For smaller RNAs, *see* **Note 3**.

Add 44 ml (4 volumes) of 100 % EtOH to buffer RPE.

1. Make up 70 % EtOH in DEPC water.
2. Prepare or thaw DNase I stock solution (1,500 Kunitz units/550 μl).
3. If homogenized RLT lysate was frozen in RLT, thaw in 37 °C water bath for 20 min. RLT precipitates when frozen and thawed. If there are precipitates, spin at 1,960 × *g* for 5 min and collect supernatant. Add an equal volume of 70 % EtOH to the homogenized lysate. Mix well by pipetting. Do not vortex.
4. Apply up to 700 μl of sample to RNeasy mini spin column. Centrifuge for 30 s at 7,800 × *g*; discard flow-through. Repeat with remainder of the sample. For higher yields can try passing flow-through a second time over the column.
5. Add 350 μl of buffer RW1 onto the mini spin column. Centrifuge for 30 s at 7,800 × *g*. Discard flow-through.
6. Add 10 μl DNase I stock solution to 70 μl buffer RDD (*see* **Note 8**). Mix gently by inverting. Do not vortex!
7. Pipet this 80-μl mix onto the spin column and let it sit at room temperature for 15 min. Add 350 μl RW1 onto spin column. Centrifuge 30 s at 7,800 × *g*. Discard flow-through.

8. Add 500 μl buffer RPE onto the mini spin column. Centrifuge for 30 s at 7,800 × *g*; discard flow-through.
9. Add 500 μl buffer RPE [make sure ethanol (4 volumes × the amount in the bottle) has been added to RPE] on the mini spin column. Centrifuge for 2 min at 15,300 × *g*. Discard flow-through.
10. Dry-spin the mini column.
    Centrifuge at 15,300 × *g* for 1 min.
11. Transfer RNeasy column into a new 1.5-ml collection tube.
    To elute, transfer RNeasy mini spin column to a new collection tube. Pipet 20 μl of RNase-free water directly onto the column membrane. Close the tube lightly, let it stand for 1 min, and spin 15,300 × *g* for 1 min.
12. Add another 20 μl of RNase-free water and spin 15,300 × *g* for 1 min. To obtain a higher total RNA concentration, this second elution step may be performed by using the first eluate. The yield might be 15–30 % less than the yield obtained using a second volume of RNase-free water, but the final concentration will be higher.
13. Mix and transfer the eluted RNA to a sterile 1.5 ml microcentrifuge tube. Remove an aliquot for quantitation and quality control (*see* **Note 7** and Fig. 1).
14. The rest of the RNA should be immediately snap-frozen and stored at −80 °C or over liquid nitrogen in a freezer until target cRNA preparation (Subheading 3.3). Keep on ice when pulled out to use.
15. Quantitate RNA on NanoDrop. Record μg/μl and 260/280 ratio. Yield will vary depending on the type and amount of starting material. The A_{260}/A_{280} ratio should be between 1.8 and 2.1 (*see* **Note 7** and Fig. 1).
16. Aliquot 1.5 μl of 100 ng/μl concentration for the BioAnalyzer. The RIN should be >8 (*see* **Note 7** and Fig. 1).
17. If the RNA is pure and not degraded, proceed to Subheading 3.3 (target cRNA preparation) followed by Subheading 3.4 (microarray hybridization).

Preparation of Homogenized RLT Lysate from Cultured Cells

Preparation of RNA from cultured cells (Subheading 3.2.2.3) is straightforward: Cells are disrupted in RLT buffer, the RLT lysate is homogenized with QIAshredder columns, and RNA is then purified using the RNeasy Mini Kit procedure. *Important*: For tissue culture cells rich in fat, such as lipid-loaded fat cells (e.g., 3T3-L1 fat cells) and brain cells, use the RNeasy Lipid Tissue Mini Kit according to manufacturer's instructions or follow the TRIzol method as described in ref. 7.

The procedure described below is for purifying RNA using the standard Qiagen RNeasy Mini Kit. *This procedure does not apply to the RNeasy Lipid Tissue Mini Kit or the RNeasy Fibrous Tissue Mini Kit.* For other kits, refer to the manufacturers' instructions provided with the kit.

1. Suspension cells: Plan on 5×10^6–1×10^7 cells/column. Centrifuge the appropriate volume of cells at $300 \times g$ in an RNase-free tube and remove all traces of medium. Loosen cell pellet thoroughly by flicking the tube. Add 700 μl of Buffer RLT plus β-ME (10 μl β-ME/ml RLT) per column to lyse the cells. Vortex or pipet to mix. Proceed to homogenization (**step 3**).
2. For cells grown as a monolayer (1×10^7 cells maximum)—Cells can either be lysed directly in the cell culture dish/flask or trypsinized and pelleted and then lysed.
   (a) To lyse in the culture dish without prior trypsinization:
      - Determine the number of cells.
      - Completely remove cell medium.
      - Add the appropriate volume of RLT:
        350 μl for <6 cm diameter dish.
        600 μl for 6–10 cm diameter dish.
      - Collect the cells with a rubber policeman.
      - Pipet the lysate into an RNase-free microcentrifuge tube.
      - Vortex or pipet to mix.
      - Ensure that no cell clumps are visible.
      - Homogenize using a Qiashredder as in **step 8** of Subheading 3.2.2.

   (b) To trypsinize cells:
      - Determine the number of cells.
      - Aspirate medium.
      - Wash with 1× PBS.
      - Aspirate PBS.
      - Add 0.1–0.25 % trypsin in PBS.
      - After cells detach, transfer to RNase-free centrifuge tube.
      - Spin $300 \times g$ 5 min.
      - Remove all traces of supernatant.
      - Disrupt cell pellet by flicking.

        Add RLT as above and homogenize with Qiashredder as above for suspension cells.

        Proceed to homogenization (step 3).
3. Homogenize cells by passing the cell suspension in QIAshredder column at maximum speed for 2 min. This is the "homogenized RLT lysate." Use immediately or store at −80 °C. The homogenized RLT lysate can be stored for several months at −80 °C.

4. Proceed to Subheading 3.2.2.2 ["Isolation of Total RNA (>200 bp) from Homogenized RLT Lysate Using the RNeasy Mini Kit, with DNase Digestion"] and follow the entire procedure (**steps 1–16**). If the RNA is pure and undegraded, proceed to Subheading 3.3 (target cRNA preparation) followed by Subheading 3.4 (microarray hybridization).

Tissue Specimens: Stabilization of Tissue in RNA*later* and Preparation of Homogenized RLT Lysates

Stabilization of Tissue in RNAlater

RNA*later*® Tissue Collection: RNA Stabilization Solution is an aqueous tissue storage reagent that rapidly permeates most tissues to stabilize and protect RNA in fresh specimens. It eliminates the need to immediately process or freeze samples; the specimen can simply be submerged in RNA*later* Solution and stored for analysis at a later date. Samples in RNA*later* Solution can be stored for extended periods under conditions where RNA degradation would normally take place rapidly.

RNA*later* Solution can be used for RNA preservation with most tissues and cultured cells but may not be effective in tissues that are poorly penetrated by the solution. Adipose tissue in particular is not compatible with RNA*later* due to its high abundance of fat, so use Allprotect Tissue Reagent (Qiagen) in place of RNA*later* to stabilize adipose tissue (*see* ref. 7).

Procedure

1. Use RNA*later* Solution with fresh tissue only; do not immerse frozen samples in RNA*later* Solution.
2. Before immersion in RNA*later* Solution, cut large tissue samples to ≤0.5 cm in any single dimension, working as quickly as possible.
3. Place the fresh tissue in 5–10 volumes of RNA*later* Solution to preserve for later extraction.
4. Do not freeze samples in RNA*later* Solution immediately. Store at 4 °C overnight (to allow the solution to thoroughly penetrate the tissue), and then either proceed with RNA extraction or move the sample to −20 or −80 °C for long-term storage. If storing the stabilized tissue at −80 °C, remove the stabilized tissue from the solution and immediately freeze the stabilized tissue.
5. Most samples in RNA*later* Solution can be stored at room temp for 1 week or at 2–8 °C for 4 weeks without compromising RNA quality, or at −20 or −80 °C indefinitely.
6. Qiagen has recently introduced Allprotect Tissue Reagent, which can also be used for stabilizing nucleic acids in tissues without the need for freezing the tissues immediately for long-term storage.

   Allprotect Tissue Reagent delivers immediate stabilization of DNA, RNA, and protein in tissue samples for long-term storage without freezing. Another advantage of this stabilization procedure is that it can be used in combination with the AllPrep DNA/RNA/Protein Mini Kit, which allows simultaneous purification of DNA, RNA, and protein from the same precious sample.

The procedure involves submerging the tissues in Allprotect Tissue Reagent to immediately stabilize DNA, RNA, and proteins. Stabilization takes place at room temperature, eliminating the use of hazardous liquid nitrogen or dry ice. Tissues stabilized in Allprotect Tissue Reagent can be stored at room temperature for up to 7 days or at 2–8 °C for up to 12 months without showing any degradation of DNA, RNA, and protein. For longer storage, stabilized tissues can be archived at –20 or –80 °C.

Preparation of RLT Lysate from Tissue

Before preparing an RLT lysate, decide which homogenization technique and which RNeasy kit is appropriate for your tissue type (*see* **Notes 2** and **9**). The protocol in this chapter applies only to the standard RNeasy Mini Kit. Refer to the manufacturer's instructions if you use any other kit.

Retrieve tissue from RNA*later* Solution with sterile forceps, quickly blot away excess RNA*later* Solution with an absorbent lab wipe or paper towel, and then submerge the sample in 1 ml of RNA isolation lysis solution (RLT buffer, supplied with the kit). Homogenize tissue promptly after placing it in RLT buffer. (The homogenization method will differ depending on tissue type; *see* **Note 9**). Pass the homogenate through a QIAshredder column at maximum speed for 2 min. This is the "homogenized RLT tissue lysate."

Use immediately for RNA preparation or store at −80 °C. The homogenized RLT tissue lysate can be stored for several months at −80 °C.

For RNA preparation, proceed to Subheading 3.2.2.2 ["Isolation of Total RNA (>200 bp) from Homogenized RLT Lysate Using the RNeasy Mini Kit, with DNase Digestion"] and follow the entire procedure (**steps 1–16**). If the RNA is pure and undegraded, the next step will be target cRNA preparation (Subheading 3.3) followed by microarray hybridization (Subheading 3.4).

Automating RNA Isolation

Large-scale gene expression screening projects require reliable high-throughput methods to isolate high-quality RNA from a large number of samples. To meet this growing need, a few automation instruments are now available of which the innovative QIAcube marketed by Qiagen appears to enable seamless integration of automated, low-throughput sample prep into the laboratory workflow. Fully automated purification of nucleic acids or proteins using QIAGEN spin kits on the QIAcube eliminates the need for tedious manual steps. Sample preparation on the QIAcube follows the same steps as the manual procedure. The QIAcube enables purification of highly pure nucleic acids or proteins, and performance is comparable to the manual procedure. The QIAcube is preinstalled with protocols for purification of plasmid DNA, genomic DNA, RNA, viral nucleic acids, and proteins, plus DNA and RNA cleanup.

3.2.3 Flow-Sorted Cells or Few Hundred Cells: Use of Ambion's RNAqueous-Micro Kit for Isolating RNA from Specialized Cell Populations

To isolate total RNA from micro-sized samples, such as flow-sorted cells or other procedures that yield only a few hundred cells, use Ambion's RNAqueous-Micro Kit. This kit employs a guanidinium-based lysis/denaturant as well as glass fiber filter separation technology. DNA-free™ DNA removal reagents are included in this kit, making the isolated RNA suitable for most downstream applications.

Since relatively few cells are used, the amount of RNA recovered will be low and thus two rounds of linear cDNA amplification are required for target preparation. Quantitation of RNA using a spectrophotometer may not be accurate due to the sensitivity limit of instrument but Agilent Picochip analysis would be able to give an approximate concentration and quality of the RNA. *See* Subheading 3.3.6.

3.3 Target-Labeled cRNA/cDNA Preparation

RNA prepared by using any of the above methods can be used for the labeling process (Fig. 1b) for hybridization onto the gene chips. The RNA must be labeled abundantly with biotin so that it can be detected after hybridizing to spots on the microarray. Depending on the type of the arrays used, cRNA or cDNA will be labeled with biotin. If the study involves application of the 3′ IVT arrays, this is accomplished by first reverse-transcribing the RNA to make cDNA and then performing second-strand cDNA synthesis (in the presence of RNase H) to form double-stranded cDNA that accurately represents the transcripts present in the original RNA preparation. After sample cleanup, the double-stranded cDNA is reverse-transcribed in the presence of biotin, forming biotin-labeled cRNA. After sample cleanup, quality control, and fragmentation to lengths spanning 35–200 bp, the biotin-labeled cRNA, along with hybridization controls, is hybridized to the microarray, which is then stained with streptavidin–phycoerythrin, washed, and scanned (Fig. 1b). Signal intensity from each "spot" is quantitated for samples and controls, and the fold-differences of genes present on the microarray can be determined.

For application on Gene 1.0; 2.0 ST and Exon ST arrays, the cRNA undergoes another reverse transcription to synthesize the sense strand DNA, which would then be fragmented and labeled with biotin prior to hybridization.

The following procedures (recommended by Affymetrix; *see* **Note 10**) (Subheadings 3.3.1–3.3.7) are used to prepare biotinylated cRNA target using 50 ng to 2 μg of purified total RNA as the starting sample, which will provide sufficient amount of cRNA target for hybridization on 3′ IVT gene chips. (For less than

50 ng total RNA, *see* **Note 11**.) Preparation of labeled cDNA for hybridization onto Gene ST and Exon ST arrays is described in Subheadings 3.3.8.1–3.3.8.7.

3.3.1 First-Strand cDNA Synthesis (IVT Express Kit)

1. Mix in a 0.2-ml PCR tube:

   50 ng to 2 μg of RNA sample (*see* **Note 11**).

   2 μl diluted poly-A RNA controls (*see* **Note 12**).

   2 μl T7-oligo(dT) primer (50 μM).

   RNase-free water to a final volume of 12 μl.
2. Flick the tube to mix and spin the tube.
3. Incubate the reaction for 10 min at 70 °C using a PCR machine to denature the RNA/T7 oligo dT primer, quickly chill the tube (by putting on ice), and keep at 4 °C or on ice at least 2 min.
4. Spin the tube briefly.
5. Prepare sufficient First-Strand Master Mix by adding:

   4 μl of 5× First-Strand Reaction Buffer

   2 μl of DTT (0.1 M)

   1 μl of dNTP (10 mM)
   for each sample. As usual, include extra if preparing a Master Mix for several samples.
6. Mix well, spin down, and transfer 7 μl of First-Strand Master Mix to each denatured sample tube (from **step 3** above). Flick to mix and spin briefly.
7. Incubate for 2 min at 42 °C.
8. Add 1 μl of SuperScript II to each RNA sample, mix by flicking, spin down, immediately incubate for 1 h at 42 °C, and then cool the sample for at least 2 min at 4 °C.
9. Spin down the tube and immediately proceed to second-strand cDNA synthesis.

3.3.2 Second-Strand cDNA Synthesis

1. Prepare sufficient Second-Strand Master Mix in a separate tube with:

   91 μl of RNase-free water

   30 μl of 5× Second-Strand Reaction Mix

   3 μl of dNTP (10 mM)

   1 μl of *E. coli* DNA ligase

   4 μl of *E. coli* DNA polymerase I

   1 μl of RNase H
   for each sample. As usual, include enough for an extra sample to account for losses during handling.
2. Add 130 μl of Second-Strand Master Mix to each first-strand synthesis sample flick to mix, spin down, and incubate for 2 h at 16 °C.

3. Add 2 μl of T4 DNA Polymerase, incubate for 5 min at 16 °C, and then hold at 4 °C.
4. Add 10 μl of 0.5 M EDTA and proceed to cleanup of double-stranded cDNA step (see below).

3.3.3 Cleanup of Double-Stranded cDNA [Sample Cleanup Module at Room Temperature (RT)]

1. Prepare cDNA wash buffer by adding 24 ml of 100 % EtOH to the wash buffer concentrate.
2. All steps are performed at room temperature; work without stopping.
3. Transfer the 162 μl of the cDNA from **step 4** of Subheading 3.3.2 into a 1.5- or 2-ml Eppendorf tube.
4. Add 600 μl of cDNA binding buffer to the tube containing the cDNA. Mix by vortexing for 3 s. The color of the mixture should be yellow. (If the color of the mixture is orange or violet, add 10 μl of 3 M sodium acetate, pH 5.0, and mix; it should turn yellow).
5. Apply 500 μl of the sample mix to the cDNA Cleanup Spin Column (purple column); centrifuge for 1 min at 15,340 × *g*.
6. Discard flow-through and reload spin column with remaining sample mixture; centrifuge for 1 min at 15,340 × *g*.
7. Discard flow-through and collection tube. Transfer column to a new 2-ml collection tube.
8. Pipet 750 μl of cDNA wash buffer to spin column; centrifuge for 1 min at 15,340 × *g*. Discard flow-through and collection tube.
9. Transfer column to a new collection tube. Cut off cap of spin column, label the side of the collection tube, and centrifuge for 5 min at max speed.
10. Transfer spin column to a 1.5-ml collection tube. Pipet 14 μl of cDNA elution buffer *directly onto the spin column membrane* (THIS IS CRITICAL). Incubate 1 min at room temp, centrifuge for 1 min at max speed. Measure amount eluted and record amount. You should retrieve 12–14 μl of eluate.
    Proceed to synthesis of biotin-labeled cRNA (IVT reaction) (Subheading 3.3.4) or freeze at −20 °C.

3.3.4 Synthesis of Biotin-Labeled cRNA by In Vitro Transcription (IVT) Reaction

1. Transfer all of the eluate (i.e., the template cDNA) from the cleanup procedure into a 0.2-ml PCR tube and add:
    8 μl RNase-free water.
    4 μl 10× IVT Labeling Buffer.
    12 μl NTP Mix.
    4 μl Enzyme Mix.
2. Carefully mix by tapping the tube and spin briefly.

3. Incubate at 37 °C for 16 h in a thermal cycler with a heated lid.
4. Biotin-labeled cRNA can be stored at −20 or −70 °C, if not used for cleanup immediately.

3.3.5 Cleanup and Quantification of Biotin-Labeled cRNA (Sample Cleanup Module at RT)

Cleanup of cRNA can be done either by columns as described below or with RNAclean XP beads (Agencourt, Beverly, MA), following manufacturer's instructions, depending on the sample throughput.

1. Save an aliquot (0.5 μl) of the unpurified IVT product for analysis by gel electrophoresis or Agilent 2100 Bioanalyzer to estimate the yield and size distribution of labeled transcripts.
2. Transfer each biotin-labeled cRNA sample to a 1.5-ml tube, add 60 μl of RNase-free water, and vortex for 3 s.
3. Add 350 μl of IVT cRNA binding buffer and vortex for 3 s.
4. Add 250 μl of ethanol (96–100 %) and mix well by pipetting. Do not centrifuge.
5. Apply sample to the IVT cRNA Cleanup Spin Column in a 2-ml collection tube and centrifuge for 15 s at ≥ 9,000 × *g*.
6. Transfer the column into a new 2-ml collection tube, pipet 500 μl of IVT cRNA wash buffer, and centrifuge for 15 s at ≥9,000 × *g*. Discard flow-through.
7. Pipet 500 μl 80 % ethanol onto the column and centrifuge for 15 s at ≥9,000 × *g*. Discard flow-through.
8. Open the cap of the column and centrifuge for 5 min at maximum speed. Discard the collection tube with the flow-through.
9. Transfer the column into a clean 1.5-ml collection tube, and pipet 11 μl of RNase-free water directly onto the membrane. Incubate for 1 min at RT and centrifuge for 1 min at maximum speed.
10. Pipet another 10 μl of RNase-free water onto the membrane and elute as above. Combined eluate (21 μl) can be stored at −20 or −70 °C if not used for quantification immediately.
11. To determine cRNA concentration and purity, check the absorbance at 260 and 280 nm of 1:100 dilution of 1 μl cRNA sample.
12. Check the quality of unfragmented samples by gel electrophoresis or with an Agilent 2100 Bioanalyzer (*see* **Note 13**).

3.3.6 Target Preparation from Flow-Sorted Cells or Few Hundred Cells

As noted in Subheading 3.2.3, total RNA yields from microscale samples such as purified cell populations and flow-sorted cells are generally low. The low amount of RNA content in these samples requires special procedures to be applied for efficient recovery of RNA. Our experience is that Ambion's RNAqueous-Micro Kit is best suited for isolating RNA from these samples.

The amount of RNA obtained from these samples is generally less than what is required for standard target preparation procedure. Two rounds of linear amplification must be performed on these samples following the manufacturer's (Arcturus/Molecular Devices) instructions. After the second round of cDNA synthesis, purified cDNA is used as the target in the synthesis of Biotin-Labeled cRNA by In Vitro Transcription (IVT) Reaction (Affymetrix IVT Labeling Kit) and cleaned up as before as described above (Subheading 3.3.4).

3.3.7 Fragmentation of Biotinylated cRNA

1. Transfer 15–20 μg of cRNA into a 0.2-ml PCR tube and add 8 μl of 5× Fragmentation Buffer and RNase-free water to a final volume of 40 μl to prepare fragmented cRNA.
2. Incubate at 94 °C for 35 min in a thermal cycler. Put on ice following the incubation.
3. Save an aliquot (1 μl) for analysis on the Bioanalyzer or gel electrophoresis to confirm the fragmentation. The size distribution should be 35–200 bp (*see* **Note 14**).
4. Store the undiluted fragmented cRNA at −70 °C until ready to perform the hybridization.

3.3.8 Preparation of Labeled cDNA for Hybridization onto Gene ST and Exon ST Arrays

The following procedure, with a minimum of 50 ng total RNA using the Whole-Transcript Sense Target Labeling and Control Reagents kit, can be used for preparing biotinylated sense DNA targets for hybridization onto Gene ST and Exon ST arrays.

Preparation of Total RNA/T7 Primers/Poly-A Controls

1. Mix Total RNA and the T7 Primers/Poly-A Controls solution in a 0.2 ml tube as follows:

Total RNA, 50 ng	Variable (1–3 μl)
T7 Primers/Poly-A Controls	2 μl (Appropriate dilution based on the input concentration)
RNase-free water	Variable
Total volume	5 μl

2. Denature sample in a heated lid thermal cycler at 70 °C.
3. Cool sample at least 2 min at 4 °C. Spin to collect condensate. Place on ice.
4. Prepare the First-Cycle, First-Strand Master Mix as follows for a single reaction (adjust as needed):

5× First-Strand Buffer	2 μl
DTT, 0.1 M	1 μl
dNTP Mix, 10 mM	0.5 μl
RNase Inhibitor	0.5 μl
SuperScript II	1 μl
Total volume	5 μl

5. Add 5 µl of First-Cycle, First-Strand Master Mix to 0.2 ml tube containing Total RNA/Poly-A Controls/T7Primer Mix for 10 µl total volume.

   Program the thermal cycler with heated lid as below. Place samples in cycler when 25 °C is reached.

25 °C	10 min
42 °C	60 min
70 °C	10 min
4 °C	Hold

   Allow reaction to cool only 2 min at 4 °C before immediately continuing to first-cycle, second-strand cDNA synthesis. Keeping the reaction at 4 °C longer than 10 min. can reduce cRNA yields.

First-Cycle, Second-Strand Synthesis

1. Make a fresh dilution of 17.5 mM $MgCl_2$ each time. Mix 2 µl of 1 M $MgCl_2$ with 112 µl RNase-free water.
2. Prepare the First-Cycle, Second-Strand Master Mix as follows for a single reaction. Add RNase H and DNA Polymerase last and immediately aliquot into reactions.

RNase-free water	4.8 µl
$MgCl_2$, 17.5 mM	4.0 µl
dNTP Mix, 10 mM	0.4 µl
DNA Polymerase I	0.6 µl
RNaseH	0.2 µl
Total volume	10 µl

3. Add 10 µl of the First-Cycle, Second-Strand Master Mix to the First-Strand cDNA reaction for a total volume of 20 µl. Flick-mix and spin down.

   Program the thermal cycle as below without heated lid.

16 °C	120 min
75 °C	10 min
4 °C	Hold

   Allow reaction to cool only 2 min at 4 °C before immediately continuing to first-cycle, second-strand cDNA synthesis. Keeping the reaction at 4 °C longer than 10 min can reduce cRNA yields.

cRNA Synthesis by IVT

1. Assemble IVT Master Mix in a separate tube as follows for a single reaction (adjust volume as necessary):

10× IVT Buffer	5.0 μl
IVT NTP Mix	20.0 μl
IVT Enzyme Mix	5.0 μl
Total volume	30.0 μl

2. Transfer 30 μl of the IVT Master Mix to each first-cycle, second-strand cDNA synthesis sample at RT° for a final volume of 50 μl.
3. Transfer each reaction to a labeled 1.5 ml microcentrifuge tube. Pipet up and down to mix.
4. Incubate the reaction at 37 °C for 16 h in the Hybridization Oven 640 (set temp. to 37.5 °C to obtain 37 °C).
5. Proceed to the Cleanup procedure for cRNA using the cRNA Cleanup Spin Column from the GeneChip Sample Cleanup Module.

   If cRNA is not going to be purified immediately, samples can be stored at −80 °C.

   NOTE: cRNA can be purified using columns, as described below, or RNAclean XP beads (Agencourt, Beverly, MA), following manufacturer's instructions.
6. Add 50 μl RNase-free water to each IVT reaction for a final volume of 100 μl. Vortex briefly at low speed.
7. Add 350 μl of cRNA binding buffer to each sample. Vortex briefly at low speed.
8. Add 250 μl of 100 % absolute ETOH to each reaction. Vortex briefly at low speed.
9. Apply each sample to the IVT cRNA Cleanup Spin Column sitting in a 2 ml collection tube.
10. Centrifuge at ≥8,000 × *g* (10,000 rpm) for 15 s. Discard the flow-through.
11. Transfer the column to a new 2.0 ml collection tube.
12. Add 500 μl of cRNA wash buffer to the column and centrifuge for 15 s. at ≥8,000 × *g* (10,000 rpm). Discard the flow-through.
13. Wash again with 500 μl of 80 % v/v ETOH. Centrifuge for 15 s at ≥8,000 × *g* (10,000 rpm). Discard flow-through.
14. Open the column cap and spin at ≤25,000 × *g* (maximum speed) for 5 min. with the caps open.
15. Transfer the IVT cRNA Cleanup Spin column to a new 1.5 ml collection tube and add 15 μl of RNase-free water directly to the membrane.

16. Incubate at RT° 5 min. Centrifuge at ≤25,000 × g (maximum speed) for 1 min.
17. Return elute back onto the column and incubate at RT° 5 min. Centrifuge at ≤25,000 × g (maximum speed) for 1 min.
18. The eluted cRNA is ~13.5 µl. Determine the cRNA yield by NanoDrop assay for RNA.

Second-Cycle, First-Strand cDNA Synthesis: ST cDNA

This procedure requires the use of the GeneChip WT cDNA Synthesis Kit.

1. Mix 10 µg cRNA with Random Primers in a 0.2 µl tube as follows for a single reaction:

cRNA, 10 µg*	Variable
Random Primers (3 µg/µl)	1.5 µl
RNase-free water	Up to 8 µl
Total volume	8 µl

2. Flick-mix and spin down tubes.
3. Program the thermal cycler as below with the heated lid.

70 °C	5 min
25 °C	5 min
4 °C	Hold

4. Cool samples at 4 °C for at least 2 min.
5. In separate tube prepare Second-Cycle, Reverse Transcription Master Mix for a single reaction as follows (adjust volume as necessary):

5× First-Strand Buffer	4.0 µl
DTT, 0.1 mM	2.0 µl
dNTP + dUTP, 10 mM	1.25 µl
SuperScript II	4.75 µl
Total volume	12.0 µl

6. Transfer 12 µl of Second-Cycle, First-Strand cDNA Master Mix for a total reaction volume of 20 µl.
7. Program thermal cycler as below:

25 °C	10 min
42 °C	90 min
70 °C	10 min
4 °C	Hold

Hydrolysis of cRNA and Cleanup of Single-Stranded ST cDNA

This can be done either by sample cleanup module as described below or by Ampure XP beads (Agencourt), following manufacturer's instructions.

1. Add 1 μl of RNase H to each sample and incubate as below:

37 °C	45 min
95 °C	5 min
4 °C	Hold

2. Proceed to the cleanup step using the cDNA Cleanup Spin Columns from the GeneChip Sample Cleanup Module.
3. Add 80 μl of RNase-free water to each sample and transfer to a labeled 1.5 ml microcentrifuge tube.
4. Add 370 μl of cDNA binding buffer to each sample and vortex 3 s.
5. Apply entire sample (total volume is 471 μl) to the cDNA spin column sitting in 2 ml collection tube.
6. Spin at ≥8,000 × *g* (10,000 rpm) for 1 min. Discard flow-through.
7. Transfer the spin column to a new 2.0 ml collection tube and add 750 μl cDNA wash buffer.
8. Spin at ≥8,000 × *g* (10,000 rpm) for 1 min. Discard flow-through.
9. Open cap of spin column and spin at ≤25,000 × *g* (maximum speed) for 5 min with the cap open.

   Discard the flow-through and transfer column to a 1.5 ml collection tube.
10. Add 15 μl of cDNA elution buffer directly to the column membrane and incubate at RT° for 1 min.
11. Spin at ≤25,000 × *g* (maximum speed) for 1 min.
12. Repeat the elution step by adding another 15 μl cDNA elution buffer to the membrane and incubating at RT° 1 min.
13. Spin at ≤25,000 × *g* (maximum speed) for 1 min.
14. The total volume of eluted single-stranded is ~28 μl. Take 1 μl from each sample and determine yield by NanoDrop.

Fragmentation of Single-Stranded DNA

This procedure requires the use of the GeneChip WT Terminal Labeling Kit.

1. Prepare single-stranded DNA in a 0.2 ml tube as follows for each reaction:

Single-stranded DNA	5.5 μg (Volume can vary)
RNase-free water	Up to 31.2 μl
Total volume	31.2 μl

2. Prepare the Fragmentation Master Mix as follows for a single reaction:

RNase-free water	10.0 μl
10× cDNA Fragmentation Buffer	4.8 μl
UDG, 10 U/μl	1.0 μl
APE 1, 1,000 U/μl	1.0 μl
Total volume	16.8 μl

3. Add 16.8 μl of the Fragmentation Mix to each sample and program thermal cycler as below:

37 °C	60 min
93 °C	2 min
4 °C	Hold (at least 2 min)

4. Mix well and spin down tubes. Transfer 45 μl to a new 0.2 μl tube for labeling.

Labeling of Fragmented Single-Stranded DNA

This procedure requires the use of the GeneChip WT Terminal Labeling Kit.

1. Prepare labeling reactions in 0.2 ml tube as follows for a single reaction:

Fragmented single-stranded DNA	45 μl
5× TdT Buffer	12 μl
TdT	2 μl
DNA labeling reagent, 5 mM	1 μl
Total volume	60 μl

2. Mix and spin down samples. Program thermal cycler as below:

37 °C	60 min
70 °C	10 min
4 °C	Hold (at least 2 min)

3.4 Microarray Hybridization, Washing, Staining, and Scanning

This procedure requires the use of the GeneChip Hybridization, Wash and Stain kit (Affymetrix 900720).

3.4.1 Preparation of Hybridization Cocktail

You will need a 65, 99, and a 45 °C heat block or use a PCR machine.

The 20× Eukaryotic Hybridization Control (EHC) (made by Affymetrix) needs to be heated at 65 °C for 5 min. Vortex EHC and briefly spin in a microcentrifuge after heating.

1. Prepare hybridization cocktail for 15–20 μg cRNA (5′ IVT arrays) in 1.5 ml tubes as described below:

Component	20 μg of cRNA	15 μg of cRNA
Fragmented, biotinylated cRNA	30 μl	40 μl
Control oligo B2 (*see* **Note 4**)	5 μl	5 μl
20× EHC (*see* **Note 4**)	15 μl	15 μl
Herring sperm DNA (10 mg/ml)	3 μl	3 μl
BSA (50 mg/ml)	3 μl	3 μl
DMSO	30 μl	30 μl
2× Hybridization solution	150 μl	150 μl
H_2O	64 μl	54 μl
Final volume	300 μl	300 μl
Volume onto chip	200 μl	200 μl

Note: (a) For convenience and reproducibility, make a Master Mix containing control oligo B2, ECH, herring sperm DNA, BSA, and DMSO; add 56 μl/tube. (b) Heat hybridization cocktail to 99 °C for 5 min followed by 45 °C for 5 min. (c) Centrifuge hybridization cocktail at max speed for 5 min.

2. Prepare hybridization cocktail for labeled, fragmented single-stranded DNA for hybridization onto Gene ST and Exon ST arrays in 1.5 ml tubes as described below:

   NOTE: The cocktail mix for the Human Gene 1.0 ST array is the 169 array format.

   The cocktail mix for the Human, Mouse, and Rat Exon 1.0 ST arrays is the 49 array format.

Component	49/64 Format	169 Format	Concentration
Fragmented and labeled DNA target	~60.0 μl	27.0 μl	~25 ng/μl
Control oligo B2 (3 nM)	3.7 μl	1.7 μl	50 pM
20× Eukaryotic hyb controls	11.0 μl	5.0 μl	1.5, 5, 25, 100 pM
2× Hybridization mix	110.0 μl	50.0 μl	1×
DMSO	15.4 μl	7.0 μl	7 %
Nuclease-free water	Up to 220.0 μl	Up to 100.0 μl	
Total volume	220 μl	100.0 μl	

3. Low speed vortex briefly and spin down.
4. Denature cocktail for 5 min at 99 °C in heat block.
5. Cool for 5 min. at 45 °C in the Hybridization Oven 640. Microcentrifuge at full speed 1 min.

3.4.2 Prehybridization of the Probe Array

Equilibrate probe array to room temperature and fill it with 1× hybridization buffer through one of the septa, using a micropipettor with another tip for venting through the other septum. Incubate array at 45 °C for 10 min. For arrays other than IVT prehybridization step is not needed.

3.4.3 Hybridization of the Probe Array

Remove the buffer from the array and cartridge and inject the appropriate amount of denatured cocktail into the array.

Array format	Hyb volume
49 (standard)	200 µl
64	200 µl
169	80 µl

Place arrays in the Hybridization Oven 640 and hybridize at 45 °C, 60 rpm, for 17 ± 1 h. Target cRNA hybridization step should be followed by washing, staining, and scanning without delay for the best results.

3.4.4 Fluidics Station Setup

1. Define file location if MAS is used by selecting Tools→Defaults →File→Locations from the menu bar, and verify that all three file locations (files for probe information, fluidics protocols, and experiment data) are set correctly. (If GCOS is used, this step is not necessary.)
2. Launch MAS 5.0 or GCOS or AGCC and enter experiment name, probe array type, and the additional information of sample name, sample type, and project if using GCOS.
3. Turn on the Fluidics Station and select Run→Fluidics from the menu bar.
4. Change the intake buffer reservoir A to Wash Buffer A and intake buffer reservoir B to Wash Buffer B and prime the Fluidics Station by selecting Protocol, Prime, and All Modules, and then clicking Run.

3.4.5 Probe Array Washing, Staining, and Scanning

1. Prepare 1,200 µl of fresh streptavidin R-phycoerythrin (SAPE) stain solution for each array by mixing:

   600 µl 2× stain buffer.

   48 µl BSA.

   12 µl SAPE.

   540 µl nuclease-free water.

   Divide into two aliquots of 600 µl each for staining **steps 1** and **3**.
2. Prepare 600 µl of fresh Antibody Solution Mix:

   300 µl 2× Stain buffer.

   24 µl BSA.

6 μl Goat IgG Stock.

3.6 μl Biotinylated antibody.

266.4 μl Nuclease-free water.

3. Remove hybridization cocktail from the probe array immediately after 16 h hybridization and completely fill with ~250 μl of Wash Buffer A. This probe array can be stored at 4 °C for up to 3 h if washing and staining is not available immediately after hybridization. Perform washing and staining of the probe array using one of either Fluidics Station 450/250 or Fluidics Station 400 following the automated protocol. Different types of chips have different washing and staining protocols; use the automated washing and staining protocol recommended by the manufacturer for your particular chip.
4. Remove the probe arrays from the Fluidics Station at the end of washing and staining and check the probe array glass window for bubbles or air pockets. If there are large bubbles, fill the probe array with array holding buffer (1× MES) without making large bubbles using Fluidics Station or manually. Keep the probe array in the dark at 4 °C until ready for scanning.
5. Warm up Affymetrix GeneChip Scanner 3000 for at least 10 min before scanning.
6. Clean the excess fluid from the probe array and apply one Tough-Spot to each of the two septa, pressing to ensure that the spots remain flat. Clean the glass window of probe arrays. Insert the cartridge into the scanner and test the auto-focus to ensure that the Tough-Spots do not interfere.
7. Select Run and Scanner from the menu bar or click the Start Scan in the tool bar Select experiment name and click Start button.
8. Open the sample door on the scanner and apply the probe array without forcing the probe array into the holder. Close the door.
9. Click OK in the Start Scanner dialog box.
10. Each complete probe array image is stored in a separate data file identified by the experiment name and is saved with a data image file (.dat) extension.

3.5 Microarray Data Analysis

This is a simplified version of analysis options using Affymetrix microarray data analysis software. The technical manuals of the analysis software have extensive descriptions for choosing the various options, and we refer the reader to those websites. Describing them here is beyond the scope of this chapter.

3.5.1 Checking Scanned Image and Converting Fluorescence Intensities to Numerical Values

1. Double-click on each *.dat file to open it.
2. Under the tab for "Image settings," choose autoscale and pseudocolor, and then click "OK."
3. Click on "Grid" tab, press "G," and select View→Grid from menu bar to superimpose grid lines on the scanned image.
4. Make sure that the corners of the scanned image fit within the grid.
5. Manually align the grid by click-dragging the double arrows on the grid perimeters.
6. Select Run→Analysis from menu bar.
7. At the end of each analysis of a *.dat file (~2 min), cell intensity data is computed and a *.cel file is created (single intensity value is computed for each probe cell).

3.5.2 File Types

1. The experiment information file (*.exp) contains information about the experiment name, sample, and probe array type. This file is not used for analysis but may be required to open other files for the designated probe array.
2. The data file (*.dat) is the image of the scanned probe array. *See* **Note 15**.
3. The cell intensity file (*.cel) is derived from a *.dat file and is automatically created when a *.dat file is opened. It contains a single intensity value for each probe cell delineated by the grid.
4. The report file (*.rpt) is a text file summarizing data quality information for a single experiment. The report is generated from the analysis output file (*.chp).

3.5.3 Preliminary Data Analysis

Each transcript in an expression array is measured by 11 probe pairs which consist of a 25-mer perfect-match oligonucleotide (PM) and a 25-mer mismatch oligonucleotide (MM) that contains a single base pair mismatch in the central position. The PM/MM design allows identification and subtraction of nonspecific hybridization and background signals at the cost of introducing more variability in the gene expression index. Affymetrix software uses the MM intensity to estimate stray signal, and this is subtracted from the PM signal. Check the Eukaryotic Hybridization Controls (*see* **Note 16**) and the poly-A control (*see* **Note 17**). Every chip in an experiment must meet strict QC controls (Fig. 1c) before its data can be included in the final analysis. Further details about this analysis can be found in the Affymetrix User's Guides.

4 Notes

1. For very lipid-rich tissues such as adipose tissue, use Allprotect (Qiagen, Valencia, CA) in place of RNA*later* for stabilization. RNA*later* can stabilize brain but not adipose tissue; Allprotect can stabilize both.
2. Use the standard "RNeasy Mini Kit" for easy-to-lyse cells and tissues that are not fibrous or lipid rich. For example, the standard RNeasy Mini Kit is routinely used to isolate RNA from non-fatty tissue culture cells and liver. The RNeasy Fibrous Tissue Mini Kit can be used to isolate RNA from RNA*later*-stabilized fibrous tissue such as muscle, heart, or aorta, but not from intestine or pancreas, since these two tissues are rich in RNases and the RNase-inactivating buffer in this kit (RLT) is diluted to permit Protease K digestion. For intestine and pancreas, try the standard RNeasy kit or follow the TRIzol method as described in Chapter 2 of this volume [7]. The TRIzol method can be used to isolate RNA from any RNA*later*-stabilized tissue but is especially useful for fibrous tissues that are difficult to lyse. For very lipid-rich tissues such as adipose tissue and brain, use the RNeasy Lipid Tissue Mini Kit according to manufacturer's instructions or use the TRIzol method as described in Chapter 2 of this volume [7]. If you are not sure which RNeasy kit to use for your particular tissue, call the manufacturer.
3. RNA purified from RNeasy Mini Kits are >200 bp. RNA species smaller than 200 bp may not be quantitatively recovered by many popular RNA isolation procedures that utilize traditional ethanol precipitation or column purification of RNAs. If recovery of smaller RNAs (<200 bp) is important for your experiment, choose a suitable RNA isolation method. For small RNAs, such as miRNAs, use TRIzol (Invitrogen) or TRI reagent (Sigma) for RNA isolation, following the procedure described in ref. 10. Special total RNA isolation kits that provide quantitative recovery of miRNAs are also available, e.g., Ambion *mir*Vana miRNA Isolation Kit (Ambion, Inc., Austin, TX). Always carefully read and follow the manufacturer's instructions for the specific small RNA that you are interested in.
4. The Eukaryotic Hybridization Controls (EHC) contain a mixture of four biotin-labeled, antisense, fragmented non-eukaryotic control cRNAs (bioB, bioC, bioD, and cre), which are spiked in staggered concentrations into the hybridization cocktail (final concentrations 1.5, 5, 25, and 100 pM, respectively). These serve as controls for effectiveness of hybridization, washing, and staining. A synthetic biotinylated control oligo (oligo B2) is also spiked into the hybridization solution to provide alignment signals for the probe arrays.

The Eukaryotic Hybridization Controls and oligo B2 are provided as concentrated stocks, ready to be added to the hybridization mix (Subheading 3.4.1). The Eukaryotic Hybridization Controls must be heated to 65 °C before being added to the hybridization cocktail (Subheading 3.4.1).

5. Reduction of globin transcripts is highly recommended when whole blood RNA is used in expression arrays if IVT express kit is used for preparing the labeled targets. This can be circumvented by using NuGEN's Ovation™ Biotin RNA Amplification and Labeling System when small amounts of starting material has to be processed.

6. The integrity of cellular RNA can be adversely affected by conventional histological staining methods preparatory to laser capture microdissection [11]. Proposed solutions to this problem include adding RNase inhibitors to aqueous staining solutions [11] and alternative staining methods, e.g., carbon-labeling of Kupffer cells in vivo [12]. Paul et al. [13] have reported successful microarray profiling of laser-captured cells from atherosclerotic lesions in mice. RNA from laser-captured cells must meet the same stringent quality controls as any other RNA sample if it is to be used for microarray analysis.

7. The quality of RNA is very critical in microarray experiments, as impurities have an adverse effect on both the amplification and the labeling efficiency. The A_{260}:A_{280} ratio for RNA samples of acceptable purity should be between 1.8 and 2.1. Sample integrity is also very important and is determined by the entire electrophoretic trace of the RNA sample using the Agilent Bioanalyzer, including the presence or absence of degradation products. The RNA integrity number (RIN) is a software tool for estimating the integrity of total RNA samples on a quantitative scale of 1 (worst) to 10 (best). The RIN for samples of acceptable integrity should be >8.

8. The DNase treatment of RNA is crucial. In addition to removing the DNA, it purifies the RNA and reduces background.

9. For easy-to-lyse tissues, addition of the RLT lysis buffer provided with the standard RNeasy Mini Kit and passage over the Qiashredder column may be sufficient to lyse the tissue and homogenize the lysate. Mechanical disruption and homogenization is, however, likely to improve the yield (see RNeasy Mini Handbook). For the RNeasy Fibrous Tissue Mini Kit, fibrous tissues such as muscle, heart, or aorta should be mechanically disrupted and homogenized in the RNeasy Fibrous Tissue Mini Kit lysis buffer using a roto-stator such as the TissueRuptor (Qiagen) or a Polytron/Ultra-Turrax (Brinkmann Instruments/Kinematica Inc.) or in bead mills such as the TissueLyser (Qiagen), which can lyse and disrupt

multiple samples simultaneously. For the RNeasy Lipid Tissue Mini Kit, lipid-rich tissues such as adipose tissue or brain should be mechanically disrupted in the Qiazol Lysis reagent using a roto-stator or bead mill. Intestine is best disrupted and homogenized in TRIzol using a roto-stator, for example, as described in Chapter 2 of this volume [7].

10. The protocols described here are minimally adapted from protocols in recent (~2010–2011) Affymetrix user manuals. Methodologies and instrumentation frequently evolve over time. If newer versions of the recommended kits and instrumentation become available, refer to the updated user manuals from the manufacturer.
11. The Affymetrix GeneChip IVT Express Kit protocol should be applied to samples with RNA concentration greater than 20 ng. For small size samples less than 20 ng, NuGEN's SPIA amplification protocol can be used. For hybridization, 15–20 μg of biotin-labeled cRNA is needed for a standard expression gene chip.
12. Poly-A hyb control: Polyadenylated transcripts (lys, phe, thr, and dap) are spiked into the labeling mixture at this step as positive controls to verify effectiveness of the procedure, from first-strand cDNA synthesis onwards.
13. The median size distribution of the cRNA should be around 400–600 nt. This, however, may not be applicable in a 2-round amplification process.
14. Fragmentation of cRNA should not be overdone as this could cause high background, low signal, and high 3′/5′ ratios.
15. A general visual inspection of the entire GeneChip probe array should be performed after scanning. B2 oligos, which are spiked into the hybridization cocktail (*see* **Note 4**), should be used for checking the quality of hybridization grid alignment.
16. The Eukaryotic Hybridization Controls (EHC) are a mixture of four biotin-labeled, antisense, fragmented noneukaryotic control cRNAs, which are spiked at staggered concentrations into the hybridization cocktail (*see* **Note 4**). Oligos complementary to these sequences are always spotted onto all gene chips arrays and the EHCs thereby serve as hybridization controls. All four transcripts should maintain a maximum 1:2 ratio of signal intensities of the 5′and 3′ probe sets. BioB should be present at least 50 % of the time while bioC, bioD, and cre should be always present, with increasing signal intensities.
17. The poly-A spike in controls added during the labeling process (i.e., added to target RNA in Subheading 3.3.1) should be a present call with increasing signal intensity in the order of lys, phe, thr, and dap.

References

1. Jison ML, Munson PJ, Barb JJ, Suffredini AF, Talwar S, Logun C, Raghavachari N, Beigel JH, Shelhamer JH, Danner RL, Gladwin MT (2004) Blood mononuclear cell gene expression profiles characterize the oxidant, hemolytic, and inflammatory stress of sickle cell disease. Blood 104:270–280
2. Chai V, Vassilakos A, Lee Y, Wright JA, Young AH (2005) Optimization of the PAXgene blood RNA extraction system for gene expression analysis of clinical samples. J Clin Lab Anal 19:182–188
3. Debey S, Schoenbeck U, Hellmich M, Gathof BS, Pillai R, Zander T, Schultze JL (2004) Comparison of different isolation techniques prior gene expression profiling of blood derived cells: impact on physiological responses, on overall expression and the role of different cell types. Pharmacogenomics J 4:193–207
4. DePrimo SE, Wong LM, Khatry DB, Nicholas SL, Manning WC, Smolich BD, O'Farrell AM, Cherrington JM (2003) Expression profiling of blood samples from an SU5416 Phase III metastatic colorectal cancer clinical trial: a novel strategy for biomarker identification. BMC Cancer 3:3
5. McPhail S, Goralski TJ (2005) Overcoming challenges of using blood samples with gene expression microarrays to advance patient stratification in clinical trials. Drug Discov Today 10:1485–1487
6. Raghavachari N, Xu X, Munson PJ, Gladwin MT (2009) Characterization of whole blood gene expression profiles as a sequel to globin mRNA reduction in patients with sickle cell disease. PLoS One 4:e6484
7. Wagner EM (2009) Monitoring gene expression: quantitative real-time RT-PCR. In: Freeman LA (ed.) Lipoproteins, 2nd edn. Humana, Totowa, NJ
8. Affymetrix (2006) Globin reduction protocol: a method for processing for whole blood RNA samples for improved array results. Affymetrix Technotes
9. Whitley PMS, Santiago J, Johnson C, Setterquist R (2005) Improved microarray sensitivity using blood RNA samples. Ambion Technotes 20–22
10. Varallyay E, Burgyan J, Havelda Z (2008) MicroRNA detection by northern blotting using locked nucleic acid probes. Nat Protoc 3:190–196
11. Wang H, Owens JD, Shih JH, Li MC, Bonner RF, Mushinski JF (2006) Histological staining methods preparatory to laser capture microdissection significantly affect the integrity of the cellular RNA. BMC Genomics 7:97
12. Gehring S, Sabo E, San Martin ME, Dickson EM, Cheng CW, Gregory SH (2009) Laser capture microdissection and genetic analysis of carbon-labeled Kupffer cells. World J Gastroenterol 15:1708–1718
13. Paul A, Yechoor V, Raja R, Li L, Chan L (2008) Microarray gene profiling of laser-captured cells: a new tool to study atherosclerosis in mice. Atherosclerosis 200:257–263

Chapter 4

Northern Analysis of Gene Expression

Lita A. Freeman

Abstract

Northern analysis detects and quantitates full-length RNA transcripts by hybridizing a labeled sequence-specific probe to RNA fractionated by size and blotted onto a membrane. The first step in Northern analysis is RNA extraction from the tissue(s) of interest, followed by RNA denaturation and electrophoresis to separate all RNA transcripts in the tissue by size. The size-fractionated RNA is then transferred (blotted) onto a membrane. The membrane is incubated with a labeled probe complementary to the sequence of interest. Stringent washing removes nonspecifically bound probe and leaves labeled probe bound specifically to the transcript(s) of interest on the membrane. Northern analysis provides information on gene transcript sizes and abundance and the presence of alternative splice variants or transcription initiation sites. Northern analysis can also confirm the presence or absence of specific exons within a given transcript. The method is medium throughput, as multiple samples can be analyzed side by side with relatively little effort. Finally, any potential PCR amplification artifacts are completely avoided using this methodology. Northern analysis offers a unique view of a gene's transcripts and provides novel insight into gene expression.

Key words RNA, Gene expression, Hybridization, Transcription, Alternative splicing, Alternative promoters, Lipoproteins

1 Introduction

Northern analysis (Northern blotting, Northern blot analysis) provides information on gene transcript sizes and abundance without the use of RT-PCR and is an invaluable tool in gene expression studies. An overview of the procedure is shown in Fig. 1.

To determine expression of a lipoprotein gene in a given tissue by Northern analysis, the first step is to isolate the tissue and extract its RNA (Fig. 1). Typical tissues of interest in lipoprotein research include liver, intestine, macrophages, adipose tissue, inflammatory cells, and brain [1, 2]. RNA is isolated using methods that preserve RNA integrity. RNA is denatured and then fractionated by size using gel electrophoresis. Size-fractionated RNA is subsequently transferred (blotted) onto a membrane, which is then incubated with a labeled probe that hybridizes to the sequence of interest

Lita A. Freeman (ed.), *Lipoproteins and Cardiovascular Disease: Methods and Protocols*, Methods in Molecular Biology, vol. 1027, DOI 10.1007/978-1-60327-369-5_4, © Springer Science+Business Media, LLC 2013

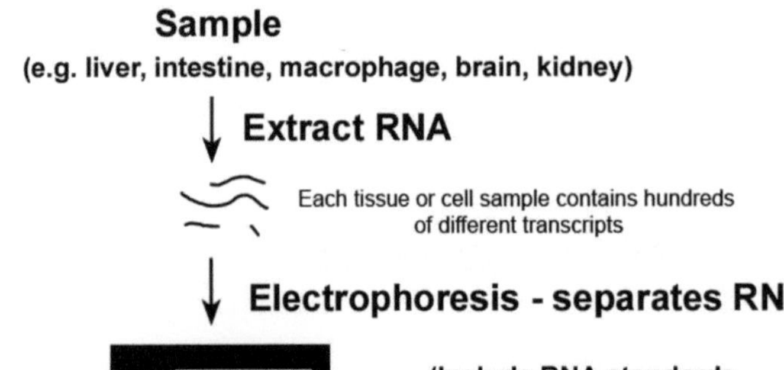

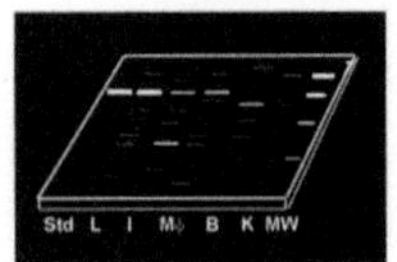

(Include RNA standards and MW markers on gel)

Each lane contains all the transcripts in the sample, separated by size

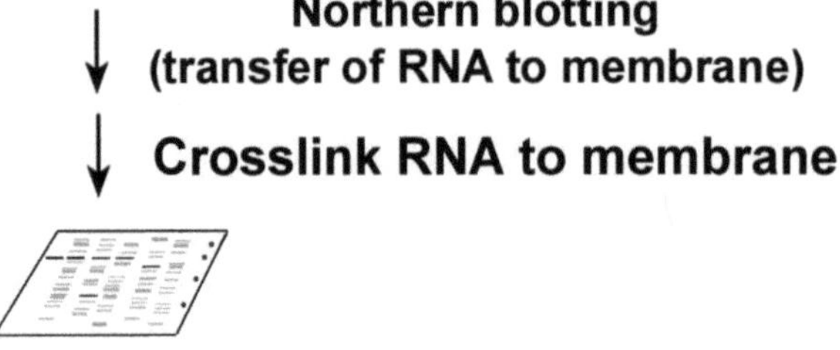

Northern blot. Each transcript originally present in the gel is now attached to the membrane.

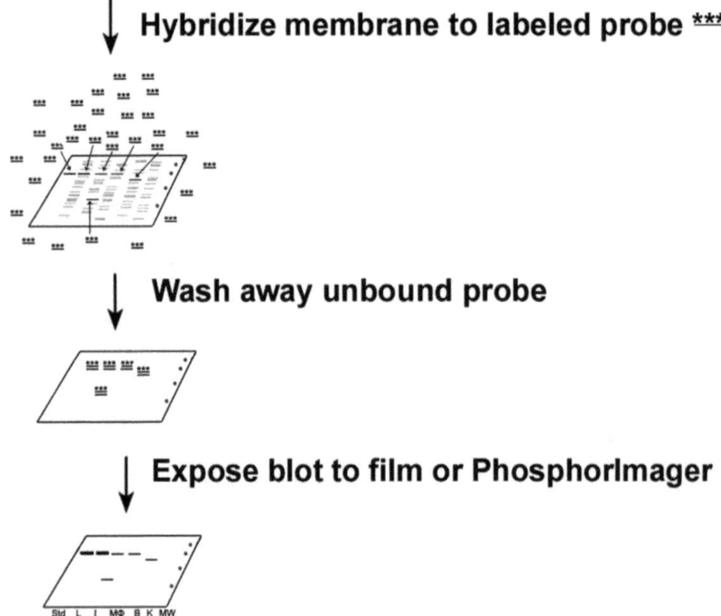

Fig. 1 Northern blotting: overview of the procedure. Tissues or cells are harvested and RNA is extracted from each sample using techniques that minimize RNA degradation. Hundreds or thousands of different transcripts, differing in size and sequence, are present in each RNA sample. For Northern analysis, the transcripts from each sample (e.g., liver, L; intestine, I; macrophages, MΦ; brain, B; and kidney, K) are separated according to size by

(Fig. 1). The membrane is washed to remove nonspecifically bound probe, leaving labeled probe bound specifically to the transcript(s) of interest on the membrane (Fig. 1). Transcript size and abundance are determined after visualizing and quantitating the signal from the probe.

Like RT-qPCR or microarray analysis, Northern analysis is used to quantitate changes in lipoprotein metabolism gene expression in specific cells and tissues in response to alterations in diets, addition of drugs, or overexpression or knockout or knockdown of a gene of interest (*see* refs. 3–9 for selected examples).

Unlike RT-qPCR or microarray analysis, however, Northern analysis probes full-length transcripts for both size and abundance. Moreover, while all three techniques can quantitate the abundance of a given sequence, only Northern analysis can specifically identify and quantitate different full-size transcript variants in one experiment. A full-length cDNA Northern probe will reveal all transcripts from a gene, within the limits of sensitivity of the procedure. Northern probes can also be designed to detect specific exons, 5′UTRs, 3′UTRs, etc., and Northern analysis is therefore very useful in confirming alternative splice variants, initiation sites, etc., present in a specific transcript. Finally, PCR artifacts are avoided, as the method does not employ PCR at any stage.

Thus, Northern analysis, in combination with sequencing of 5′ and 3′ RACE products, supplies unique qualitative and quantitative information on a gene's full-length transcripts present in a given tissue or cell type.

Here we describe a Northern analysis protocol that avoids the use of formaldehyde, is sensitive and specific, and produces very clean blots with essentially no background. Moreover, large mRNAs, such as the 6.78 kb ABCA1 transcript, are readily transferred and detected. This method can be used alone or in combination with previous methods described in this volume to assess lipoprotein-related gene expression in different tissues under differential genetic, nutritional, pharmaceutical, or environmental conditions and in healthy vs. diseased states.

Fig. 1 (continued) electrophoresis on an agarose gel, along with RNA standards (Std) and molecular weight markers (MW). A hypothetical gel demonstrating the presence of different transcripts in different tissues is shown here; in reality, 28S rRNA and 18S rRNAs are the major species visible on an RNA gel stained with ethidium bromide. All other transcripts in the sample appear as a smear up and down the lane and their identity cannot be distinguished by ethidium bromide staining. Following electrophoresis, all RNA molecules are transferred from the gel to a membrane and then cross-linked to the membrane. To visualize a particular transcript of interest, a DNA (or RNA) molecule complementary to that transcript is labeled and then incubated with the membrane. The labeled probe hybridizes specifically to its complementary RNA molecule affixed to the blot. The membrane is thoroughly washed to remove unbound probe and then exposed to autoradiographic film or a PhosphorImager to visualize transcripts complementary to the probe. Note that transcript sizes and the presence of alternative transcripts, as well as relative transcript quantities, are readily visualized by Northern analysis

2 Materials

Prepare all materials using RNase-free technique (*see* **Note 1**).

2.1 RNA Preparation

2.1.1 Tissue Isolation and Stabilization

RNA*later* (Ambion Life Technologies) or, for adipose tissue, Allprotect (Qiagen, Valencia, CA).

2.1.2 RNA Isolation

Caution: All RNA isolation kits contain strong denaturants. Wear gloves and eye protection during use.

1. RNA isolation reagents or kits. *See* Subheading 3.1.2 for guidance as to the proper RNA isolation method for your tissue or RNA species of interest.
   (a) TRIzol reagent (Invitrogen, Carlsbad, CA).
   (b) QIAzol (Qiagen).
   (c) UltraSpec^{R} (Biotecx, Houston, TX).
   (d) The RNeasy line of RNA isolation kits from Qiagen (*see* **Note 2**). Note that Qiagen has developed a robotic workstation, the QIAcube, for automated RNA purification using RNeasy spin-column kits.
   (e) For small RNAs, such as miRNAs, purification options (*see* **Note 3**) include:
      - TRIzol (Invitrogen) or TRI reagent (Sigma, St. Louis, MO). Use as described in ref. 10.
      - The *mir*Vana miRNA Isolation Kit (Ambion Life Technologies). Efficiently purifies all RNA larger than 10 nucleotides (nt). Also includes procedures for isolating RNA fractions specifically enriched or depleted in small RNA species.
      - The miRNeasy Mini Kit (Qiagen). Yields total RNA larger than approximately 18 nt. Can be automated using the QIAcube (Qiagen).

        Avoid column-based procedures that result in loss of small RNA species (such as regular RNeasy kits, in which RNA species <200 bp are lost). These cannot be used for small RNA isolation [10].

   Store all reagents according to manufacturer's instructions.
2. RNase-free water.
3. Common laboratory reagents such as ethanol, isopropanol, chloroform, or 2-mercaptoethanol may need to be supplied by the user; read the manufacturer's instructions carefully and plan ahead before proceeding.
4. (Optional) Mid- to high-throughput RNA isolation systems, e.g., QIAcube (Qiagen).

2.1.3 (Optional) PolyA+ RNA

1. PolyA+ RNA isolation from purified RNA: Kits and reagents are commercially available from many suppliers. The polyA Spin mRNA Isolation Kit from New England Biolabs (Ipswich, MA) is one of many good choices. Store at 4 °C.
2. PolyA+ RNA isolation directly from a crude lysate (without prior total RNA purification): Use either of the following:
   (a) Oligotex Direct mRNA kits (Qiagen). Store at room temperature.
   (b) Dynabeads mRNA purification kits (Invitrogen). Store at 4 °C.

2.1.4 RNA Quantitation

1. Instrument and reagents for quantitating RNA as described in Chapters 2 and 3 of this volume [1, 2].
2. (Optional) RNase-free glycogen (UltraPure™ Glycogen, Invitrogen).

2.2 Electrophoresis

2.2.1 Pouring the Agarose Gel

1. A 500 ml RNase-free flask (*see* **Note 1**). Disposable polycarbonate sterile 500-ml capped Erlenmeyer flasks (Corning, Inc., Corning, NY) are convenient.
2. DEPC (diethyl pyrocarbonate) (Sigma). DEPC inactivates RNases. Use in fume hood.
3. DEPC-treated H_2O. Add 1.0 ml DEPC per 1.0 L water in a glass flask cover and stir overnight. Autoclave on liquid cycle. DEPC-treated water is also commercially available.
4. NorthernMax^R-Gly 10× Gel Prep/Running Buffer (Ambion Life Technologies). This solution is supplied RNase-free. Store at 4 °C.
5. 1.5 g Agarose, e.g., SeaKem GTG Agarose (Lonza Rock-land Inc., Rockland, ME). If possible, reserve a bottle for RNA use only.
6. Heat-proof gloves.
7. Horizontal gel electrophoresis apparatus, e.g., Wide Mini Sub-Cell GT with 7 cm × 10 cm gel tray and a 20-well adjustable comb (0.75 thick) and accompanying comb holder (*see* **Note 4**) (BioRad Laboratories, Hercules, CA). Multichannel-pipette-compatible combs are also available for high-throughput gel loading.
8. (Optional) For microRNA, use a denaturing polyacrylamide gel, not an agarose gel, for electrophoresis [10].

2.2.2 Preparing Samples and Standards

1. Heating block prewarmed to 50 °C for denaturing samples.
2. RNA molecular weight markers (e.g., ssRNA Ladder, catalog # N0362S; New England Biolabs). Store at −80 °C.
3. (Optional) Quantitation standards, for example an in vitro-transcribed RNA of known concentration, complementary to the probe, for absolute quantitation; or, for relative quantitation,

such as normalizing sample intensities between blots, a massive RNA prep from a tissue that expresses the gene of interest. Divide quantitation standards into aliquots and store frozen at –80 °C.

4. RNA sample (up to 30 μg RNA). Store at –80 °C.
5. Glyoxal Sample Loading Dye (Ambion). *Caution*: contains ethidium bromide, a potential carcinogen. Wear gloves! Store at –20 °C.

2.2.3 Running the Gel

1. Horizontal gel apparatus (*see* Subheading 2.2.1 **item** 7, above).
2. 10× NorthernMax^R-Gly Gel Prep/ Running Buffer (Ambion).
3. 1× NorthernMax^R-Gly Gel Prep/Running Buffer. Dilute 100 ml 10× NorthernMax^R-Gly Gel Prep/Running Buffer to a final volume of 1.0 l with RNase-free water.

2.2.4 Viewing and Imaging the Gel

1. UV transilluminator.
2. Fluorescent ruler (#FRK-100) (Research Products International Corp., Mount Prospect, IL).
3. UV goggles.
4. Camera and film or a digital imaging system with an ethidium bromide filter.
5. Molecular biology grade water.
6. 10× TAE: 0.4 M Tris base, 0.01 M EDTA, pH'd to 8.3 with glacial acetic acid.
7. 2× TAE: Dilute 400 ml 10× TAE to a final volume of 2.0 l with molecular biology-grade water. Part of this will be used in the next section as transfer buffer.

2.3 Blotting and Cross-Linking

2.3.1 Sponge Method

1. Zeta-Probe^R GT (Genomic-Tested) (BioRad Laboratories).
2. Thick sponge larger than the gel (e.g., Spontex Professional Grout Sponge or Large Heavy Duty Sponge (Spontex USA, Columbia, TN; spontexusa.com) (*see* **Note 5**).
3. Two glass baking dishes or Tupperware dishes larger than the gel and sponge.
4. Six pieces of Whatman 3 MM paper (GE Healthcare, Piscataway, NJ) precut to the same size as the membrane.
5. 2× TAE.
6. Plastic wrap or parafilm.
7. Membrane marking pen (Whatman # 10499001) (GE Healthcare).

8. Whatman grade GB004 or GB003 blotting sheets (formerly Schleicher & Schuell GB004 or GB003) (GE Healthcare).
9. RNase-free water or 10 mM Tris–HCl, pH 7.4.
10. Ethidium bromide (Invitrogen). *Caution*: Possible carcinogen! Wear gloves and dispose as toxic waste. Stock solution is supplied as 10 mg/ml; for staining RNA, dilute to a final concentration of 0.5 μg/ml.

2.3.2 TurboBlotter (Upwards Blotting)

IMPORTANT: Wear gloves or use forceps when handling the membrane!

1. TurboBlotter™ (Whatman/GE Healthcare)—available in two sizes, small and large.
2. Zeta-Probe® GT TurboBlotter (Genomic-Tested) (BioRad Laboratories).
3. (Optional) TurboBlotter refills (Whatman/GE Healthcare)—contain blotting paper and wicks. Also contains Nytran membrane, if the user prefers to use this membrane rather than Zeta-Probe GT.
4. Transfer buffer (2× TAE).
5. Twenty sheets of dry GB004 or GB003 blotting paper.
6. Four sheets of dry Whatman 3 MM paper.
7. Four sheet of Whatman 3 MM paper, precut to the size of the gel.
8. Plastic wrap or Parafilm.
9. Precut "buffer wick" (can use the wick provided in the refill pack as a template), wide enough so that the blotting stack is completely covered by the short dimension of the wick and long enough for the long ends of the wick to dip into opposite sides of the buffer tray.
10. Membrane marking pen.
11. RNase-free water or 10 mM Tris–HCl pH 7.4.
12. Ethidium bromide (Invitrogen). *Caution*: Possible carcinogen! Wear gloves and dispose as toxic waste. Stock solution is supplied as 10 mg/ml; for staining RNA, dilute to a final concentration of 0.5 μg/ml.

2.3.3 Cross-Linking

1. Whatman 3 MM paper (GE Healthcare).
2. UV light source for cross-linking RNA to blot (e.g., Stratalinker UV Crosslinker, Stratagene, La Jolla, CA).

2.4 Working with ^{32}P

1. Training and approval by your radiation safety department is required before working with ^{32}P.

2. Plexiglass shield (VWR, West Chester, PA) covered with absorbent bench paper.
3. (Optional) Tray (e.g., cafeteria tray).
4. Thick plexiglass tube holders for microcentrifuge tubes (VWR).
5. Thick plexiglass tube holders for 50-ml centrifuge tubes (VWR).
6. Geiger counter, preferably with a pancake probe for highest sensitivity.
7. Radioactive label tape.
8. Solid ^{32}P waste container (VWR).
9. Liquid ^{32}P waste container (VWR).
10. Count-Off cleaner (*see* **Note 6**).

2.5 Preparing the Probe

2.5.1 ^{32}P-Labeling of the DNA Fragment

1. 25–50 ng gel-purified double-stranded DNA fragment containing the sequence of interest.
2. 25–50 ng gel-purified double-stranded DNA fragment containing the sequence of a loading control appropriate for the tissue(s) being analyzed.
3. [$\alpha^{32}P$]dCTP 3,000 Ci/mmol, 10 mCi/ml (PerkinElmer). *Caution*: Use maximum radiation safety precautions when working with the source vial! *See* Subheading 3.4. [$\alpha^{32}P$]dCTP can be purchased in either non-stabilized (must be stored at –20 °C and thawed before each use) or stabilized (can be stored at 4 °C; no need to thaw before use) form. [$\alpha^{32}P$]dCTP should be used no later than 10 days past the reference date for best performance (*see* **Note 7**). Prevent unauthorized access to radioactivity: do not leave [$\alpha^{32}P$]dCTP ^{32}P unattended in an unlocked room and store [$\alpha^{32}P$]dCTP shielded, in a locked freezer or refrigerator. A permanent record must be kept of all ^{32}P usage.
4. GE Healthcare Amersham Ready-To-Go DNA Labeling Beads (-dCTP) (#27-9240-01).
5. 95–100 °C heating block or water bath.
6. H_2O, molecular biology grade.
7. 37 °C heating block.

2.5.2 Removing Unincorporated Nucleotides with ProbeQuant™ G-50 Micro Columns

1. ProbeQuant G-50 Micro Columns (GE Healthcare #28-9034-08) (formerly #27-5335-01; *see* **Note 8**).
2. Scintillation tubes.
3. Liquid scintillation fluid.
4. Scintillation counter.

2.6 Prehybridization, Hybridization, and Washing

2.6.1 Prehybridization

1. Shaking water bath (or hybridization oven, *see* **Note 9**) preheated to hybridization temperature (usually 65 °C).
2. Glass baking dishes.
3. RNase-free water or 10 mM Tris–HCl, pH 7.4 (~500 ml for rewetting blot).
4. Salmon testes DNA (salmon sperm DNA) (10 mg/ml in dH_2O, sonicated) (Sigma, Catalog #D7656).
5. PerfectHyb™ Plus Hybridization Buffer (PerfectHyb™) (Sigma, Catalog #H7033).
6. Whatman 3 MM paper.
7. Heat-sealable plastic bag (VWR) (or screw-capped glass hybridization tubes if using a hybridization oven; *see* **Note 9**).
8. Bag sealer, e.g., Fuji Desktop Poly Sealer, Ampac* (Kapak) (VWR) (unnecessary if using a hybridization oven).

2.6.2 Hybridization

1. Shaking water bath (or hybridization oven), preheated.
2. Scissors.
3. 10-ml pipettes.
4. Prewarmed heat sealer (unnecessary if using a hybridization oven; *see* **Note 9**).
5. Dry waste container for ^{32}P.
6. Liquid waste container for ^{32}P.
7. Paper towels, full box.
8. PerfectHyb™ Plus Hybridization Buffer (Sigma).
9. 100 μl salmon sperm DNA (10 mg/ml) (Sigma).
10. 25×10^6 cpm (or more) of radiolabeled, column-purified probe (prepared in Subheading 3.5).
11. 95 °C heating block or water bath.

2.6.3 Washes

1. 20× SSC: 3.0 M sodium chloride, 0.3 M trisodium citrate; adjust pH to 7.0 with HCl.
2. 10 % SDS.
3. 1.0 liter Low-stringency wash buffer:

   100 ml 20× SSC.

   10 ml 10 % SDS.

   Molecular biology grade or DEPC-treated water to a final volume of 1.0 L.

   Use at room temperature.
4. Two glass baking dishes.
5. 50-ml disposable centrifuge tube with cap.

6. Thick plexiglass holder for 50-ml tube centrifuge tubes (VWR).
7. Scissors.
8. Forceps.
9. Paper towels (full box).
10. High-stringency wash buffer:

    25 ml 20× SSC.

    10 ml 10 % SDS.

    Molecular biology grade or DEPC-treated water to a final volume of 1.0 L.

    Prewarm to 65 °C before use.
11. Ultrahigh-stringency wash buffer:

    5.0 ml 20× SSC.

    100 ml 10 % SDS.

    Molecular biology grade or DEPC-treated water to a final volume of 1.0 L.
12. Whatman 3 MM paper.
13. Plastic food wrap.

2.7 Detection

2.7.1 Film

1. Autoradiography cassette.
2. Two autoradiography screens (e.g., Kodak BioMax MS Intensifying Screens Cat# 851-8706, Carestream Health Molecular Imaging, New Haven, CT).
3. Autoradiography film (e.g., Kodak BioMax MS film, Carestream Health Molecular Imaging).

2.7.2 Storage Phosphor/PhosphorImager Technology

Typhoon or Storm imaging system (GE Healthcare).

2.8 Probe Stripping

1. Shaking water bath set to 95 °C.
2. Heat-proof gloves.
3. Pyrex baking dish.
4. Stripping buffer: 0.1× SSC, 0.5 % SDS.

3 Methods

3.1 RNA Preparation

3.1.1 Tissue Isolation and Stabilization

RNA is very susceptible to degradation but must remain intact for Northern analysis. Thus, tissues and cells must be harvested under conditions that minimize RNA degradation (*see* Appendix for general guidelines and refs. 1 and 2 for specific procedures). Samples that are not immediately processed for RNA extraction should be immersed immediately into a stabilization reagent (RNAlater or Allprotect) that permeates tissue/cells and protects RNA from degradation (*see* **Note 10**). For specific tissue isolation and stabilization protocols, *see* refs. 1 and 2.

3.1.2 RNA Isolation

RNA isolation methods that yield RNA suitable for Northern analysis include TRIzol reagent (Invitrogen), Qiazol (Qiagen), UltraSpec[R] (Biotecx), and RNeasy kits (*see* **Note 2**). If small RNAs (smaller than ~200 bp) are to be investigated by Northern analysis, special RNA isolation methods are required, e.g., TRIzol (Invitrogen), TRI reagent (Sigma, Ambion), or the *mir*Vana™ RNA Isolation Kit (Ambion) (*see* **Note 3**). The miRNeasy Mini Kit (Qiagen) enriches miRNA and RNAs <200 nt and the procedure can be automated on the QIAcube. Avoid methods that have not been optimized for miRNA or other small RNA isolation, as small RNA species may be lost.

For mid- to high-throughput applications, several companies have developed instruments that deliver automated walkaway preparation of nucleic acids, including RNA. Interested users should certainly give these a try. The QIAcube (Qiagen), a robotic workstation for automated purification of DNA, RNA, or proteins, processes up to 12 samples per run using Qiagen spin-column kits, yielding high-quality RNA.

Certain tissues commonly used for lipoprotein studies pose special challenges for RNA isolation. Adipose tissue must be stabilized with Allprotect rather than RNA*later* due to its high lipid content, and RNA from fatty tissues such as adipose tissue or brain should be isolated using the TRIzol method [1] or the RNeasy Lipid Tissue Kit. Intestine must be cleaned out as described in ref. 1 and homogenized using a rotostator in a strong denaturant such as TRIzol, following the methodology described in ref. 1 or using the UltraSpec kit according to manufacturer's instructions, as described in ref. 6. The RNeasy Fibrous Tissue Kit is appropriate for fibrous tissues that are not RNase rich, such as aorta and heart, but should not be used for RNase-rich tissues such as intestine and pancreas. For RNA purification from whole blood or peripheral blood mononuclear cells (PBMCs), *see* ref. 2.

DNase treatment of RNA is not required prior to Northern analysis but will not interfere with the procedure as long as RNA integrity is preserved. Thus, RNA isolated for RT-qPCR or microarray analysis and quality-controlled by methods described in this volume can be used for Northern analysis as long as sufficient RNA is available. In general, any method that produces high-quality RNA in reasonable abundance (20–30 μg RNA per sample) is more than sufficient for all but the most scarce transcripts; 5 μg or even 2 μg RNA may give a detectable signal, depending on the transcript.

When planning your experiment, determine whether you will need quantitation standards. If you are only interested in comparing relative signal intensities of samples run side by side on a single blot, quantitation standards are not strictly necessary, but may be helpful in determining the linear range of signal intensity. RNA transfer and hybridization efficiency does vary between blots, so if multiple blots are to be compared, then quantitation/normalization standards must be included on each blot. The standard may be an in vitro-transcribed RNA complementary to the probe or aliquots from a single large RNA prep from a tissue expressing the transcript of interest (*see* **Note 11**). Load the same amount of the RNA standard on each of the gels, and use its intensity to normalize sample intensities between different blots. Loading several different concentrations of a quantitation standard can also help to assess linearity of signal intensities (*see* **Note 11**). If absolute quantitation is desired, serial dilutions of quantitated RNA prepared by in vitro transcription can be used as absolute quantitation standards.

Always include positive and negative controls.

3.1.3 PolyA+ RNA

If your transcript is very difficult to detect due to low abundance, and it is polyadenylated, polyA+ RNA isolation (capturing polyA+ transcripts using beads coupled to polydT) can provide increased sensitivity. *Note that polyA+ RNA does not contain transcripts that are not polyadenylated* (e.g., miRNA or other noncoding RNAs. Replication-dependent histone mRNAs are also not polyadenylated).

In our experience, isolating polyA+ RNA is not usually necessary for Northern analysis. Simply using a comb with wider teeth or pouring a thicker gel to accommodate more RNA in the sample well is usually sufficient to improve signal intensity. However, polyA+ RNA isolation can be beneficial in some cases and, of course, is useful for determining the polyadenylation status of specific transcripts.

If you have already isolated RNA and wish to purify polyA+ RNA from the existing RNA prep, a number of commercially available reagents and kits (e.g., the PolyA Spin mRNA Isolation Kit from New England Biolabs) are available. Alternatively, you can prepare a tissue or cell lysate and purify polyA+ mRNA directly from the crude lysate, without the extra steps involved in isolating total RNA first. DynaBeads (Invitrogen) and Oligotex Direct mRNA kits (Qiagen) are designed to isolate polyA+ RNA from a crude lysate, without prior total RNA isolation, and can be used to purify polyA+ RNA from total RNA as well. Again, this is usually not required for Northern analysis, will result in loss of non-polyadenylated transcripts, and becomes necessary only for enrichment of very scarce polyadenylated transcripts.

3.1.4 RNA Quantitation and Determination of Sample Volume

After RNA isolation, save an aliquot and quantitate as described in Chapter 2 or 3 of this volume [1, 2]. Determine the volume of each RNA sample required for Northern analysis (20–30 μg/well is standard, but as little as 5 μg or even 2 μg RNA per well may still give an acceptable signal). Ensure that the total sample volume (RNA + Glyoxal Sample Loading Dye; *see* **step 5** below) will not exceed the volume of the wells (see below). The same number of micrograms must be loaded in each well if the samples are to be directly compared; if the most dilute sample has only enough RNA to load 7 μg, for example, then only 7 μg RNA should be loaded on the gel for all the other samples as well. If necessary, RNA can be concentrated by ethanol or isopropanol precipitation (using RNase-free glycogen (UltraPure™ Glycogen, Invitrogen) as a carrier to reduce losses) or column binding followed by resuspension in a smaller volume. Quantitate the RNA again if concentrated in this manner.

3.2 Electrophoresis

During electrophoresis, RNA must be maintained in a denatured state, since native RNA has secondary structure which will affect its electrophoretic mobility. In the procedure described below, RNA is denatured by incubation in sample buffer containing glyoxal, which prevents secondary structure from forming and keeps the RNA denatured during electrophoresis. The gel itself and the gel running buffer do not contain glyoxal.

3.2.1 Pouring the Agarose Gel

1. Combine in a 500 ml RNase-free flask:

   135 ml DEPC-treated H_2O.

   15.0 ml NorthernMax^{R}-Gly 10× Gel Prep/Running Buffer.

   1.5 g agarose.

   Cover flask but not tightly—steam must be allowed to escape or explosive boiling will occur after cap removal.
2. Microwave on high with occasional swirling with until agarose is completely dissolved. *Caution*: Solution is extremely hot and can overboil or cause steam burns. Handle with heat-proof gloves!
3. Allow solution to cool on lab bench with occasional swirling until flask is still hot but can be touched (55–65 °C).
4. Pour into a 10-cm-long gel tray to a depth of 0.6–1.2 cm (*see* **Note 12**) and insert comb (*see* **Note 4**).
5. While the agarose gel is hardening, prepare the samples and standards.

3.2.2 Preparing Samples and Standards

Before beginning:

Prewarm a heating block to 50 °C for glyoxal denaturation of samples.

Thaw RNA molecular weight markers (e.g., ssRNA Ladder, New England Biolabs) and denature along with the experimental samples.

Include positive and negative controls, when feasible.

If using quantitation standards, denature along with the samples and RNA molecular weight markers.

Again, ensure that the total volume of sample plus loading dye will not exceed the volume of the wells.

1. Combine:

   1 Volume RNA sample (up to 30 μg RNA) (for lower RNA concentrations *see* **Note 13**), MW marker, or quantitation standard.

   1 Volume Glyoxal Sample Loading Dye (contains ethidium bromide, a potential carcinogen; *see* **Note 14**. Wear gloves!).

2. Incubate 30 min at 50 °C in a heating block (*see* **Note 15**).
3. Place on ice until ready to load gel. Alternatively, samples may be stored at −20 °C for several days before electrophoresis.

3.2.3 Running the Gel

1. Remove the comb carefully.
2. Place the gel (in the gel mold) into the gel apparatus.
3. Pour 1× NorthernMax^{R}-Gly Gel Prep/Running Buffer (Ambion) into the gel apparatus until the gel is submerged.
4. Briefly centrifuge glyoxal-denatured samples and standards from **step 3**, above, at 4 °C to collect liquid at the bottom of the tube.
5. Load size standards, samples, and optionally quantitation/normalization standards into the wells.
6. Run gel at 5 V/cm until the bromophenol blue dye front (~500 nt) has migrated approximately ¾ the length of the gel.
7. While the gel is running, prepare for transfer (Subheading 3.3, below). The probe can also be prepared during this time (Subheading 3.5; Read "Working with ^{32}P" Subheading 3.4 before preparing the probe below).

3.2.4 Viewing and Imaging the Gel

The Glyoxal Gel Sample Loading Dye from Ambion contains ethidium bromide. Thus, RNA is pre-stained during sample preparation and the gel can be photographed immediately without further staining. If a different glyoxal sample buffer without ethidium bromide was used, or if the RNA is not visible after electrophoresis, stain the gel in 0.5 μg/ml ethidium bromide in water for 20–30 min and then destain with two 10-min washes in water. Either traditional film or a digital imager can be used for imaging.

1. Place gel on UV transilluminator (*see* **Note 16**). The agarose above the wells may be cut away with a razor blade now or at any later time before blotting, if desired.
2. Position a fluorescent ruler alongside the molecular weight marker. *Wear UV goggles* and turn on the transilluminator. The ruler should be placed so that "0 cm" is aligned with the origin (i.e., the gel well). Ensure that the ruler is straight.

3. Photograph gel with camera and film or capture images with a digital imaging system using an ethidium bromide filter (*see* **Note 17**).
4. Wash gel 1×, ~10 min with gentle shaking in water (molecular biology grade) (no need to use DEPC water; *see* **Note 18**).
5. Wash gel 2×, ~10 min/wash with gentle shaking in 2× TAE.
6. Gel is now ready for transfer.

3.3 Blotting and Cross-Linking

After electrophoresis and photography, RNA in the gel is transferred to a membrane by capillary action and then cross-linked to the membrane prior to hybridization.

The traditional "upwards blotting" method, in which liquid is wicked upwards through the gel and then through the membrane, carrying up the RNA from the gel and blotting it onto the membrane [11], is widely used. However, transfer efficiency in this traditional "upwards blotting" method is suboptimal, as considerable weight is placed on top of the gel stack, crushing the gel and preventing complete transfer of RNA out of the gel.

We describe here the "sponge method" of blotting [12] (Fig. 2), an "upwards blotting" method that facilitates wicking, reduces gel crushing (the pressure of the gel stack is partially transmitted to the sponge at the bottom of the stack), and improves transfer efficiency. Zeta-Probe[R] GT (Genomic Tested) blotting membranes work very well in this procedure when used with radioactive probes (*see* **Note 19**).

Transfer efficiency is also improved by "downwards blotting," which minimizes the weight placed over the gel and prevents crushing. The TurboBlotter™ system (Fig. 3) is a convenient and very effective method for downwards transfer. We have modified the procedure slightly, using Zeta-Probe[R] GT (Genomic Tested) blotting membranes (*see* **Note 19**) in place of the supplied membrane and 2× TAE as the transfer buffer.

3.3.1 Sponge Method (Adapted from Zeta-Probe GT Instruction Manual, BioRad)

IMPORTANT: Wear gloves or use forceps when handling the membrane!

Figure 2 summarizes the steps involved in this procedure.

1. Cut the membrane (Zeta-Probe GT, BioRad) (*see* **Note 19**) to the size of the gel. Place the cut membrane in distilled water for at least 5 min before use. Also cut six pieces of Whatman 3 MM paper the same size as the membrane, and keep dry until ready to assemble the gel stack.
2. Place a thick sponge larger than the gel (*see* **Note 5**) in the bottom of a deep dish (e.g., a glass baking dish or Tupperware container).
3. Place enough 2× TAE in the dish to immerse the sponge a little more than halfway up its side.

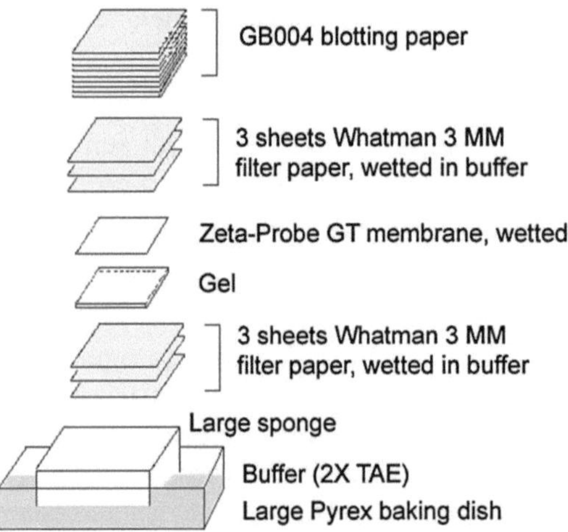

Fig. 2 Upwards transfer (sponge method). A large sponge is placed in a large Pyrex baking dish containing transfer buffer (2× TAE) and saturated with buffer. Three sheets of Whatman 3 MM paper wetted in transfer buffer are place on top of the sponge, followed by the gel and the prewetted membrane (Zeta-Probe GT recommended for ^{32}P-labeled probes; manufacturer-recommended membrane for nonradioactive probes). Plastic wrap or Parafilm (not shown) is placed under the edges of the gel to ensure that buffer flows only through the gel. This prevents "short-circuiting" of transfer buffer around the gel and preserves transfer efficiency. Three more sheets of Whatman 3 MM paper wetted in transfer buffer are placed on top of the membrane and a thick stack of dry GB004 or GB003 blotting paper (~50–100 sheets) is place on top of the stack. Transfer buffer is drawn up by capillary action from the sponge through the Whatman paper, gel, and membrane towards the dry blotting paper. RNA is carried upwards by the transfer buffer from the gel to the membrane, where it binds noncovalently. Using a sponge as the base enhances flow through the blotting stack and decreases pressure on the gel

4. Prepare a second dish containing 2× TAE.
5. Wet a piece of precut Whatman 3 MM paper with transfer buffer, place on top of the sponge, and debubble by deftly "ironing" the surface with a disposable 2- or 10-ml plastic pipette (*see* **Note 20**). Repeat with two more pieces of precut Whatman 3 MM paper.
6. Without losing track of the sample orientation in the gel, flip the gel over so that the smooth side of the gel (originally in contact with the bottom of the gel mold) is facing up (*see* **Note 21**), and place the gel on top of the Whatman 3 MM paper stack. Debubble with a plastic pipette as in **step 5**, above.
7. Spread plastic wrap over the gel stack, making sure the 3 MM paper and sponge are completely covered. Cut a "window" around the gel with a razor blade so that transfer buffer will flow only through the gel. Parafilm can be used in place of plastic wrap; just slip it under the gel's edges.
8. Remove the precut Zeta-Probe GT membrane from the distilled water, dip it into 2× TAE, and place it over the gel, avoiding bubble formation (*see* **Note 20**). Debubble with a plastic

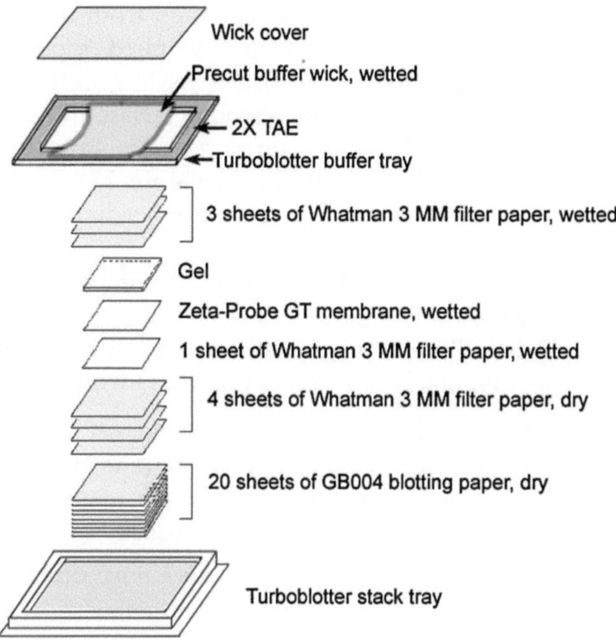

Fig. 3 Downwards transfer (TurboBlotter method). A dry TurboBlotter stack tray is placed on the lab bench. Twenty sheets of dry GB004 or GB003 blotting paper, four sheets of dry Whatman 3 MM filter paper, one sheet of 3 Whatman 3 MM paper prewetted in transfer buffer, and then the membrane, precut to the size of the gel and prewetted, are stacked into the tray. The agarose gel is then added to the stack, followed by three sheets of Whatman 3 MM paper prewetted in transfer buffer. The buffer tray is attached and filled with buffer. A precut buffer "wick" is then placed across the stack, completely covering it, with both ends of the long dimension placed in the buffer tray. Transfer buffer is drawn by capillary action from the buffer tray and wick downwards through the Whatman paper, gel, and membrane towards the dry blotting paper. RNA is carried downwards by the transfer buffer from the gel to the membrane, where it binds noncovalently. Downwards blotting reduces the weight of material above the gel and prevents crushing of gel pores during transfer, improving transfer efficiency

pipette as above. Try not to let the membrane slide around during debubbling, or else a small amount of RNA may transfer and faint "ghost bands" may show up on the blot.

9. Using a membrane marking pen, mark the location of the wells and/or other landmarks (e.g., needle pokes showing positions of MW markers and rRNA bands, or top and bottom of gel).
10. Place three more sheets of Whatman 3 MM paper wetted in transfer buffer sequentially on top of the gel stack, debubbling after each sheet is added.
11. Place ~50–70 pieces of dry Whatman grade GB004 or GB003 blotting sheets on top of the stack (optional: and a light weight on top of the blotting sheets). Make sure the stack is stable and will not tip over.
12. Transfer overnight.
13. After transfer, disassemble the stack. Mark the side of the membrane that was in contact with the gel (the "RNA side") and

keep the RNA side facing up from this point on. The damp blot can be quickly photographed or otherwise imaged (using an EtdBr filter) to visualize RNA on the membrane (*see* **Note 22**) but should be placed in RNase-free water or 10 mM Tris–HCl pH 7.4 as soon as possible. Place the gel into ethidium bromide (0.5 μg/ml) (*see* **Note 23**). Immediately proceed to cross-linking of the blot (Subheading 3.3.3).

3.3.2 TurboBlotter (Downwards Blotting)

IMPORTANT: Wear gloves or use forceps when handling the membrane!

TurboBlotters are available in small or large sizes.

Figure 3 summarizes the steps involved in this procedure.

1. Cut the membrane (Zeta-Probe GT) (*see* **Note 19**) to the size of the gel. Place the cut membrane in distilled water for at least 5 min before use. Then soak membrane in transfer buffer (2× TAE) for 5 min before use.
2. Place the TurboBlotter "stack tray" on the bench. Make sure it is level.
3. Place 20 sheets of dry GB004 or GB003 blotting paper in the stack tray.
4. Place four sheets of dry Whatman 3 MM paper on top of the GB004 or GB003 paper.
5. Wet one sheet of Whatman 3 MM paper; precut to the size of the gel, with transfer buffer (2× TAE), and place on top of stack. Remove any bubbles by "ironing" with a 2- or 10-ml plastic disposable pipette (*see* **Note 20**).
6. Place the precut, prewetted membrane (Zeta-Probe GT) (from **step 1**, above) on top of stack. Remove any bubbles by "ironing" with a 2- or 10-ml plastic disposable pipette (*see* **Note 20**).
7. Place gel on top of the Zeta-Probe GT membrane. Remove bubbles as in **step 6**, above. Keep track of the sample orientation in the gel and ensure that the smooth side of the gel (originally in contact with the bottom of the gel mold) is the side that contacts the membrane (*see* **Note 21**).
8. Cover any exposed Whatman paper (i.e., any area not covered by the gel) with plastic wrap or Parafilm to ensure that buffer flows only through the gel.
9. Wet the top of the gel with transfer buffer.
10. Wet three sheet of Whatman 3 MM paper; precut to the size of the gel, with transfer buffer (2× TAE); and place one by one on top of stack, debubbling between each sheet.
11. Attach the TurboBlotter "buffer tray" to the bottom tray. Use the circular alignment buttons to align both trays.

12. Fill the buffer tray with transfer buffer (2× TAE) (125 ml for up to 11 cm × 14 cm transfers; 200 ml for larger transfers up to 20 cm × 25 cm).
13. Soak a precut "buffer wick" in transfer buffer (2× TAE) and place across the stack so that the blotting stack is completely covered by the short dimension of the wick and the long ends of the wick dip into opposite sides of the buffer tray.
14. Cover the stack with the wick cover. Make sure the ends of the wick are immersed in transfer buffer.
15. Transfer for at least 3 h. More time may be required for thicker gels or longer transcripts.
16. After transfer, disassemble the stack. Place the gel into ethidium bromide (0.5 μg/ml) (*see* **Note 23**). Mark the side of the membrane that was in contact with the gel (the "RNA side") and keep the RNA side facing up from this point on. The damp blot can be quickly photographed or otherwise imaged (using an EtdBr filter) to visualize RNA on the membrane (*see* **Note 22**) but should be placed in RNase-free water or 10 mM Tris–HCl pH 7.4 as soon as possible. Immediately proceed to cross-linking of the blot (Subheading 3.3.3).

3.3.3 Cross-Linking

1. *Briefly* blot the membrane on filter paper (e.g., Whatman 3 MM) until liquid is no longer visible on surface (i.e., until membrane is no longer shiny) (≤2 min). Do not allow the membrane to completely dry out.
2. Cross-link RNA to membrane by irradiating the damp membrane (RNA side facing the irradiation source) with 1,200 μJ of UV light. The Stratalinker UV Crosslinker is designed for this task.
3. Dry membrane thoroughly at room temperature before beginning prehybridization.
4. Store membrane dry between pieces of filter paper, wrapped in Saran Wrap, until use (*see* **Note 24**).
5. Before proceeding to prehybridization and hybridization, the blot must be rewetted in RNase-free water or 10 mM Tris–HCl pH 7.4 (Subheading 3.6.1 **step 2**).

3.4 Working with ^{32}P

Do not work with ^{32}P or any other radioisotope until you have received proper training and certification by your radiation safety department.

Working with ^{32}P seems intimidating at first but becomes routine with practice. Careful adherence to safety guidelines is essential. Always change gloves frequently, have plenty of paper towels and bench paper laid down everywhere to catch any spills, shield everything, and monitor obsessively. All waste should be disposed of in liquid or solid ^{32}P waste containers.

Specific radiation safety tips for this chapter include:

Work behind a plexiglass shield with plentiful absorbent bench paper underneath as well as over the bottom of the shield. A tray may be used inside this area as extra protection in capturing spills.

Use thick plexiglass tube holders for all tubes (microcentrifuge tubes and 50-ml tubes) containing radioactive material.

Wear a lab coat and protective lenses at all times.

Wear a double layer of gloves and change gloves constantly.

Keep a Geiger counter on nearby at the most sensitive setting and monitor hands/gloves and workplace constantly. Pancake probes are most effective for monitoring. Don't move the probe along swiftly while monitoring surfaces or small radioactive spots will be missed.

Monitor the floor and the bottom of your feet as well as the rest of your body.

Monitor sinks, faucet handles, door handles, centrifuges, heating blocks, refrigerator handles, phones, water baths, anything that may have been contaminated, and, especially, items that you are sure have not been contaminated!

Label EVERYTHING that touches radioactivity with radioactive tape with your name, date, isotope (^{32}P), and total amount of radioactivity (in cpm or uCi) in the tube.

Dedicate as much equipment as possible (pipetmen, centrifuges, racks, sealer, water bath, heating block, etc.) as ^{32}P-only.

Keep a solid and liquid waste container handy nearby (preferably behind the plexiglass shield) so that radioactive materials do not have to be carried far.

When working with highly concentrated ^{32}P, small but concentrated droplets can spill but go unnoticed during monitoring. After working with concentrated ^{32}P, clean all surfaces and discard of all disposables, paper towels, bench paper, etc., as radioactive waste, even if no contamination was noted.

Store radioactive materials in capped, unbreakable, shielded containers. No glass!

Do not leave ^{32}P unattended in an unlocked laboratory and use a locked freezer or refrigerator for long-term storage.

No eating, no drinking, no gum chewing, etc., is allowed in any laboratory containing radioactive material.

Perform a final monitoring, including swipe tests, when done.

All rules and regulations pertaining to use and disposal of radioactivity in your institute must be strictly obeyed.

3.5 Preparing the Probe

^{32}P-labeled double-stranded DNA fragments are easy to prepare and to use as probes in hybridization experiments. A DNA fragment containing the target sequence is gel-purified, labeled with ^{32}P in a single incubation step, and purified away from unincorporated nucleotides by passage through a spin column. ^{32}P-labeled single-stranded RNA probes can be used as an alternative to ^{32}P-labeled double-stranded DNA probes (*see* **Note 25**).

Full-length cDNA probes are best for general use, as their length ensures a strong signal. Moreover, since a full-length cDNA probe contains all exons present in the original gene transcript, alternative splice variants transcribed from the same gene and containing one or more exons present in the full-length probe will be detected. Splice variants can be investigated further with exon-specific probes, if desired. Exon-specific probes are shorter, may require lower hybridization temperatures, and may give weaker signals than full-length DNA probes; however, by altering labeling or hybridization conditions or even the type of probe used, short probes can be used successfully for Northern analysis (*see* **Note 26**).

For Northern analysis of miRNA, use locked nucleic acids (LNA) as probes [10].

Nonradioactive detection techniques are gaining in popularity. Somewhat more handling is required, but avoiding radioactive work is undeniably a plus. Manufacturers such as Ambion and Roche have very detailed online methods and troubleshooting guides for nonradioactive detection, which we recommend to anyone contemplating nonradioactive Northerns.

In our experience, first-time users have better success with radioactive Northerns, as nonradioactive Northerns often require some optimization. Also, the chemiluminescent signal from nonradioactive Northern blots may change in a complex fashion over time, thus compromising signal quantitation. Using ^{32}P-labeled probes and quantitating the number of counts per band using PhosphorImager technology (*see* Subheading 3.7.2 below) is direct, sensitive, and most importantly, more quantitative than chemiluminescent detection methods. We therefore focus on the use of ^{32}P-labeled probes in this chapter.

3.5.1 ^{32}P-Labeling of the DNA Fragment

The protocol below describes ^{32}P-labeling of a double-stranded, gel-purified DNA fragment containing the sequence of interest. The gel-purified DNA fragment is first denatured and then added to a "bead" [GE Healthcare/Amersham Ready-To-Go DNA Labeling Beads (-dCTP)], a lyophilized mixture of random-sequence oligodeoxynucleotides, a DNA polymerase (Klenow fragment), and deoxynucleotides (minus dCTP). The oligonucleotides in the bead anneal to random sites on the denatured DNA fragment and act as primers for the DNA polymerase. [$\alpha^{32}P$]dCTP is added and, along with the other deoxynucleotides in the bead, is incorporated into newly synthesized DNA fragments as DNA polymerization takes place. Unincorporated nucleotides are removed by size exclusion chromatography using a spin column; the large DNA fragment elutes into the void volume and is retained, while the small dNTPs, including [^{32}P]-dCTP, remain in the beads and are discarded. Labeling efficiency is determined and the probe is then ready for hybridization.

1. Thaw [α^{32}P]dCTP 3,000 Ci/mmol, 10 μCi/μl, if stored at –20 °C. *Caution*: Use maximum radiation safety precautions when working with the source vial!

   Once thawed, open the vial carefully, making sure no radioactive liquid flies away from the lid. Pipette up and down a few times with a 200-μl pipette to mix. Do not vortex. Take every precaution to make sure no drops of liquid, however small, are spilled onto the work area. Dispose of pipette tip in a shielded waste container. Change gloves and monitor both inside and outside of the working area each time the source vial is handled.
2. Tap the Ready-To-Go DNA Labeling Bead (-dCTP) tube against the lab bench to bring the bead to the bottom of the tube. Keep the bead at room temperature.
3. Denature the DNA (25–50 ng in ≤45 μl) by heating at 95–100 °C for 3–5 min (*see* **Note 27**). Immediately place on ice for 2 min, and then briefly centrifuge at 4 °C to collect contents at the bottom of the tube. Keep on ice thereafter.
4. To the tube containing the Ready-To-Go Labeling Bead, taking maximum radiation safety precautions while adding the [α^{32}P]dCTP, add:

   Denatured DNA (25–50 ng) ≤45 μl.

   [α^{32}P]dCTP, 3,000 Ci/mmol ×μl (enough for 50 uCi; *see* **Note 7**).

   H_2O, molecular biology grade, to a final volume of 50 μl.

   Incubate 1 h to overnight at 37 °C (*see* **Note 28**) in a heating block. Cover the tube with a plexiglass shield (e.g., the cover for a plexiglass microcentrifuge tube holder).

   Caution: The labeling reaction is VERY RADIOACTIVE. Be sure the tube is labeled with radioactive tape and completely shielded to protect others in the lab.
5. Remove unincorporated dNTPs (including free [α^{32}P]dCTP, which may cause background) by purification with ProbeQuant™ G-50 Micro Columns (Subheading 3.5.2).

3.5.2 Removing Unincorporated Nucleotides with ProbeQuant™ G-50 Micro Columns

1. Vortex to resuspend resin in column.
2. Loosen cap ¼ turn and remove bottom closure.
3. Place column in the supplied collection tube or 1.5 ml screw-cap tube.

4. Pre-spin column in a microcentrifuge at room temp, 735 × *g* (*see* **Note 29**), for 1 min. Do not pulse as the recommended speed may be surpassed, damaging the column. Proceed to the next step immediately.
5. Place column in new 1.5-ml microcentrifuge tube.
6. Slowly apply the labeling reaction (50 μl) to the CENTER of the column to ensure complete [^{32}P]-dCTP removal. Be sure not to poke the resin with the pipette tip while applying the sample.
7. Centrifuge 2 min at room temperature, 735 × *g*.
8. Monitor the eluate (containing the labeled DNA probe) and the column (containing unincorporated dNTPs) with a Geiger counter—both should be offscale on the least sensitive setting.
9. Carefully dispose of the column as dry radioactive waste, making sure no bead material spills out. Retain the eluate.
10. Place 1 μl of the eluate in a scintillation vial containing 10 ml scintillation fluid, and count in a scintillation counter. Usually >10^6 cpm/μl obtained from a 25–50 ng labeling reaction is sufficient (*see* also **Note 30**).
11. ^{32}P decay within the DNA strand causes strand breakage. In addition, the half-life of ^{32}P is 14.3 days. Thus, the performance of the probe decreases with time. Use the probe within 5 days of preparation; sooner is better; immediate use is best.

3.6 Prehybridization, Hybridization, and Washing

We provide a detailed protocol for prehybridizing and hybridizing the probe to the membrane in heat-sealed plastic bags in a shaking water bath. This is a classic technique but messy and somewhat tricky to execute properly. A convenient and popular alternative method is to prehybridize and hybridize membranes in screw-cap tubes placed in a "rotisserie" inside a heated oven (termed a "hybridization oven") (*see* **Note 9**). The two methods differ only in the use of plastic bags vs. screw-capped tubes for prehybridization and hybridization; otherwise all steps are identical. Following **Notes 32**, **35,** and **38** in the procedure below will enable the reader to use hybridization ovens rather than heat-sealed plastic bags. Both methods give equivalent results.

3.6.1 Prehybridization

The prehybridization solution coats the surface of the membrane and blocks nonspecific binding of the radiolabeled probe over the entire membrane. Adding salmon sperm DNA to the prehybridization solution decreases nonspecific hybridization of the labeled probe to non-target transcripts and prevents nonspecific background up and down gel lanes.

1. Preheat shaking water bath to 65 °C. (If using a hybridization oven, *see* **Note 31.**) Half-fill a glass baking dish with water and

place inside water bath. The water level in the water bath should be as high as the water level in the dish to maintain the temperature inside the dish.

2. Rewet blot in a second dish containing RNase-free water or 10 mM Tris–HCl, pH 7.4 for at least 5 min at room temperature with shaking before prehybridizing.
3. Denature 150 μl salmon testes DNA (salmon sperm DNA) (10 mg/ml in dH_2O, sonicated) by heating to 95–100 °C 5–10 min in a heating block or water bath (*see* **Note 27**). Immediately transfer tube to ice and cool for at least 2 min. Spin tube briefly at 4 °C to collect contents at bottom of tube. Keep on ice until use.
4. Add denatured salmon sperm DNA to 15.0 ml PerfectHyb™ Plus Hybridization Buffer (PerfectHyb™) and mix by gentle pipetting. Avoid introducing bubbles. Do not vortex.
5. Remove excess liquid from the membrane by very briefly blotting with Whatman 3 MM paper. *Do not allow membrane to dry.*
6. Place damp membrane in heat-sealable plastic bag (or screw-capped tube if using a hybridization oven; *see* **Note 32**). Immediately add the salmon sperm DNA/PerfectHyb™ mix from **step 4**, above, to the plastic bag containing the damp membrane. Seal the bag using a bag sealer, avoiding bubbles. Check for leaks, and then place the bag into the dish in a 65 °C shaking water bath, guiding bubbles away from membrane (*see* **Note 33**).
7. Prehybridize at 65 °C for at least 1 h with shaking (or rotation on the rotisserie).

3.6.2 Hybridization

Labeled probe, along with additional salmon sperm DNA, is denatured and added to the hybridization bag, where it hybridizes to RNA on the blot. The denatured probe should not be added directly to the bag or tube containing the blot, as this can lead to "hot spots" on the blot. Instead, the denatured probe and salmon sperm DNA are added to 10 ml prewarmed PerfectHyb, mixed, and then added to the hybridization bag or tube containing the blot and prehybridization buffer.

Before starting, place 2–3 layers of absorbent bench paper, scissors, pipettes, warmed heat sealer, a radioactive dry waste container, and a box of paper towels behind the plexiglass radiation shield.

1. 15–30 min before use, place 10 ml PerfectHyb™ into a 50-ml disposable centrifuge tube. Screw the top on firmly, place in a rack, and prewarm to 65 °C in a water bath.
2. 10 min before use, place into separate tubes:

   100 μl salmon sperm DNA (10 mg/ml).

   25×10^6 cpm (or more) of radiolabeled, column-purified probe (from Subheading 3.5.2, above).

   Heat each tube at 95 °C for 10 min (*see* **Note 27**).

3. Immediately transfer tubes to ice and cool for 2 min. Spin briefly at 4 °C to collect contents at bottom of tube. Return to ice immediately after removing tube from centrifuge. Shield this tube and keep cold until use (*see* **Note 34**).
4. Remove the plastic bag containing the membrane from the 65 °C bath. Cut off a small piece from one corner of the plastic bag.
5. Remove the 50-ml tube containing 10 ml PerfectHyb™ from the 65 °C bath (from Subheading 3.6.2 **step 1**), and add the denatured salmon sperm DNA and ^{32}P-labeled probe (from Subheading 3.6.2 **step 3**). Pipette the radioactive mixture up and down with a 10-ml pipette 3–4 times to mix, avoiding bubble formation, and then with the same pipette, add this radioactive mixture to the plastic bag containing the prehybridized membrane and prehybridization solution. (If using a hybridization oven, *see* **Note 35**.)
6. Place several paper towels under the cut corner of the plastic bag and squeeze out any bubbles through this corner. Then heat-seal the corner. With care, only a little radioactive solution is extruded, and it can be contained on the paper towels. Dispose of the paper towels as radioactive waste.
7. Place clean paper towels under the bag. Knead the solution in the bag to thoroughly mix contents. Check the seals at this point by applying pressure to the bag near the seals/seams and checking for leaks. Double seals are a good safeguard, even if the bag doesn't leak. Wipe the outside of the bag with a damp paper towel to remove any radioactive material from the outside of the bag.
8. Return the bag to the dish in the 65 °C shaking water bath. Make sure that the bag is covered with water and that the level of water outside the dish is at least as high as inside the dish. The glass dish can be covered with a plate with a lead weight on top for extra shielding and/or to prevent the dish from sliding around excessively in the water bath. Hybridize at 65 °C overnight (*see* **Note 36**), with shaking. Check periodically for bubbles during the first hour of hybridization—bubbles frequently appear as the hybridization heats up. The bubbles can be kneaded to the top of the bag, which can be propped up to isolate the bubbles away from the membrane or even resealed to form a barrier between the bubbles and the membrane.
9. Unplug the heat-sealer and clean thoroughly with Counts-Off and paper towels. Place several layers of bench paper underneath to catch drips during cleaning. If radioactivity can no longer be removed and the sealer is still hot, label it with radioactive tape and store it completely shielded.
10. Discard all radioactive waste, clean up the working area, ***and monitor everywhere for ^{32}P contamination.***

3.6.3 Washes

Prepare wash buffers in advance and preheat the high- and ultra-high-stringency buffers to 55–65 °C before use by submerging in a water bath or microwaving (*see* **Note 37**).

Caution: Have extra layers of bench paper and lots of paper towels handy, especially while taking out the membrane and removing the hybridization fluid for the first wash. Be extra-vigilant in avoiding spills and contamination. Perform a dry run and make sure everything is in place before beginning the wash steps. Dispose of all wash and hybridization solutions as liquid radioactive waste and all bags, bench paper, paper towels, pipettes, etc., as dry radioactive waste.

1. Prepare 1.0 liter low-stringency wash buffer at room temperature.
2. Fill two glass baking dishes with 500 ml low-stringency wash buffer and place one behind the plexiglass radiation shield (*see* **Note 38**).
3. Behind the radiation shield, place a 50-ml disposable centrifuge tube (preferably in a plexiglass 50-ml tube holder, for shielding), dry waste container, scissors, forceps, and lots of paper towels.
4. Remove the hybridization bag from the 65 °C water bath, shielding yourself and others from radioactivity. Place the bag on a stack of paper towels behind the radiation shield and quickly dry it. Uncap the 50-ml centrifuge tube.
5. Cut off a corner of the plastic bag and carefully drain the radioactive contents into the 50-ml centrifuge tube. Cap the 50-ml tube IMMEDIATELY to avoid spills. Cut away one or two sides of the plastic bag, remove the blot with forceps (or gloves if necessary), and immediately place into the first dish containing low-stringency wash buffer. *Do not allow the blot to dry, even briefly, until after the very last wash.* Even a little dryness will result in non-removable background. Change gloves.
6. Incubate the blot in low-stringency wash buffer ~2–5 min at room temperature, shaking by hand, to briefly rinse the membrane.
7. Remove the blot and place into the second glass dish containing low-stringency wash buffer. Incubate ≥30 min at room temperature, shielded, on a shaker.
8. While the blot is washing, clean up the working area behind the plexiglass shield. As small droplets of used radioactive hybridization solution may easily go undetected, it is best at this point to dump all liquid waste into the liquid waste container, wrap everything else up and discard as dry radioactive waste, put on clean gloves, and place clean bench paper down behind the shield. Carefully monitor for radioactive spills.

9. Using a paper towel to catch drips, pour off the low-stringency wash buffer into liquid radioactive waste. IMMEDIATELY add 500 ml high-stringency wash buffer, prewarmed to ~55–65 °C, to the dish containing the blot. Incubate ≥30–45 min in a 65 °C water bath, with shaking.
10. Perform a second wash in 500 ml high-stringency wash buffer at 65 °C, as above.
11. Perform two 65 °C washes, ≥30–45 min each, in ultrahigh-stringency wash buffer prewarmed to ~55–65 °C, with shaking.
12. Remove membrane with forceps and briefly blot on Whatman 3 MM paper until no longer shiny, then wrap in plastic food wrap. The blot should be kept damp if it is to be stripped and reprobed later; otherwise it is OK if the blot becomes dry after this point.

3.7 Detection

3.7.1 Film

1. Prepare an autoradiography cassette containing two autoradiography screens: one affixed to the bottom of the cassette and the second stuck to the lid.
2. In a darkroom, place the blot (RNA side up) on the bottom screen, cover with one piece of film, close cassette, and place in −80 °C freezer (*see* **Note 39**).
3. Expose overnight and develop film.
4. Take additional longer or shorter exposures as necessary.
5. Bands can be quantitated by densitometric scanning.

3.7.2 Storage Phosphor/PhosphorImager Technology

Using storage phosphor technology in place of film improves sensitivity and especially quantitation of Northern blots. Storage phosphor technology, also known as PhosphorImager technology, captures high-energy radiation (beta particles, X-rays, UV light, gamma rays) on a screen by causing molecules in the screen to enter an excited state. This forms a "latent image" on the screen. To visualize the image, the screen containing the excited molecules is exposed to light with a specific wavelength to return the molecule to the ground state. This liberates a photon, which is captured digitally to form an image. The most important advantage of this technology is that it counts photons directly and is consequently linear over a wide dynamic range, unlike autoradiography film, which saturates quickly and is also nonlinear at very low density (*see* Fig. 9 of ref. 13 for comparison of storage phosphor vs. film for ^{32}P). GE Healthcare now sells several state-of-the art storage phosphor systems (Typhoon, Storm) that deliver high-resolution imaging and quantitation of ^{3}H, ^{14}C, ^{125}I, ^{32}P, ^{33}P, as well as a number of fluorophores. Some systems also enable direct chemiluminescent imaging as well. These systems are expensive, but their sensitivity, flexibility, and in particular their wide linear dynamic range are unmatched by film.

3.8 Probe Stripping

Zeta-Probe GT blots can be stripped and reprobed multiple times with different probes. At the very least, blots should be reprobed with an appropriate loading control probe. To strip the probe from a Zeta-Probe GT membrane [12]:

1. The membrane should not have dried out completely after the previous hybridization. Strip as soon as possible after autoradiography/phosphorimaging.
2. Place the blot in a dish containing RNase-free water. Incubate on a shaker at room temperature for at least 5 min to completely rewet the damp blot.
3. Wash the membrane twice, 20 min per wash, in 500 ml 0.1× SSC/0.5 % SDS *in a glass baking dish* at 95 °C, preferably in a shaking water bath. *Caution*: *Hot!* Wear heat-proof gloves and handle hot dishes and solutions with care. You can begin each stripping step at a cooler temperature (e.g., 65 °C), incubate the blot in stripping buffer with shaking inside the water bath until the temperature reaches 95 °C, incubate with shaking at 95 °C for 15–20 min, and then let the water bath cool a bit before removing the dish. *Stripping the blots by boiling on a hot plate is not recommended*, since the heated glass dish will explosively shatter if accidentally placed in contact with even a few drops of cold water on the lab bench.
4. After stripping the blot, expose to film (or a storage phosphor screen) for the appropriate length of time to ensure that the probe has been removed.

4 Notes

1. RNA is highly susceptible to degradation by RNases present in the environment and in cells and tissues. Tips for preventing RNA degradation during and after RNA isolation from cells and tissues are presented in the Appendix.
2. Qiagen sells RNeasy kits for general use (RNeasy Mini Kit), for lipid-rich tissues (RNeasy Lipid Tissue Mini Kit), and for fibrous tissues that do not have high levels of RNases (RNeasy Fibrous Tissue Mini Kit). The Qiagen website provides useful guidance in choosing the appropriate RNeasy kit for your sample. The maximum binding capacity for RNeasy mini-spin columns provided in the RNeasy "Mini" Kits is 100 μg RNA. Overloading these columns is not recommended. For isolation of >100 μg RNA, use multiple mini-spin columns or an RNeasy Midi Kit, or use alternative purification methods.
3. For very small RNAs such as microRNAs, many traditional RNA isolation methods will not give good recoveries. For example, very small RNAs may not be quantitatively recovered

in procedures involving traditional ethanol precipitation of RNAs. If using the TRIzol or TRI reagent or kits such as the *mir*Vana miRNA Isolation Kit or the miRNeasy Mini Kit, follow the specific instructions for the class of small RNAs that you are interested in. In addition, very small RNAs are not well resolved on the 1 % agarose gels described in this chapter, and preparing and hybridizing probes for very small RNAs presents a challenge. We refer the reader to ref. 10 for Northern analysis of miRNAs and other small RNAs.

4. In general, the thinner the comb, the sharper the band. The 0.75-mm-thick 20-well comb from BioRad yields very crisp bands when used with the Wide Mini Sub-Cell GT system. Using a 1.5-mm-thick comb will decrease resolution and may cause minor transcripts slightly smaller or larger than the predominant transcript to go unnoticed. 0.75- and 1.5-mm-thick combs hold ~30 and 60 μl, respectively, in a 1-cm-deep gel. Using an adjustable-height comb with ~2 mm between the comb's teeth and the floor of the gel mold allows more sample to be loaded.
5. Place two or more smaller straight-edged sponges side by side if a large sponge is not available. Rinse sponge thoroughly with distilled water to remove any detergent, if it is fresh out of a new package, and then rinse once or twice with transfer buffer before using.
6. 409 Household cleanser, available in supermarkets, drugstores, etc., can be used in place of Counts-Off in a pinch.
7. ^{32}P decays with a half-life of 14.3 days. As radioactive decay occurs, decay products that inhibit labeling efficiency build up. For optimal results, have a fresh batch of $\alpha^{32}P$-dCTP (3,000 Ci/mmol, 10 mCi/ml) shipped to you the day it is prepared by the supplier and prepare your probe with 5.0 μl of fresh $\alpha^{32}P$-dCTP the day it arrives. This unfailingly results in a good hot probe. Ten days—or at the very maximum 2 weeks—past the reference date is about as long as you can use $\alpha^{32}P$-dCTP for probe labeling. Increase the volume of $\alpha^{32}P$-dCTP to compensate for the decrease in activity due to decay according to the following radioactive decay equation:

$$N = N_o e^{-(0.693t/T_{1/2})}.$$

N_o : the original quantity of radioactive material.

e : the natural number 2.71828.

t : elapsed time.

$T_{1/2}$: half-life (14.3 days for ^{32}P).

Decay tables are available online and are also supplied with $\alpha^{32}P$-dCTP.

8. Some online literature from the manufacturer may still refer to this product by its previous catalog number (#27-5335-01) but it must be ordered as #28-9034-08.
9. As a convenient alternative to placing membranes in sealed plastic bags in a shaking water bath, membranes can be prehybridized and hybridized in screw-topped glass tubes that are rotated in a special hybridization oven (e.g., Hybaid Hybridization Ovens, Thermo Scientific, Waltham, MA). *See* **Notes 32**, **35,** and **38** for further details. However, even if a hybridization oven is used, a water bath, preferably a large, shaking one, is useful for prewarming hybridization and wash solutions.
10. RNA*later* can be used to stabilize all tissues except adipose tissue, due to its high lipid content. All protect must be used to stabilize adipose tissue but can be used to stabilize other tissues as well. Our experience in isolating RNA has been with fresh tissues and cells or RNA*later*-stabilized tissues.
11. Since RNA transfer, hybridization, and washing efficiencies vary between blots, Northern hybridization has been traditionally regarded as only a semiquantitative method, particularly when film densitometry is used for quantitation (Subheading 3.7). Quantitation can, however, be improved by using quantitation/normalization standards, which allow signals to be compared between blots and can be used to assess linearity, and PhosphorImagers, which give a linear response over a wider dynamic range than film. To compare samples on more than one blot, load an RNA normalization standard on each blot and divide the signal for each sample by the signal of the normalization standard on its blot. Normalized signal intensities can then be compared between blots. A normalization standard can be a sample from the experiment, or RNA from a large RNA prep that has been aliquotted and frozen specifically to be used for a standard, or even an in vitro-transcribed RNA that hybridizes to your probe. Ideally, three or more concentrations spanning the lowest-, intermediate- and highest-expected signal intensities would be used on each blot to assess linearity. Absolute quantitation—determining the number of pg of each transcript—is almost never performed with Northern blots, but if desired, an in vitro transcript can be made and quantitated for use as a quantitative standard. Dilutions spanning the expected range of signal intensities can be loaded on the gel(s) alongside the samples to be quantitated. Again, quantitation using storage phosphor technology rather than film is strongly recommended if strict quantitation is desired.

    Densitometric analysis of images on film is an acceptable method of determining relative signal intensities within a blot and also between blots as long as quantitative standard(s) are used to normalize signal intensities between blots, as described above. For best results, the signal should not saturate the film but should not be too weak, either. Linearity is rarely achieved by densitometric scanning of film—exposing the blot twice as long is unlikely to double the quantitated signal intensities for all samples, especially with a wide range of signal intensities between samples. Densitometric scanning of autoradiographic film is appropriate for identifying increases and decreases in gene expression under different experimental conditions, but the quantitative fold-differences will likely differ from the fold-differences obtained by other methods.

12. Some protocols recommend a maximum gel thickness of 6 mm. However, we routinely obtain good transfer out of a 12 mm deep agarose gel, so if RNA concentration is low, a deeper gel allows more RNA to be loaded.
13. If a 1:1 ratio of sample volume + Glyoxal Sample Loading Dye exceeds the gel well volume, 0.5 volume Glyoxal Sample Loading Dye can be added to 1 volume RNA sample. In this case the sample should be incubated for 1 h rather than 30 min at 50 °C.
14. The Glyoxal Sample Loading Dye contains not only glyoxal but also bromophenol blue and ethidium bromide. Since ethidium bromide binds RNA and remains bound throughout electrophoresis, the gel does not have to be stained later to visualize the RNA after electrophoresis.
15. A water bath may be used instead. However, a dry heating block is cleaner and is less likely to result in RNase contamination and degradation of the sample.
16. Do not overexpose RNA in the gel to UV light. Turn off the UV light source when the gel is not being photographed.
17. Obtain exposures with MW markers clearly visible next to the ruler markings. If the ruler markings are too bright compared to the MW markers, image short exposures to see the ruler markings and then longer exposures to see the MW markers, both clearly showing the position of the gel well. You can poke holes next to the gel well and MW markers with a 16-gauge needle to mark their positions at this point and then later, when setting up the blot, mark the position of the holes/MW markers onto the membrane, using a permanent membrane marker. Digital annotation of markers or other landmarks is also possible with some imaging systems. Finally, obtain good exposures of the 28S and 18S RNA bands, which should be visible as sharp bands with little downwards smearing throughout the gel if the RNA is undegraded. The 18S band should be about half as intense as the 28S band if the samples are undegraded. 28S and 18S RNA band intensities are also useful for comparing sample loading between lanes. Always keep track of orientation, for example, by cutting off a corner of the gel.

18. Strict RNase-free precautions are no longer necessary after running the gel. In fact, for very long transcripts, a small amount of nicking may increase the migration out of the gel onto the membrane. We have not found this necessary for transcripts as long as ABCA1 (6.78 kb), but nicking might be required for complete transfer of extremely long RNAs, such as apoB mRNA. RNA can be nicked in-gel with 10 mM NaOH treatment for 10–15 min, followed by neutralizing with 0.5 M Tris–HCl, pH 7.4, for 30 min at room temperature, then proceeding with blotting as usual [12].
19. We have used Zeta-Probe GT, a quaternary amine-derivatized nylon membrane, in over 100 successful radioactive Northerns and Southerns with excellent sensitivity and no background and recommend it highly. If you are having difficulties with your current membrane, try this one, particularly in combination with Sigma's PerfectHyb hybridization buffer. *IMPORTANT*: For nonradioactive Northern blotting, use the membrane recommended by the manufacturer (e.g., BrightStar-Plus membrane (Ambion) when using Ambion's psoralen–biotin nonisotopic probes), and follow manufacturer's instructions precisely.
20. Bubble formation can be almost completely avoided by holding the wetted blotting paper or membrane by the sides, with the center dipping down slightly. Touch the center of the blotting paper or membrane to the center of the gel and smoothly lay down the rest of the paper from the center outwards. Bubbles tend to roll out as the paper or membrane is smoothed along the gel. It is still important to debubble after each step by "ironing" the blotting paper, gel, or membrane with a 2- or 10-ml plastic disposable pipette.
21. After the gel is poured and solidifies in the gel mold, the "air" side of the gel is not smooth and will not make uniform contact with the membrane. Placing the membrane against the smooth side of the gel (the side that solidified against a smooth plastic surface) gives bands that are less "marbled" and improves uniformity.
22. 28S and 18S RNA should be visible on the blot if total RNA was loaded on the gel but will be absent if polyA+ RNA was used. 28S and 18S RNA are not polyadenylated.
23. The gel can be left in ethidium bromide (0.5 μg/ml in water or 2× TAE) for 1 h to overnight and should be photographed or otherwise imaged to determine whether any RNA is left in the gel. No RNA should be visible. Save the image or photo of the gel after transfer for later troubleshooting, if necessary.
24. Store blots short term at room temperature for up to a few weeks but at −20 °C long term. Protect blots against scratching or other damage during storage.

25. ^{32}P-labeled single-stranded RNA probes can be used as an alternative to ^{32}P-labeled double-stranded DNA probes, if desired. Clone a double-stranded DNA fragment containing the sequence of interest downstream of a prokaryotic RNA polymerase promoter (e.g., T7 or T3 RNA polymerase) in reverse orientation, linearize downstream, and then transcribe in vitro with T7 or T3 polymerase in the presence of a ^{32}P-rNTP such that RNA complementary to the transcript of interest is transcribed from the promoter. Alternatively, attach a T7 or T3 promoter to a DNA fragment by PCR and transcribe the fragment in vitro with T7 or T3 polymerase in the presence of a ^{32}P-rNTP, optionally without cloning [11]. Working from a cloned and sequenced fragment is more rigorous and is recommended. ^{32}P-labeled single-stranded RNA probes can give a stronger signal than ^{32}P-labeled double-stranded DNA probes, since RNA–RNA duplexes on the blot are more stable than DNA–DNA duplexes. However, any RNA molecule is at risk for degradation by contaminating RNases. Also, the increased stability of RNA–RNA duplexes can have negative consequences, such as very stable stem-loop formation in the probe that decreases its availability for hybridization and increased difficulty in stripping blots. ^{32}P-labeled double-stranded DNA fragments are easy to prepare and use, and thus we focus on this procedure.
26. Probes shorter than 200 bp have significantly lower melting temperatures than longer probes and hybridization and washing conditions must be altered to ensure that the probe binds to its target during hybridization and remains bound throughout washing [11]. The melting temperature (T_m) is the temperature at which 50 % of the base pairs in a duplex have been denatured [11] and is used to estimate appropriate hybridization and wash temperatures. The following guidelines [11] are useful as a starting point:

For DNA–RNA duplexes (when a short DNA fragment is used as a Northern probe)

$$T_m(\text{in}\,^\circ\text{C}) = 67 + 16.6\log_{10}\left(\frac{[\text{Na}^+]}{1+0.7[\text{Na}^+]}\right) + 0.8(\%[\text{G}+\text{C}]) - 500/n$$

For RNA–RNA duplexes (when a short RNA fragment is used as a Northern probe)

$$T_m(\text{in}\,^\circ\text{C}) = 78 + 16.6\log_{10}\left(\frac{[\text{Na}^+]}{1+0.7[\text{Na}^+]}\right) + 0.7(\%[\text{G}+\text{C}]) - 500/n$$

Na^+ = concentration of sodium +.

n = length of duplex.

P = temp correction for % mismatch = 1 °C per 1 % mismatch.

Hybridize at least 5–10 °C below the melting temperature (T_m) using high concentrations (0.1–1 pmol/ml) of probe for at least 3–4 h [11]. Wash briefly first under conditions of low stringency (decreased temperature and/or increased salt concentration), then under conditions of stringency equal to that used for hybridization [11], and then expose the blot. Increase the washing temperature and/or time if the background is too high and decrease if the signal is too low. For further details on working with short probes, in particular oligonucleotide probes, *see* ref. 11. The specific activity (cpm/μg) of any probe can also be increased by performing labeling reactions (random priming or in vitro transcription) in the presence of all four radioactive dNTPs (random priming) or rNTPs (in vitro transcription).

27. When using a heating block to denature the fragment, use a tube that fits snugly inside the holes in the heating block. If the fit is not snug the fragment will not reach the correct temperature and may not fully denature.

28. The incubation time can be shortened to 15–30 min but the probe may not be maximally labeled. The labeling reaction can be left overnight (16 h) at 37 °C with no ill effects.

29. rpm = $(1{,}000)\,(657/r)^{1/2}$.

    r = radius from center of spindle to bottom of rotor bucket.

30. The efficiency of incorporation can also be determined by determining acid-precipitable counts as described in the manufacturer's protocol.

31. A 65 °C water bath is useful even if you prehybridize and hybridize in a hybridization oven, as it can be used to prewarm solutions.

32. If using a hybridization oven, preheat the hybridization oven to 65 °C. Place the damp membrane lengthwise into the recommended screw-capped glass tube, add the PerfectHyb plus denatured salmon sperm DNA from Subheading 3.6.1 **step 3** using a pipette, smooth out bubbles with a 10-ml pipette, screw the cap onto the tube, and insert the tube into the hybridization oven's rotisserie. Close the door, turn on the rotisserie, and watch the flow of liquid in the tube. *The entire surface of the blot must remain covered with prehybridization (and later, hybridization) buffer throughout the rotation cycle of the tube.* If necessary, adjust the tube position, the position of the membrane in the tube, or the volume of hybridization buffer to keep the entire membrane in contact with fluid.

33. Prehybridization blocks ^{32}P-labeled DNA or dNTPs from permanently binding to the membrane. If a part of the membrane

contacts a bubble rather than prehybridization fluid, that part of the membrane will not be blocked. It will irreversibly bind ^{32}P during the hybridization step and end up as a dark spot on the autoradiogram. Thus, bubbles should be avoided during prehybridization. To minimize bubbles in the hybridization bag while heat-sealing, first make sure paper towels are close at hand. Place the damp membrane in the bag and seal all sides of the bag. Avoid wrinkling the bag. Cut off one corner of the bag, making a hole big enough to fit a pipette. Then mix the denatured salmon sperm DNA with PerfectHyb without forming bubbles—pipette up and down gently but thoroughly, without introducing air into the solution, and transfer the solution into the bag, using the pipette. Hold the bag at an upwards angle with the opening of the bag near the heat sealer. Stroke the bag from the bottom to the top to eliminate bubbles, catching excess prehybridization solution on paper towels, and immediately heat-seal the top of the bag. If bubbles are still present, cut off an even smaller corner of the bag again, nudge bubbles out of the small opening, and reseal the bag. It may be helpful to lay the bag flat against a large glass or plastic plate and smooth out bubbles with a pipette, sealing the bag immediately. After sealing, wipe off the outside of the bag and check for leaks along the seals by pressing hard on the middle of the bag and looking for liquid oozing out of the seams. Reseal any leaky seams.

34. Do not leave the denatured probe at room temperature after denaturation, as the two single strands may re-hybridize to each other to form double-stranded DNA, which cannot hybridize to its target RNA on the membrane. If there is any possibility of probe renaturation, denature it a second time by reheating again.

35. If using a hybridization oven, remove the screw-capped hybridization tubes from the oven and place behind the plexiglass shield. Do not remove the prehybridization fluid. Add the hybridization fluid (PerfectHyb plus denatured salmon sperm DNA and radioactive probe) with a 25-ml pipette and mix the prehybridization plus hybridization solutions together by pipetting up and down repeatedly (against the side of the tube—avoid scratching the membrane with the pipette), without introducing bubbles. Iron out any bubbles with the pipette and dispose of the pipette as radioactive waste. Recap the tube. Return the tube to the rotisserie, check that the entire blot surface is coated with hybridization fluid throughout the rotation cycle as in **Note 32**, and rotate the tube at 65 °C overnight.

36. If maximizing sensitivity is not an issue, shorter hybridization times can be used—for example, 2 h for an actin loading control hybridization.
37. Since microwaving does not heat solutions evenly, mix the solution well before measuring buffer temperature. To be conservative, raise the temperature to no more than 55–60 °C. It is better to add a cooler solution and let it warm up during the wash than to overheat and inadvertently strip the probe from the blot.
38. For blots hybridized in hybridization tubes, our experience is that blots are cleanest when washed in a shaking water bath following the method in Subheading 3.6.3 rather than in hybridization tubes, likely due to increased wash solution volume and more vigorous agitation. Some Hybaid models conveniently include a platform shaker within a shielded, air-heated cabinet; when using these shakers ensure that the high-stringency and ultrahigh-stringency wash buffers are close to 65 °C, since buffers will not warm up as quickly in air as in water.
39. Kodak BioMax MS (maximum sensitivity) film in combination with two Kodak BioMax MS Screens will maximize sensitivity.

References

1. Wagner EM (2010) Monitoring gene expression: quantitative real-time RT-PCR. In: Freeman LA (ed.) Methods in molecular biology: lipoproteins, 2nd edn (Walker JM, Series ed.). Humana, Totowa, NJ
2. Raghavachari N (2010) Microarray technology: basic methodology and application in clinical research for biomarker discovery in vascular diseases. In: Freeman LA (ed.) Methods in molecular biology: lipoproteins, 2nd edn (Walker JM, Series ed.). Humana, Totowa, NJ
3. Basso F, Freeman L, Knapper CL, Remaley A, Stonik J, Neufeld EB, Tansey T, Amar MJ, Fruchart-Najib J, Duverger N, Santamarina-Fojo S, Brewer HB Jr (2003) Role of the hepatic ABCA1 transporter in modulating intrahepatic cholesterol and plasma HDL cholesterol concentrations. J Lipid Res 44:296–302
4. Joyce CW, Wagner EM, Basso F, Amar MJ, Freeman LA, Shamburek RD, Knapper CL, Syed J, Wu J, Vaisman BL, Fruchart-Najib J, Billings EM, Paigen B, Remaley AT, Santamarina-Fojo S, Brewer HB Jr (2006) ABCA1 overexpression in the liver of LDLr-KO mice leads to accumulation of pro-atherogenic lipoproteins and enhanced atherosclerosis. J Biol Chem 281:33053–33065
5. Berge KE, Tian H, Graf GA, Yu L, Grishin NV, Schultz J, Kwiterovich P, Shan B, Barnes R, Hobbs HH (2000) Accumulation of dietary cholesterol in sitosterolemia caused by mutations in adjacent ABC transporters. Science 290:1771–1775
6. Wu JE, Basso F, Shamburek RD, Amar MJ, Vaisman B, Szakacs G, Joyce C, Tansey T, Freeman L, Paigen BJ, Thomas F, Brewer HB Jr, Santamarina-Fojo S (2004) Hepatic ABCG5 and ABCG8 overexpression increases hepatobiliary sterol transport but does not alter aortic atherosclerosis in transgenic mice. J Biol Chem 279:22913–22925
7. Basso F, Freeman LA, Ko C, Joyce C, Amar MJ, Shamburek RD, Tansey T, Thomas F, Wu J, Paigen B, Remaley AT, Santamarina-Fojo S, Brewer HB Jr (2007) Hepatic ABCG5/G8 overexpression reduces apoB-lipoproteins and atherosclerosis when cholesterol absorption is inhibited. J Lipid Res 48:114–126
8. Freeman LA, Kennedy A, Wu J, Bark S, Remaley AT, Santamarina-Fojo S, Brewer HB Jr (2004) The orphan nuclear receptor LRH-1 activates the ABCG5/ABCG8 intergenic promoter. J Lipid Res 45:1197–1206
9. Gonzalez-Navarro H, Nong Z, Freeman L, Bensadoun A, Peterson K, Santamarina-Fojo S (2002) Identification of mouse and human macrophages as a site of synthesis of hepatic lipase. J Lipid Res 43:671–675

10. Varallyay E, Burgyan J, Havelda Z (2008) MicroRNA detection by northern blotting using locked nucleic acid probes. Nat Protoc 3:190–196
11. Sambrook J, Russell D (2001) Molecular cloning: a laboratory manual, 3rd edn. Cold Spring Harbor Laboratory Press, Cold Spring Harbor, NY
12. Laboratories BR (2009) Zeta-Probe® GT (genomic tested) blotting membranes instruction manual, bulletin # LIT292, pp 1–39. BioRad Laboratories, Hercules, CA
13. Johnston RF, Pickett SC, Barker DL (1990) Autoradiography using storage phosphor technology. Electrophoresis 11:355–360

Chapter 5

Laser Capture Microdissection for Analysis of Macrophage Gene Expression from Atherosclerotic Lesions

Jonathan E. Feig and Edward A. Fisher

Abstract

Coronary artery disease, resulting from atherosclerosis, is the leading cause of death in the Western world. Most previous studies have subjected atherosclerotic arteries, a tissue of mixed cellular composition, to homogenization in order to identify the factors in plaque development, thereby obscuring information relevant to specific cell types. Because macrophage foam cells are critical mediators in atherosclerotic plaque advancement, we reasoned that performing gene analysis on those cells would provide specific insight in novel regulatory factors and potential therapeutic targets. We demonstrated for the first time in vascular biology that foam cell-specific RNA can be isolated by laser capture microdissection (LCM) of plaques. As expected, compared to whole tissue, a significant enrichment in foam cell-specific RNA transcripts was observed. Furthermore, because regression of atherosclerosis is a tantalizing clinical goal, we developed and reported a transplantation-based mouse model. This involved allowing plaques to form in apoE−/− mice and then changing the plaque's plasma environment from hyperlipidemia to normolipidemia. Under those conditions, rapid regression ensued in a process involving emigration of plaque foam cells to regional and systemic lymph nodes. Using LCM, we were able to show that under regression conditions, there was decreased expression in foam cells of inflammatory genes, but an up-regulation of cholesterol efflux genes. Interestingly, we also found that increased expression of chemokine receptor CCR7, a known factor in dendritic cell migration, was required for regression. In conclusion, the LCM methods described in this chapter, which have already lead to a number of striking findings, will likely further facilitate the study of cell type-specific gene expression in animal and human plaques during various stages of atherosclerosis, and after genetic, pharmacologic, and environmental perturbations.

Key words apoE, Atherosclerosis, Gene expression, LCM, CCR7

1 Introduction

Macrophage foam cells are critical in the development of atherosclerosis [1, 2]. Therefore, a better understanding of the gene expression changes in foam cells during disease progression and regression has become an important goal in order to develop potential therapies and interventions [3–5]. The study of macrophage foam cell gene expression in arterial lesions, however, is hampered by the cellular heterogeneity of arterial tissue, which

Lita A. Freeman (ed.), *Lipoproteins and Cardiovascular Disease: Methods and Protocols*, Methods in Molecular Biology, vol. 1027, DOI 10.1007/978-1-60327-369-5_5, © Springer Science+Business Media, LLC 2013

besides macrophages, also contain lymphocytes, smooth muscle cells, endothelial cells, and fibroblasts. To overcome these technical obstacles, we describe here a method for the use of laser capture microdissection (LCM) to selectively procure macrophage foam cells from arterial lesions (as identified by the macrophage-specific marker CD68/Macrosialin) [6–9]. RNA extracted from the microdissected foam cells exhibited a 30-fold enrichment in CD68 mRNA levels as compared to homogenization of the whole artery [10], suggesting that LCM provides a method to identify gene expression changes specific to a particular cell type and that may be regulated by disease stage (e.g., progression vs. regression) or by different genetic and pharmacologic manipulations.

We previously developed a transplantation-based mouse model of atherosclerosis regression by allowing plaques to form in apoE−/− mice. Then, by grafting their aortic segments into a wild-type recipient, the plaque's plasma environment is changed from hyperlipidemia to normolipidemia [11, 12]. As a control, transplants are performed into apoE−/− recipients. At various time points after transplantation, LCM was used to isolate the foam cells from the grafts and RNA was extracted. Using real-time quantitative reverse transcription polymerase chain reaction (qRT-PCR), we were able to show that gene expression levels of inflammation markers such as MCP-1 and VCAM-1 were significantly down-regulated, and cholesterol efflux genes such as LXRα, ABCA1, and SR-BI levels were increased. Notably, CCR7 (a migratory factor for a number of leukocyte types, including dendritic cells) was up-regulated (Fig. 1) [13]. This was a crucial finding, as this gene turned out to be functionally required for the depletion of foam cells observed during regression [13]. In short, LCM makes possible the quantitative analysis of gene expression in macrophage foam cells and adds a powerful dimension to the study of atherosclerosis. It is important to note that while not described here, a similar approach can be taken for the molecular analysis of vascular smooth muscle cells, except that the identifying marker would be smooth muscle-alpha actin instead of CD68.

2 Materials

2.1 Animals and Tissue Processing

1. apoE−/− mice (The Jackson Laboratory, Bar Harbor, ME).
2. Ketamine/Xylazine anesthesia working solution (Sigma-Aldrich, St. Louis, MO).
3. 1× Phosphate-buffered saline (PBS, store at room temperature).
4. 1× PBS containing 10 % Sucrose (w/v) and an RNase inhibitor (store at room temperature). The RNase inhibitor should be added fresh before use.

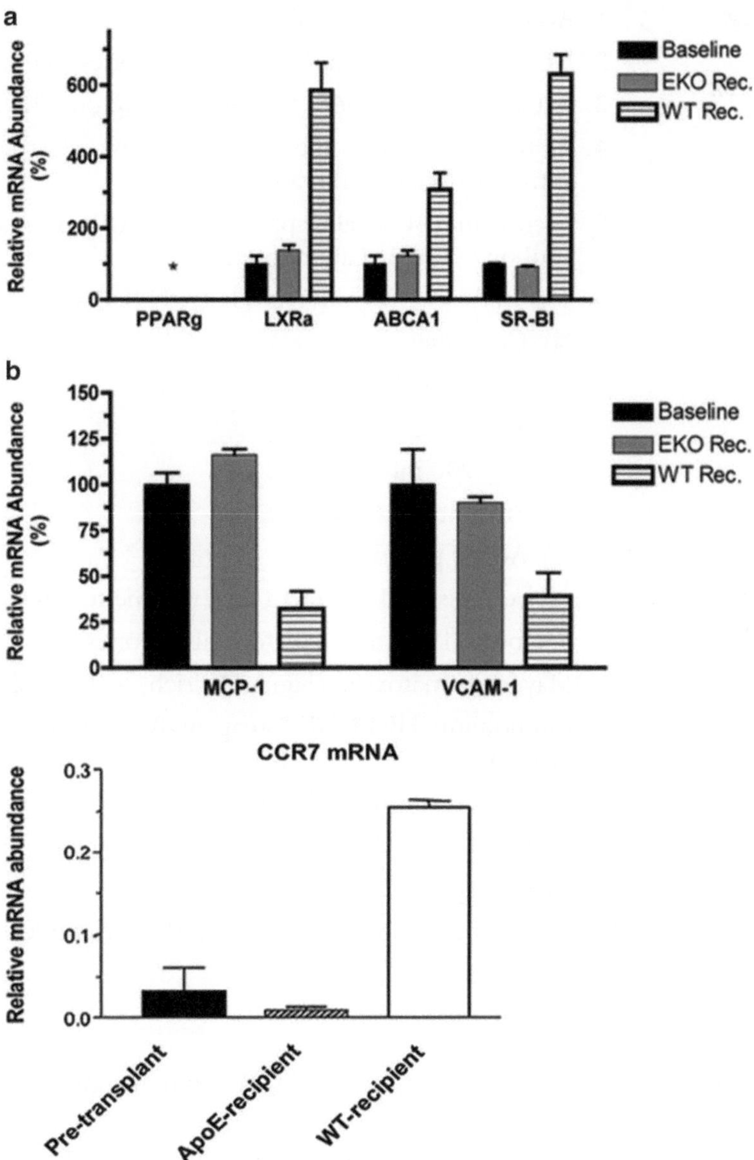

Fig. 1 Changes in the expression of genes associated with cholesterol efflux, inflammation, and migration in foam cells after the reversal of dyslipidemia

5. ProtectRNA RNase Inhibitor 500× concentrate (Sigma-Aldrich).
6. OCT cryoembedding medium (VWR, San Diego, CA).
7. Superfrost Plus slides 75 × 25 mm (Fisher Scientific, Pittsburgh, PA).
8. RNase Away reagent (Molecular Bioproducts, San Diego, CA).
9. Diethyl pyrocarbonate (DEPC, Sigma-Aldrich): working solution contains 0.1 % v/v DEPC in distilled water (*see* **Note 1**).

2.2 Histological Detection of Macrophage Foam Cells for LCM

1. Acetone (Fisher Scientific).
2. Normal rabbit serum (Vector Laboratories, Burlingame, CA).
3. Rat anti-mouse CD68 (Serotec, Kidlington, UK).
4. Biotinylated rabbit anti-rat IgG mouse-adsorbed secondary antibody (Vector Laboratories).
5. Vectastain ABC alkaline phosphatase (AP) enzymatic detection antibody (Vector Laboratories).
6. Tris Buffer—100 mM Tris–HCl, pH 8–8.4.
   (a) Combine:
      - 6.3 g Tris–HCl
      - 7.4 g Tris base
      - 900 ml distilled H_2O.
   (b) Stir until dissolved.
   (c) Adjust pH to 8–8.4.
   (d) Add distilled H_2O to a final volume to 1.0 l.
7. Vector Red substrate (Vector Laboratories).
8. Mayer's hematoxylin (Sigma-Aldrich, *see* **Note 2**).
9. Ammonium Hydroxide, Reagent ACS (Fisher Scientific, *see* **Note 3**).
10. Eosin Y (Harleco, Gibbstown, NJ).
11. 100 % Ethanol, Anhydrous, ACS/USP grade.
12. Xylenes (Fisher Scientific).
13. Permount Mounting Medium (Fisher Scientific).
14. Slide rack.
15. Staining dish.

2.3 Laser Capture Microdissection and RNA Isolation

1. Arcturus PixCell II LCM System (Arcturus, Mountain View, CA).
2. CapSure Macro LCM caps (Molecular Devices, Sunnyvale, CA).
3. PicoPure RNA isolation kit (Molecular Devices).
4. RNase-free DNase set (Qiagen, Valencia, CA).

2.4 Real-Time Quantitative RT-PCR

1. 7300 Real-Time PCR System (Applied Biosystems, Foster City, CA).
2. MicroAmp Fast Optical 96-Well Reaction Plate (Applied Biosystems).
3. The MicroAmp Optical Adhesive Film (Applied Biosystems).
4. Gene-specific Taqman primers and probes (Table 1).
5. iScript One-Step RT-PCR Kit for Probes (Bio-Rad, Hercules, CA).

Table 1
Oligonucleotide primer sequences for quantitative RT-PCR

Peroxisomal proliferator activated receptor γ (PPARγ)	
Fwd	5′-CATTCTGGCCCACCAACTTC-3′
Rev	5′-AAGGAATGCGAGTGGTCTT-3′
Probe	5′-FAM-TCAGCTCTGTGGACCTCTCCGTGATG - BHQ1-3′
Liver-X-receptor α (LXRα)	
Fwd	5′-GGAGGCAACACTTGCATCCT-3′
Rev	5′-AGGGCTGTAGGCTCTGCTGA-3′
Probe	5′-FAM-AGGGAGGAAGCCAGGATGCCCC-TAMRA-3′
ATP-binding cassette transporter A1 (ABCA1)	
Fwd	5′-ATTGCCAGACGGAGCCG-3′
Rev	5′-TGCCAAAGGGTGGCACA-3′
Probe	5′-FAM-CCAGCTGTCTTTGTTTGCATTGCCC-TAMRA-3′
Scavenger receptor BI (SR-BI)	
Fwd	5′-GATGATGACCTTGGCGCTG-3′
Rev	5′-TCACCAACTGTGCGGTTCATA-3′
Probe	5′-FAM-TCACCATGGGCCAGCGTGCTT-TAMRA-3′
CCR7	
Fwd	5′-CACGCTGAGATGCTCACTGG-3′
Rev	5′-ATCTGGGCCACTTGGATGG-3′
Probe	5′-FAM- CAGTGCCCAAGTGGAGGCCTTGATC-TAMRA- 3′
MCP-1	
Fwd	5′-TTCCTCCACCACCATGCAG-3′
Rev	5′-CCAGCCGGCAACTGTGA-3′
Probe	5′-FAM-CCCTGTCATGCTTCTGGGCCTGC-TAMRA-3′
VCAM-1	
Fwd	5′-CCCCAAGGATCCAGAGATTCA-3′
Rev	5′-ACTTGACCGTGACCGGCTT-3′
Probe	5′-FAM-TTCAGTGGCCCCCTGGAGGTTG-TAMRA-3′

All primers were provided by Biosearch Technologies (Novato, CA)

3 Methods

3.1 Animals and Tissue Processing

All reagents were maintained under RNase-free, sterile conditions. Transplantation was performed as previously described [12]. In brief, apoE−/− mice, a standard model of human atherosclerosis [14], were fed a high-fat diet for 16 weeks. Mice were then sacrificed via exsanguination (under general anesthesia with ketamine/xylazine) by intravascular perfusion with PBS. The thorax was opened and a 21-gauge cannula inserted into the left ventricle. The right atrium was incised to allow efflux of blood. Perfusion was at physiological pressure (100 mmHg) with PBS. The aortic arch (graft) was removed and transplanted into the descending aorta of a recipient mouse in the region just above the iliac bifurcation [12]. After 3 days, the recipient mouse was sacrificed, perfused with a PBS/sucrose/RNA inhibitor solution and the graft was harvested. The tissue (which was cut in half cross-sectionally) was positioned in a semi-filled (with OCT) embedding mold to ultimately generate cross sections using a cryostat. The embedding mold was then fully filled with OCT and frozen on dry ice. The specimens were stored at −80 °C until sectioning, immunostaining, and LCM were performed. Eventually, the tissue was cryo-sectioned at a thickness of 6 μm, collected on Superfrost Plus slides, and maintained cold at all times after collection. All procedures were approved by the Institutional Animal Care and Use Committee of NYU School of Medicine.

3.2 CD68 Immunodetection of Macrophages

Using immunohistochemistry, we stained every fifth slide to identify the macrophage-specific marker CD68/Macrosialin. This served as a guide slide for the remaining non-immunostained slides. The following protocol assumes the staining of 15–20 slides.

1. Use 250 ml ice-cold acetone to fix cells on slides in slide rack (10 min).
2. Rinse in 250 ml, 1× PBS for 2 min (2×).
3. Block in 4 % normal rabbit serum diluted in 1× PBS for 10 min.
4. Primary antibody: Dilute the rat anti-mouse CD68 antibody 1:250 in the 4 % rabbit serum. Incubate slides in primary antibody for 1 h.
5. 45 min into the primary antibody incubation, prepare the VECTASTAIN ABC-AP by adding two drops of reagent A and two drops of reagent B into 5 ml, 1× PBS, making sure to vortex well after each drop. Allow this solution to incubate at room temperature until ready for use.
6. Rinse slides in 1× PBS (3 dips).
7. Secondary antibody: Dilute the biotinylated rabbit anti-rat antibody 1:200 in the 4 % rabbit serum. Incubate slides in secondary antibody for 10 min.
8. Rinse slides in 250 ml 1× PBS (3 dips).
9. Incubate slides for 5 min with the VECTASTAIN ABC-AP solution.
10. During the 5 min incubation, prepare the Vector Red AP substrate. To 5 ml Tris buffer add two drops of each reagent, making sure to vortex well after adding each reagent.
11. Rinse in 250 ml 1× PBS (3 dips).
12. Pipette enough Vector Red solution to cover each tissue section on the slide. Develop in the dark until desired stain intensity develops. Monitor intensity under the microscope.

13. Place slides in 250 ml distilled water to stop staining reaction.
14. Counterstain with Mayer's hematoxylin for 1 min.
15. Place slides in 250 ml distilled water for 30 s.
16. Place slides in a bluing solution for 1 min. Bluing solution is prepared by adding 500 μl of ammonium hydroxide to 300 ml of distilled water.
17. Place slides in 250 ml 70 % ethanol for 30 s.
18. Place slides in 250 ml 80 % ethanol for 30 s.
19. Place slides in 250 ml 95 % ethanol for 1 min.
20. Place slides in 250 ml 100 % ethanol for 1 min.
21. Place slides in 250 ml xylene for 3 min.
22. Allow to air dry in hood until xylenes have evaporated (5 min or less depending on air flow and number of slides).
23. Once dry, mount with a resin mounting medium and coverslip the slides.

3.3 Hematoxylin and Eosin Y (H&E) Stain

These instructions follow a traditional H&E stain, with several modifications to significantly reduce the effects of RNases on the integrity and yield of RNA (*see* **Note 4**). H&E staining is conducted on the remaining slides that are not immunostained and are to be used for laser capture microdissection. The following protocol assumes the staining of 15–20 slides.

1. Take slides out of the −80 °C freezer and leave at room temperature for 2 min.
2. Place in staining tray containing 250 ml 70 % ethanol for 30 s.
3. Place in 250 ml RNase inhibitor-treated DEPC-treated water (DEPC water with ProtectRNA RNase Inhibitor diluted to 1× final concentration) for 1 min.
4. Place in 250 ml RNase inhibitor-treated Mayer's hematoxylin (Mayer's hematoxylin with ProtectRNA RNase Inhibitor diluted to 1× final concentration) for 1 min.
5. Place in 250 ml DEPC-treated water for 15 s.
6. Place in 250 ml PBS for 15 s.
7. Place in 250 ml RNase inhibitor-treated DEPC water for 15 s.
8. Place in 250 ml 70 % ethanol for 30 s.
9. Place in 250 ml 95 % ethanol for 30 s.
10. Dip in 250 ml eosin Y twice.
11. Place in 250 ml, 95 % EtOH for 15 s.
12. Place in separate 250 ml, 95 % EtOH for 30 s.
13. Place in 250 ml 100 % ethanol for 30 s.

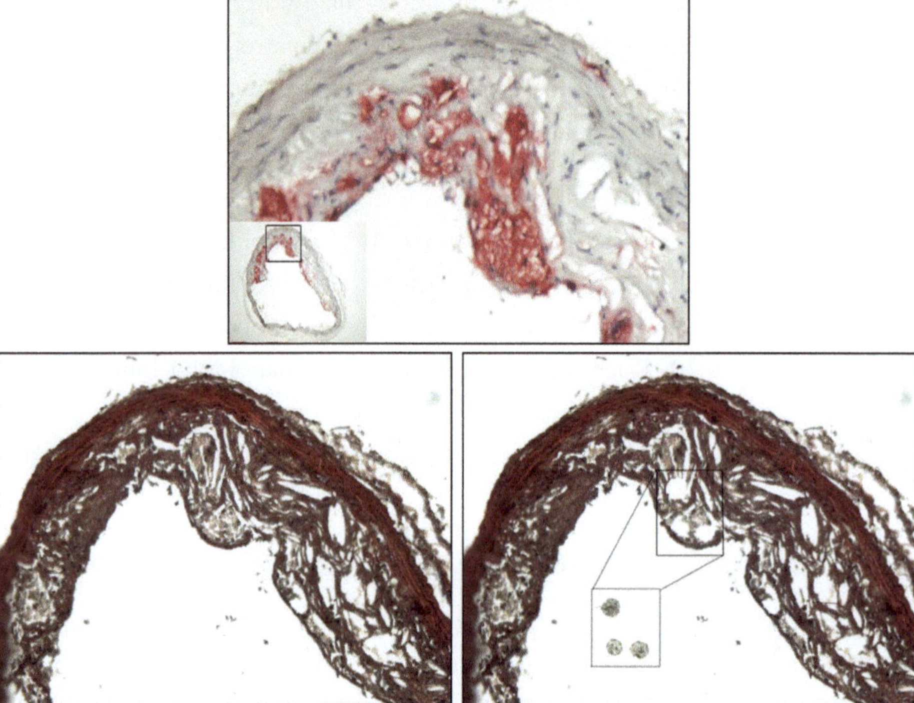

Fig. 2 CD68 (*top panel*) and H&E staining (*lower panel*: before LCM, *bottom left*; after LCM, *bottom right*) of representative cross sections of an aortic arch containing atherosclerotic plaque. The *bottom right panel* shows microdissected H&E-stained tissue with the insert displaying the microdissected macrophage foam cells

14. Place in 250 ml xylenes for 30 s.
15. Place in separate 250 ml xylenes for 3 min.
16. Allow to air dry in hood until xylenes have evaporated (5 min or less depending on air flow and number of slides).
17. Proceed directly to LCM.

3.4 Laser Capture Microdissection

The following instructions assume the use of an Arcturus, PixCell II LCM System but common to all instruments, there is direct visualization of the sections with subsequent capture of the desired cells. Identification of foam cells from the atherosclerotic plaque was guided by the staining of CD68 (Fig. 2). These slides were used as "guide slides" for the preceding and subsequent slides, which were H&E stained. In the PixCell II System, the capture of cells is achieved by placing a specially made cap, lined with a thermolabile film, onto a thinly cut section of tissue, with "pickup" completed by activating the film with a near-infrared laser diode pulse. This melting of the thermoplastic film causes the cells of interest to adhere to the film, which are then isolated from the surrounding tissue when the cap is lifted away from the slide. The

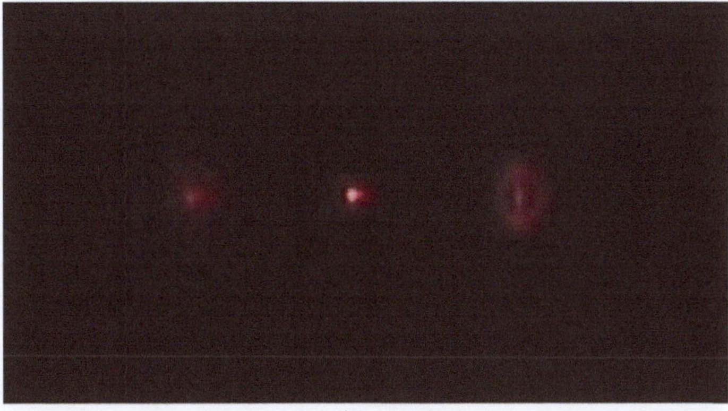

Fig. 3 Laser focusing. A focused laser will have an intensely *bright red* appearance with defined edges (*middle*). An unfocused laser will appear hazy (*left*) or as a ring containing a center line (*right*)

isolation of quality RNA was completed using the Arcturus PicoPure RNA isolation kit.

1. Set the PixCell II LCM System to the following parameters: 7.5 μm laser spot size, 40 mW power, 3.0 ms duration, 100 mV target, 0.2 ms delay between pulses.
2. Place a CD68 positively stained slide onto the microscope stage and locate the stained foam cells. Pressing the "Map" button, take several pictures at 4×, 10×, and 20× magnifications to use as guides to help locate foam cells on the H&E slides that follow.
3. Replace the CD68 slide with an H&E stained slide and activate the vacuum to prevent the slide from shifting during capture.
4. Using the rotating cap arm, collect a cap from the loading area and carefully lay the cap over the tissue section.
5. Enable the laser and focus if necessary (Fig. 3).
6. Before collecting the macrophage foam cells, attempt a few pulses in a tissue-free area to ensure proper wetting of the cap. Adjust if necessary (Fig. 4, *see* **Note 5**).
7. When ready, bring the tissue into the microscope's field of view and begin collecting foam cells. Continue collecting until the desired number of cells are obtained or until it is no longer possible to capture cells (*see* **Note 6**).
8. Once finished, lift and rotate the cap arm to bring the cap to the unloading area.
9. Into a 0.5 ml microfuge tube, included with the PicoPure kit, pipette 50 μl of extraction buffer.
10. Place the cap into the tube and invert so that the buffer floods the captured tissue.

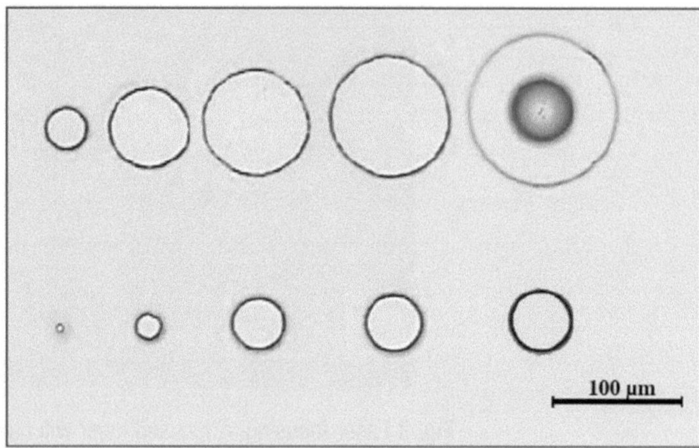

Fig. 4 Pulse wetting. *Upper row, left to right*: Result of changing power from 20 to 40, 60, 80, 100 mW. Duration, size, and intensity were kept constant at 10 ms, 7.5 μm, and 100 mV, respectively. At 100 mW, the thermoplastic film burned. *Bottom row, left to right*: Result of changing duration from 3.0 to 5.0, 10, 20, 30 ms. Power, size, and intensity were kept constant at 20 mW, 7.5 μm, and 100 mV, respectively. Note that the size of the pulse area does not become larger after 20 ms. Bar = 100 μm

11. Incubate at 42 °C for 30 min.
12. Continue RNA isolation using manufacturer's protocol, making sure to include the optional DNase treatment (*see* **Note 7**).

3.5 Analysis of Macrophage Foam Cell Gene Expression by Real-Time Quantitative RT-PCR

A very sensitive measurement method of low-abundance transcripts is qRT-PCR, and unlike Northern or RNase protection assays, it requires only a very small amount of total RNA (typically 100 pg to 1 ng). We have previously used qRT-PCR to demonstrate the selective enrichment from mouse plaques of lesional macrophage RNA by LCM [10]. The following instructions assume the use of a 7300 Real-Time PCR System from Applied Biosystems along with the iScript One-Step RT-PCR Kit for Probes from Bio-Rad (*see* **Note 8**). The protocol for the iScript kit has been slightly modified to reduce the amount of RNA sample needed for analysis. Keep reagents and samples on ice at all times.

1. Obtain an aliquot of each RNA sample and dilute in RNase-free water to a concentration of 20 pg/μl. Prepare enough sample to be used in 5 μl triplicates. Total RNA per reaction will be 100 pg.
2. Prepare a standard curve using an RNA sample (liver for all genes except CCR7, in which case we use spleen) of good quality and known concentration by making serial dilutions ranging from 20 to 0.0002 pg/μl.

3. Prepare the master mix for each gene of interest using the following setup, taking into account the necessary amount of wells per gene. Final total volume per reaction will be 25 μl. Always prepare an extra volume of master mix to account for any possible error in pipetting.
   (a) 2× RT-PCR Reaction Mix for Probes, 1× final concentration.
   (b) Forward primer, 300 nM final concentration.
   (c) Reverse primer, 300 nM final concentration.
   (d) Probe, 100 nM final concentration.
   (e) ROX reference dye, 1× final concentration.
   (f) iScript Reverse Transcriptase for One-Step RT-PCR, 0.5 μl per reaction.
   (g) Nuclease-free H_2O, to 25 μl final volume.
4. Into the 96-well plate, arrange and pipette 20 μl of the relevant master mix into its designated well.
5. Pipette 5 μl of the RNA sample into its corresponding well containing master mix. Mix well by pipetting.
6. When finished loading the wells, cover plate with the optical adhesive film and place into the 7300 Real-Time PCR System.
7. Input the following settings:
   (a) cDNA synthesis: 10 min at 50 °C.
   (b) RT activation: 5 min at 95 °C.
   (c) PCR cycling and detection (40 cycles): 15 s at 95 °C, 30 s at 55 °C (data collection); for some mRNAs of interest, the annealing temperature setting may need adjustment, as for any PCR assay.
8. Based on the generated standard curve, values for relative amounts of the gene of interested are calculated by the software provided with the 7300 Real-Time PCR System.

4 Notes

1. To prepare DEPC water, combine 0.1 % (v/v) DEPC with distilled water in an autoclave-safe container. Shake vigorously and leave for 24 h at room temperature. Autoclave and leave it for another 24 h at room temperature. DEPC-treated water is now ready to use. DEPC treatment inactivates RNases.
2. Mayer's hematoxylin must be filtered before use. Solution is toxic if swallowed and is also a known irritant to eyes, skin, and the respiratory system.

3. Ammonium hydroxide causes eye and skin burns. It also causes digestive and respiratory tract burns.
4. Eosin Y may be fatal or cause blindness if swallowed. It may cause damage to liver, kidneys, GI tract, and cardiovascular system. Wear eye protection and gloves. In addition, it is important to prepare all alcohol and xylene solutions fresh for the dehydration steps in order to achieve efficient capture during the LCM procedure.
5. Wetting of the cap refers to the melting of the thermolabile film. To obtain proper or optimal wetting, adjust the power and/or duration. Size and intensity of the pulse will vary depending on the parameters; however, keep in mind that high-power pulses will "burn" the film.
6. When calculating the number of pulses needed for your application, remember that the typical yield of total RNA from a cell is approximately 10 pg. It has been observed that approximately 10,000 pulses of macrophage foam cells, using the parameters 40 mW, 10 ms, 7.5 μm, and 100 mV (power, duration, size, and intensity), were enough to produce approximately 80 ng of total RNA. As the film becomes saturated with cells, it will become increasingly difficult to collect more, at which point it may be a sufficient amount of material to proceed to gene expression analyses.
7. It is important to eliminate genomic DNA with the RNase-free DNase set, especially when your downstream applications include RT-PCR, PCR, Microarray, qPCR, or bioanalysis. Presence of genomic DNA may affect results.
8. The probe, labeled at the 5′ and 3′ ends with 6-carboxyfluorescein (6-FAM) reporter and tetramethyl-6-carboxyrhodamine (TAMRA) quencher, respectively, is hydrolyzed by the 5′ exonuclease activity of Taq DNA polymerase, causing an increase in fluorescent signal that is measured in "real-time" after each cycle of PCR amplification. Standard curves were constructed by plotting $\log_{10}$ RNA starting quantity vs. cycle threshold. On the basis of appropriate serially diluted standard RNA, the amount of input standard RNA yielding the same amount of PCR product measured from an unknown sample was calculated.

Acknowledgments

This work was supported by NIH grant HL-084312 (E.A.F.) and by an NIH predoctoral fellowship AG-029748 (J.E.F.).

References

1. Glass CK, Witztum JL (2001) Atherosclerosis: the road ahead. Cell 104:503–516
2. Smith JD, Trogan E, Ginsberg M, Grigaux C, Tian J, Miyata M (1995) Decreased atherosclerosis in mice deficient in both macrophage colony-stimulating factor (op) and apolipoprotein E. Proc Natl Acad Sci USA 92:8264–8268
3. Chong PH, Bachenheimer BS (2000) Current, new and future treatments in dyslipidaemia and atherosclerosis. Drugs 60:55–93
4. Brewer HB Jr (2000) The lipid-laden foam cell: an elusive target for therapeutic intervention. J Clin Invest 105:703–705
5. Plutzky J (1999) Atherosclerotic plaque rupture: emerging insights and opportunities. Am J Cardiol 84:15J–20J
6. Emmert-Buck MR, Bonner RF, Smith PD, Chuaqui RF, Zhuang Z, Goldstein SR et al (1996) Laser capture microdissection. Science 274:998–1001
7. Bonner RF, Emmert-Buck M, Cole K, Pohida T, Chuaqui R, Goldstein S, Liotta LA (1997) Laser capture microdissection: molecular analysis of tissue. Science 278:1481–1483
8. Ramprasad MP, Fischer W, Witztum JL, Sambrano GR, Quehenberger O, Steinberg D (1995) The 94- to 97-kDa mouse macrophage membrane protein that recognizes oxidized low density lipoprotein and phosphatidylserine-rich liposomes is identical to macrosialin, the mouse homologue of human CD68. Proc Natl Acad Sci USA 92:9580–9584
9. Ramprasad MP, Terpstra V, Kondratenko N, Quehenberger O, Steinberg D (1996) Cell surface expression of mouse macrosialin and human CD68 and their role as macrophage receptors for oxidized low density lipoprotein. Proc Natl Acad Sci USA 93:14833–14838
10. Trogan E, Choudhury RP, Dansky HM, Rong JX, Breslow JL, Fisher EA (2002) Laser capture microdissection analysis of gene expression in macrophages from atherosclerotic lesions of apolipoprotein E-deficient mice. Proc Natl Acad Sci USA 99:2234–2239
11. Reis ED, Li J, Fayad ZA, Rong JX, Hansoty D, Aguinaldo JG, Fallon JT, Fisher EA (2001) Dramatic remodeling of advanced atherosclerotic plaques of the apolipoprotein E-deficient mouse in a novel transplantation model. J Vasc Surg 34:541–547
12. Chereshnev I, Trogan E, Omerhodzic S, Itskovich V, Aguinaldo JG, Fayad ZA, Fisher EA, Reis ED (2003) Mouse model of heterotopic aortic arch transplantation. J Surg Res 111:171–176
13. Trogan E, Feig JE, Dogan S, Rothblat GH, Angeli V, Tacke F, Randolph GJ, Fisher EA (2006) Gene expression changes in foam cells and the role of chemokine receptor CCR7 during atherosclerosis regression in ApoE-deficient mice. Proc Natl Acad Sci USA 103:3781–3786
14. Breslow JL (1996) Mouse models of atherosclerosis. Science 272:685–688

Part II

Sequencing

Chapter 6

Sequencing PCR-Amplified DNA in Lipoprotein and Cardiovascular Disease Research

Victoria Youngblood and James G. Taylor VI

Abstract

The discovery of novel genetic variants and mutations in lipoprotein and cardiovascular disease research requires DNA sequencing. Large-scale genomics facilities will increasingly accomplish this with a combination of "next-generation" DNA sequencing methodologies. However, laboratories with limited access to these emerging technologies can still support focused genomic studies with the use of automated Sanger sequencing. Here, we describe two robust methods for medium-throughput DNA sequencing from PCR-amplified fragments of genomic DNA.

Key words PCR, DNA sequencing, Automated Sanger sequencing, Single-nucleotide variant, Mutation, Genetic variant, Automation

1 Introduction

The first publicly available assemblies of human genomic sequence in 2002 generated anticipation for rapid advances in deciphering the genetic basis of human diseases [1, 2]. The completed Human Genome Project was quickly followed by the development of a Human Haplotype Map (www.hapmap.org), an effort to assemble known single-nucleotide variants (SNVs) into haplotypes from pedigrees in major world populations. The HapMap is now in its third generation, including data for more than two million genetic markers [3]. The utility of these genetic tools is now being demonstrated with an increasing number of genome-wide association (GWA) and resequencing studies, which have identified susceptibility loci and haplotypes for a variety of human conditions, including lipoprotein and cardiovascular diseases [4–9]. However, GWA studies are designed to localize disease susceptibility to a defined chromosomal region using genetic markers, and thus it is unlikely that SNVs used in these studies actually represent the etiologic variant attributable to the disease state.

Lita A. Freeman (ed.), *Lipoproteins and Cardiovascular Disease: Methods and Protocols*, Methods in Molecular Biology, vol. 1027, DOI 10.1007/978-1-60327-369-5_6,

An essential step to identifying the actual disease-causing variants underlying apparent genetic associations is to comprehensively sequence entire loci or even entire genomes from single individuals [10]. Sequence-based studies also allow for consideration of both common and rare alleles contributing to complex traits [4, 9]. Indeed, sequence analysis represents a flexible genotyping platform, allowing an investigator to query almost all nucleotide sites within the human genome. Large-scale genomics facilities will increasingly accomplish this with a combination of "next-generation" DNA sequencing methodologies where substantial bioinformatic infrastructure is essential. Such efforts have recently resulted in reports of entire genomic sequences for three individuals [10–12]. However, it is notable that one whole-genome sequencing study determined that nearly 84 % of 181 nonsynonymous SNV candidates identified with "next-generation" sequencing were not, in fact, variant sites when validated with standard automated Sanger sequencing [10]. Thus, laboratories with limited access to these facilities can support more focused and highly accurate studies of complex traits with the use of robust methods for automated Sanger sequencing.

PCR fragments amplified from human genomic DNA templates can be sequenced in a rapid, cost-effective manner using automated dideoxy DNA sequencing in a laboratory equipped with either a 16- or 96-channel capillary DNA sequencer using the workflow outlined in Fig. 1. This method requires a minimum of personnel and is amenable to the development of a medium-throughput production pipeline with robotic automation, even in small laboratories. The author is routinely able to generate 1 megabase of sequence reads in an average month with 2 laboratory staff members [13, 14].

2 Materials

2.1 Genomic DNA

1. Purified genomic DNA (*see* **Notes 1–3**). Store at −20 °C (short-term storage) or −80 °C (long-term storage).
2. DNA quantification method: UV light spectrophotometer (an alternative is the Quant-iT™ PicoGreen® dsDNA reagent quantification kit; Invitrogen, Carlsbad, CA).

2.2 PCR DNA Amplification Reactions in 96- or 384-Well Formats

1. AmpliTaq Gold DNA polymerase: 1 U/μL (Applied Biosystems, Santa Clara, CA) (*see* **Note 4**). Store at −20 °C.
2. Gold Reaction buffer (10×): 150 mM Tris–HCl, pH 8.0, 500 mM KCl (Applied Biosystems, Santa Clara, CA). This reagent is provided with AmpliTaq Gold. Store at −20 °C.
3. Deoxynucleotides or dNTPs (10 mM mixture of dATP, dTTP, dGTP, and dCTP; Fermentas, Glen Burnie, MD). Store at −20 °C (*see* **Note 5**).

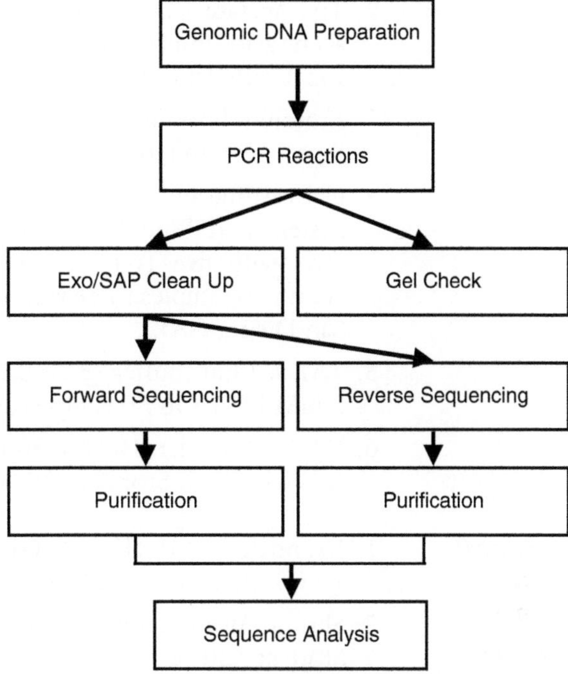

Fig. 1 Overview of a medium-throughput PCR-automated sequencing workflow. In the text, two different options for purification of sequencing reactions are presented

4. 25 mM $MgCl_2$. This reagent is provided with AmpliTaq Gold. Store at –20 °C.
5. A sterile solution of 50 % glycerol (prepared with molecular biology grade water). Store at room temperature.
6. A pair of specific primers (one forward and one reverse), each at a concentration of 10 μM (*see* **Notes 6** and 7).
7. 96-well or 384-well microtiter PCR plates (various vendors).
8. 50-mL sterile disposable reagent reservoirs (various vendors).
9. DNase/RNase-free polyolefin amplification tape (Nalge Nunc International, Rochester, NY).
10. Thermal cycler(s) fitted with sample blocks corresponding to PCR plate size.
11. Pipette tips, 10 μl, filter barrier (USA Scientific, Ocala, FL) (*see* **Note 8**).
12. Twelve multichannel Finnpipette (5–50 μl; VWR Scientific, West Chester, PA).
13. Centrifuge for PCR plates.

2.3 Agarose Gel Electrophoresis

1. Ultra Pure Agarose (Invitrogen, Carlsbad, CA). Store dry at room temperature.

2. Tris Acetate EDTA buffer (TAE) (50×): 2 M Tris, 0.358 M acetic acid, and 50 mM EDTA, pH 7.6–7.8 (Quality Biological, Inc, Gaithersburg, MD). Store at room temperature.
3. Ethidium bromide (Invitrogen), 10 mg/mL solution in water. Store at room temperature.
4. Gel casting and running apparatus. RunOne Electrophoresis System and Blue MultiCaster Long UVT (UV-transparent) trays with five 16+2 combs (tray stock number EP-1019) allows 96 samples to be run on a single agarose gel (Embi Tec, San Diego, CA).
5. TAE loading buffer (5×): 0.2 M Tris, 0.0358 M acetic acid, 5 mM EDTA in 50 % glycerol. Store at room temperature.
6. DNA size ladder, 100 bp (500 μg/mL) (New England Biolabs, Ipswich, MA). Store at −20 °C.

2.4 Exonuclease I/ Shrimp Alkaline Phosphatase PCR Cleanup Reactions

1. Exonuclease I (Exo) (GE Healthcare, Piscataway, NJ) 10 U/μL. Store at −20 °C.
2. Shrimp Alkaline Phosphatase (SAP) (GE Healthcare) 1 U/μL. Store at −20 °C.
3. SAP/Exo 10× reaction buffer: 200 mM Tris–HCl, pH 8.0, 100 mM $MgCl_2$ (*see* **Note 9**). Store at −20 °C.
4. Clean 96-well or 384-well microtiter PCR plates.
5. Amplification tape.
6. Centrifuge for PCR plates.

2.5 BigDye Terminator Sequencing Reactions

1. BigDye Terminator version 3.1 reaction mix (Applied Biosystems, Santa Clara, CA). Store in the dark at −20 °C.
2. BigDye Terminator 5× dilution buffer (Applied Biosystems) (*see* **Note 10**). Store at −20 °C.
3. Sterile molecular biology grade water.
4. SAP/Exo-treated PCR DNA template.
5. Clean 96-well or 384-well microtiter PCR plates.
6. Amplification tape.
7. Centrifuge for PCR plates.

2.6 Option 1: Purification of Sequencing Reactions by Sephadex Gel Filtration

1. Multiscreen HV filter plates (product # MAHVN4550; Millipore, Billerica, MA): 1 screen per 96-well sequencing plate.
2. Sephadex G-50 (Sigma Chemical Company, St. Louis, MO). Store at room temperature.
3. Sterile deionized, distilled water.
4. Sephadex filter plate loading device (Millipore).
5. Completed sequencing reactions in 96-well or 384-well PCR plates.

6. 12 multichannel Finnpipette (50–300 μl; VWR Scientific).
7. Centrifuge for PCR plates.
8. Manual heat sealer apparatus (*see* **Note 11**).
9. Foil seals for use with manual heat sealer.

2.7 Option 2: Automated Dye Terminator Magnetic Bead Purification

1. Agencourt CleanSEQ® reagent (sold as 8 or 50 mL kits; Agencourt Biosciences, Beverly MA). Store at 4 °C.
2. Agencourt SPRIPlate® 96R—Ring Magnet Plate (or 384-well version). Two to four magnetic plates are recommended.
3. Beckman Coulter Biomek® FX Laboratory Automated Workstation equipped with Beckman Software V3.2 and V.3 CleanSEQ 96 script (batch method) (*see* **Note 12**).
4. 85 % ethanol solution (*see* **Note 13**). Store at room temperature.
5. Vortex for PCR plates.
6. Centrifuge for PCR plates.
7. Manual heat sealer apparatus.
8. Foil seals for use with manual heat sealer.

2.8 DNA Sequencer and Sequence Analysis

1. DNA sequencer (*see* **Note 14**).
2. DNA sequence analysis software (*see* **Note 15**).

3 Methods

3.1 DNA Preparation (Again Notes 2 and 3)

1. Genomic DNA is quantified in nanograms per microliter (ng/μL) by light optical density (OD) measurement at 260 nm. An alternative method is the PicoGreen DNA quantification kit (Invitrogen, Carlsbad, CA).
2. Genomic DNA should be diluted with water or TE buffer to a working concentration of 10 ng/μL in a 96-well or 384-well plate. Our lab typically leaves two blank wells for control sequencing reactions (always in wells A01 and H12), and thus a 96-well plate can accommodate DNA from 94 different individuals.

3.2 PCR Amplification from Genomic DNA

1. Enter the following cycling program parameters into a thermal cycler:
   (a) Hold at 95 °C for 10 min.
   (b) 35 cycles: denature at 94 °C for 15 s, anneal at 60 °C for 30 s, and extension at 72 °C for 45 s (*see* **Note 16**).
   (c) 10 min at 72 °C.
   (d) Hold at 4 °C forever.

PCR Reaction				
Gene:	*gene*			
Population:	*pop*			
Samples:	*samples*			
Primer Name:	*primer*		master mix (uL)	
Date:	*date*		% overage:	7%
PCR Program:	*pcr program*	1X (uL)	# reactions:	96
	50% Glycerol	2.4	156.2	
	10X GOLD Reaction Buffer	1.5	97.6	
	MgCl2 (25mM)	1.2	78.1	
	dNTP (10mM)	0.3	19.5	
	primer Fwd. Primer (10 pmol/ul)	0.3	19.5	
	primer Rev. Primer (10 pmol/ul)	0.3	19.5	
	TaqGold (1 U/uL)	0.2	13.0	
	Sterile Water	7.8	507.5	
	DNA template (10 ng/uL)	1.0	65.1	
	Total reaction volume (uL):	15.0	976.0	

Fig. 2 Example of a spreadsheet template for the PCR reaction setup, including composition for a mastermix for 96 reactions. The suggested PCR reaction volume for an individual reaction is 15 μL, although this volume can be adjusted accordingly to between 10 and 20 μL

2. Prepare a reaction mastermix in a sterile 50 mL reagent reservoir by combining the following reagent volumes for each reaction well in the following order (*see* **Note 17**; Fig. 2):
   (a) 7.8 μL sterile water.
   (b) 2.4 μL 50 % sterile glycerol.
   (c) 1.5 μL 10× Gold Reaction buffer.
   (d) 1.2 μL $MgCl_2$ (from 25 mM stock solution).
   (e) 0.3 μL dNTP mix (from 10 mM stock solution).
   (f) 0.3 μL forward primer (from 10 μM stock solution).
   (g) 0.3 μL reverse primer (from 10 μM stock solution).
   (h) 0.2 μL Taq polymerase (1 U/μL).

   Mix completely by pipetting mixture up and down.
3. To each well of a PCR plate, add 14 μL of mastermix and 1 μL (~10 ng) genomic DNA.
4. Seal the plate with amplification tape using a roller, centrifuge at maximum speed for 1 min and place in thermal cycler.
5. Run the thermal cycling program entered in Subheading 3.2, **step 1**.
6. After completion of the thermal cycling program, remove the reaction plate from the thermal cycler and centrifuge at maximum speed for 1 min to collect any condensation into the bottom of the reaction wells.
7. Remove amplification tape and analyze 5 μL from each reaction well on a TAE agarose gel.

3.3 TAE Agarose Gel Electrophoresis

1. Mix 1 g of agarose (for 1 % gel) with 100 mL of 0.5× TAE. Melt agarose in microwave and add 2.5 μL ethidium bromide/100 mL of melted agarose. Allow the melted agarose to cool at room temperature for several minutes.
2. Assemble gel casting apparatus, pour the melted agarose into the form, and insert combs. Leave on the bench top until hardened.
3. Remove the combs and place in gel running apparatus and submerge in 0.5× TAE buffer.
4. Mix 5 μL of PCR sample with 3 μL of TAE sample buffer.
5. Centrifuge at maximum speed for 1 min and load 5 μL into each lane on the agarose gel.
6. Load 1 μL of a DNA size ladder in appropriate well.
7. Electrophorese at 100 V until the bromophenol blue is near the next well lane (~1.5 cm if you are using Embi Tec Blue MultiCaster Long UVT trays with five 16+2 combs) or until the dye front has migrated about half the distance of the gel.
8. Expose gel to ultraviolet light to visualize bands. The gel should show a prominent band for each 400–800-base-pair PCR product. If these bands are not seen on the gel, do not proceed to cleanup or sequencing steps until the PCR reaction has been optimized.

3.4 Exonuclease I/ Shrimp Alkaline Phosphatase PCR Cleanup Reactions

1. Enter the following cycling program parameters into a thermal cycler:
   (a) 37 °C for 60 min.
   (b) 72 °C for 15 min.
   (c) Hold at 4 °C forever.
2. Prepare Exo/SAP reaction mastermix in a sterile 50 mL reagent reservoir by combining the following reagent volumes for each reaction well in the following order (*see* **Note 17**; Fig. 3):
   (a) 4.3 μL sterile water.
   (b) 0.25 μL SAP.
   (c) 0.20 μL Exonuclease I.
   (d) 0.25 μL 10× SAP buffer.

   Mix well by pipetting up and down.
3. Pipette 5 μL of the Exo/SAP mastermix into each well of a new 96-well PCR reaction plate using a 12-multichannel pipette.
4. Pipette 4 μL of the remaining PCR product from Subheading 3.2 into the PCR plate containing the Exo/SAP reaction mixture. Mix well by pipetting up and down.

Shrimp Alkaline Phosphatase/Exonuclease I Pretreatment Reaction				
Gene:	*gene*			
Amplicon:	*amplicon*			
Population:	*pop*			
Samples:	*samples*		**master mix (uL)**	
Date:	*date*		% overage:	7%
PCR Program:	*pcr program*	**1X (uL)**	# reactions:	96
	SAP (1 unit/ul)	0.25	25.7	
	Exonuclease I (10 U/ul)	0.20	20.5	
	10X SAP buffer	0.25	25.7	
	Sterile Water	4.30	441.7	
	PCR product (template)	4.00		
	Total reaction volume (uL):	9.0		

Fig. 3 Example of spreadsheet template used for PCR/Shrimp Alkaline Phosphatase/Exonuclease I reaction setup, including composition for a mastermix for 96 reactions

5. Seal the plate with amplification tape using a roller, centrifuge at max speed for 1 min and place the sealed 96-well plate in a thermal cycler.
6. Run the thermal cycling program entered in Subheading 3.4, **step 1**.
7. After completion of the thermal cycling program, remove the reaction plate from the thermal cycler and centrifuge at maximum speed for 1 min to collect any condensation into the bottom of the reaction wells.
8. Remove the amplification tape and transfer 2 μL of the PCR SAP/Exo reaction product into each sequencing reaction on a fresh 96-well reaction plate using a multichannel pipette (*see* Subheading 3.5 or 3.6—Option 1 or 2).
9. Seal the remaining PCR SAP/Exo reaction products in the 96-well SAP/Exo reaction plate with adhesive foil. This product may be stored for several months at −20 to −80 °C for additional sequencing reactions in the future, if necessary.

3.5 Option 1: BigDye Terminator Sequencing Reactions and Sephadex Purification

1. Set up the thermal cycler with the following profile:
   (a) Hold at 96 °C for 2 min.
   (b) 25 cycles: 96 °C for 10 s, anneal at 50 °C for 10 s, and extension at 60 °C for 4 min.
   (c) Hold at 4 °C forever.
2. Prepare two separate (forward and reverse) BigDye Terminator (BDT) sequence mastermix by combining on ice in the following order per well (*see* **Note 18**):
   (a) 2 μL sterile water.
   (b) 2 μL 5× sequencing dilution buffer.
   (c) 2 μL BDT reaction mix (ver 3.1).

(d) 1 μL M13F primer (from a 3.2 μM stock solution for forward reactions) *or* M13R (from a 3.2 μM stock solution for reverse reactions) (*see* **Note 19**).

Mix well by pipetting up and down.

3. To each well of a PCR plate, add 7 μL of the forward BDT sequence reaction mastermix and 2 μL of PCR/SAP/Exo product from Subheading 3.4, **step 8**.
4. In a separate 96-well reaction plate, add 7 μL of the reverse BDT sequence reaction mastermix to each well and transfer 2 μL of the PCR/SAP/Exo product from Subheading 3.4 to each reaction well.
5. Seal the plate with amplification tape.
6. Centrifuge at max speed for 1 min to collect any condensation into the bottom of the wells, and place in thermal cycler.
7. Run the thermal cycling program entered in Subheading 3.5, **step 1**.
8. After completion of the thermal cycling program, remove the reaction plate from the thermal cycler and centrifuge at maximum speed for 1 min to collect any condensation into the bottom of the reaction wells (*see* **Note 20**).
9. To begin sequencing reaction purification, add Sephadex to Sephadex filter plate loading device. Remove any excess Sephadex.
10. Transfer Sephadex from loading device to multiscreen HV filter plates by aligning the filter plate on top of the device and inverting.
11. To each well of the multiscreen HV filter plates, add 300 μL of water to hydrate the Sephadex matrix.
12. Incubate hydrated Sephadex matrix in the multiscreen filter plate for 4 h at room temperature (or at 4 °C overnight).
13. After a minimum of 4 h hydration, remove the excess water by centrifuging the Sephadex multiscreen plate at 1,273 × *g* for 5 min (*see* **Note 21**). The multiscreen plate fits easily on the top of a 96-well reaction plate, where the excess water will be collected during centrifugation.
14. Remove the amplification tape from the completed sequencing reaction plate, and transfer each reaction well content to the center of the Sephadex matrix at the corresponding position on the multiscreen plate (i.e., transfer the contents of reaction well A01 to position A01 on the Sephadex multiscreen plate) (*see* **Note 22**).
15. Place a new 96-well reaction plate below the Sephadex multiscreen plate loaded with the sequencing reaction (*see* **Note 23**). Centrifuge this Sephadex multiscreen 96-well reaction plate assembly at 1,273 × *g* for 5 min.

BigDye Terminator Sequencing Reaction				
Gene:	gene			
Amplicon:	amplicon			
Population:	pop			
Samples:	samples			
Primer Name:	M13F		master mix (uL)	
Date:	date		% overage:	7%
PCR Program:	program	1X (uL)	# reactions:	96
	ABI 5X seq diluation buffer	2.00	205.4	
	BDT reaction mix (ver 3.1)	0.25	25.7	
	M13F Primer (3.2 pmol/ul)	1.00	102.7	
	Sterile Water	3.75	385.2	
	PCR/SAP product (template)	2.00		
	Total reaction volume (uL):	9.0		
	Repeat reaction plate as above with M13R primer.			

Fig. 4 Example of spreadsheet template used for the Option 2 Dilute BigDye Terminator (BDT) sequencing reaction setup, including composition for a mastermix for 96 reactions. This template can be adjusted accordingly for the standard BDT sequencing reaction protocol described under Subheading 3.5

16. Approximately 10 μL is now in each well of the reaction plate. The Sephadex multiscreen can be discarded. Heat-seal the reaction plate with foil seals and the manual heat sealer.
17. Purified sequencing reactions sealed in the reaction plate can be stored at −20 to −80 °C until loaded on a DNA sequencer. Samples may be stored for up to 1 month before analysis on the sequencer.

3.6 Option 2: Diluted BigDye Terminator Sequencing Reactions and Magnetic Bead Purification

1. Set up the thermal cycler with the cycling profile listed under Subheading 3.5, **step 1**.
2. Prepare two separate (forward and reverse) BDT sequence mixture by combining on ice in the following order per well (*see* **Note 17**; Fig. 4):
   (a) 3.75 μL sterile water.
   (b) 0.25 μL 5× sequence dilution buffer.
   (c) 2 μL BDT reaction mix (version 3.1).
   (d) 1 μL M13F primer (from a 3.2 μM stock solution for forward reactions) *or* M13R (from a 3.2 μM stock solution for reverse reactions).

   Mix well by pipetting up and down.
3. To each well of a PCR plate, add 7 μL of the forward BDT sequence mixture and 2 μL of PCR/SAP/Exo product from Subheading 3.4, **step 8**.
4. Follow remaining steps listed under Subheading 3.5, **steps 5–8**, including reaction plate centrifugation and thermal cycling.
5. To begin magnetic bead sequencing reaction purification, manually add 10 μL Agencourt CleanSEQ® reagent directly

into each well containing a completed sequencing reaction plate using a multichannel pipette (*see* **Note 24**). CleanSEQ can be poured into a sterile 50-mL reagent reservoir for easy transfer into the reaction plate.

6. Add 42 μL 85 % ethanol to each sequencing reaction/ CleanSEQ mixture and mix well by pipetting up and down ten times. Seal the reaction plate with adhesive foil.
7. Agitate the sealed plates with vigorous vortexing for 2 min and incubate at room temperature for 5 min.
8. Centrifuge the sealed reaction plate at maximum speed for 1 min to pellet the beads.
9. Place each 96-well sample plate on an Agencourt SPRIPlate 96R magnetic plate holder for 5 min (*see* **Note 25**).
10. Place the sample plates in the magnetic SPRIPlates at appropriate locations on the Beckman Coulter Biomek® FX Laboratory Automated Workstation stage. The automated FX script for CleanSEQ will aspirate the supernatant out of each reaction well (*see* **Note 26**). The robot will wash each reaction well by adding 100 μL 85 % ethanol for 30 s to each reaction well and then aspirating the ethanol. This step is repeated for a total of two wash steps.
11. Remove the plates from the magnets, and let air dry for no more than 10 min (*see* **Note 27**).
12. Elute the purified sequencing products from the magnetic beads by manually pipetting 30 μL of sterile water into each reaction well containing dried Agencourt CleanSEQ beads. Seal plate with foil, vortex for 2 min, and centrifuge the sealed plate at maximum speed for 1 min to pellet the beads.
13. Remove the foil seal and return the plate to the Agencourt SPRIPlate 96R magnetic plate holder for 5 min on the stage of the FX Workstation.
14. Place a clean 96-well reaction plate at the transfer position on the stage of the FX Workstation. One new 96-well reaction plate is required for each individual 96-well plate with sequencing reactions purified using this method (*see* **Note 23**).
15. After the 5 min incubation on the magnetic plate holders, the FX Workstation script will transfer 10 μL from each sequencing plate in the SPRIPlate 96R magnetic plate holder to clean 96-well reaction plates at the transfer position on the stage of the Workstation.
16. Remove the 96-well plates containing 10 μL CleanSEQ-purified sequencing reactions from the Workstation.
17. Heat-seal these new sequencing reaction plates with foil seals using the manual heat sealer.

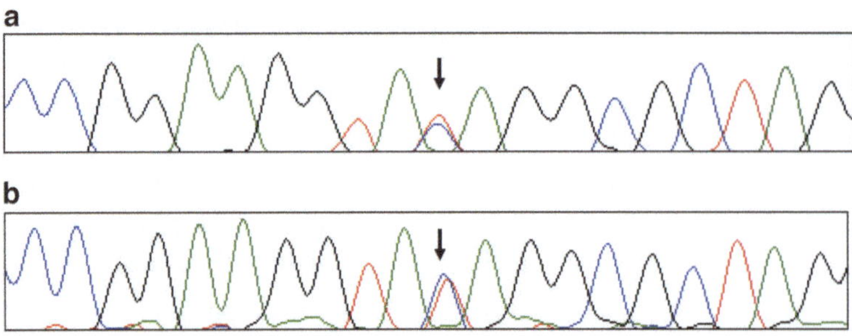

Fig. 5 Typical sequence data for ten nucleotides on either side of rs42409 generated using the conditions described for option 2. (**a**) Forward sequence data for SNV rs42409 (*arrow*). (**b**) Reverse sequence data for SNV rs42409 (*arrow*)

18. Purified sequencing reactions sealed in the reaction plate can be stored at –20 to –80 °C until loaded on a DNA sequencer. Samples may be stored for up to 1 month before analysis on the sequencer.

3.7 Analysis of Sequencing Reactions on a Capillary DNA Sequencer

1. Heat-sealed 96-well plates containing sequencing reactions are now ready for loading onto a DNA sequencer (*see* **Note 28**).
2. Once sequencing reaction products have been resolved on a capillary DNA sequencer (*see* **Note 28**), results are analyzed using DNA sequence analysis software (*see* **Note 15**). Representative sequence tracings flanking the SNV rs42409 (Fig. 5) were generated using this chapter's automated sequencing protocol with Option 2 magnetic bead purification.

4 Notes

1. Individual DNA samples should be stored in a 96- or 384-well format to make use of multichannel pipettes and/or multichannel robotic systems.
2. Access to high-quality genomic DNA from standardized human populations is often helpful for quality control and for establishing the prevalence of newly discovered genetic variants. Recommended control DNA populations include a panel of 200 healthy Caucasians (HD200CAU), 100 healthy African Americans (HD100AA), and a panel of 102 individuals comprising four major populations (the SNP500 Cancer; SNP500V). These and other prepared genomic DNA panels can be purchased from the National Institutes of General Medical Sciences (NIGMS) Human Genetic Cell Repository (Coriell Cell Repositories; http://ccr.coriell.org).
3. High-quality genomic DNA can be prepared from whole blood samples using one of the following kit systems according to the manufacturer's directions: QIAamp DNA Blood

Maxi Kit (Qiagen, Valencia, CA) or Agencourt Genfind kit (Agencourt, Beverly, MA). We strongly recommend these kits for high-yield recovery of large-fragment genomic DNA for downstream use with various genomic applications, including high-throughput oligonucleotide array genotyping or DNA sequencing. Further discussion of recovery and purification of DNA is covered in Chapter 11 by Boris Vaisman.

4. An alternative product for a hot-start DNA polymerase is Hot-Start Taq (Denville Scientific, Metuchen, NJ). This DNA polymerase and its reaction buffer can be used interchangeably with AmpliTaq Gold in these algorithms.
5. Multiple freeze–thaw cycles and long periods at room temperature degrade dNTPs. Preparation of small dNTP aliquots in microfuge tubes for single use or keeping this reagent on ice during reaction preparation prevents degradation.
6. Several recent studies have published PCR primer pairs corresponding to protein-coding exons for the Consensus Coding Sequence (CCDS) genes, a set of the best-annotated protein-coding genes, and the Reference Sequence (RefSeq) genes, another collection of annotated gene sequences which is thought to include nearly all human genes [15–17]. Sequences for these pre-designed and validated primer sets can be downloaded as supplemental text files included with these publications [15, 17] or from the Oncogenomics Section, Pediatric Oncology Branch, National Cancer Institute website (registration is free for access to the primer database; http://ntddb.abcc.ncifcrf.gov/cgi-bin/Primers.pl). Custom primer pairs for noncoding sequence and other genomic regions can also be designed with Primer3 (http://frodo.wi.mit.edu/cgi-bin/primer3/primer3_www.cgi), and each pair should amplify a fragment of 400–800 base pairs. M13 universal sequencing tags added to the 5′ end of each primer facilitates rapid setup of sequencing reactions (forward primers have M13 forward added to their 5′ end: TGTAAAACGACGGCCAGT; reverse primers have M13 reverse added to their 5′ end: CAGGAAACAGCTATGACC).
7. Long homopolymer regions with more than 6 T's or A's in a row can cause problems for automated sequencing with BigDye chemistry sequencing due to "slippage" in the region of the repeated base. This results in abrupt loss of sequencing signal at the homopolymer sequence. It is therefore recommended that target sequences are examined for stretches of T's or A's, especially when using primers from large databases [15–17]. This can be rapidly assessed using the UCSC Genome Browser's in silico PCR (http://genome.ucsc.edu/cgi-bin/hgPcr?command=start). In silico PCR will map the target region for a pair of PCR primers onto a build of the human genome. The target sequence can be visually inspected or

downloaded as a text file to search for poly T and poly A repeats. BigDye chemistry sequencing is not recommended for these homopolymer regions, and consultation with a DNA sequencing facility is recommended to identify a suitable alternative for sequencing such problematic regions.

8. Pipette tips used for genomic DNA handling must be filtered to minimize the risk of sample cross-contamination.
9. This buffer can be purchased from GE Healthcare; however, we routinely use a homemade version for SAP/Exo reactions (10× buffer: 200 mM Tris–HCl, pH 8.0, 100 mM $MgCl_2$).
10. If you do not wish to buy BigDye Terminator 5× dilution buffer (Applied Biosystems), the following homemade dilution buffer works well with BDT version 3 reactions: 175 mM Tris–HCl pH 9.0 and 1.25 mM $MgCl_2$.
11. We recommend the ALPS 25 manual heat sealer (Thermo Scientific, Waltham, MA) for sealing reaction plates with purified sequencing reactions prior to resolving them on a sequencer.
12. Although a manual version of the magnetic bead purification method is described by the vendor, our experience has been that the use of the robotic method is essential for optimizing use of laboratory personnel and quality of sequence reads.
13. 85 % ethanol should be made fresh and stored for no more than 24 h for use with this protocol. Twenty-five milliliters is required for each 96-well plate.
14. We use Applied Biosystems sequencers including models 3100 and 3730.
15. Our lab utilizes Sequencher software (Gene Codes Corporation) and SeqScape (Applied Biosystems) for reading sequence tracings. A freeware alternative for academic users is Phred/Phrap/Autofinish/Consed (http://www.phrap.org).
16. We have standardized all PCR thermal cycling programs such that they differ only in (1) annealing temperature or (2) extension time. All PCR primers are designed and tested to work at an annealing temperature of 60 °C, although occasionally this parameter is changed to between 61 and 67 °C. The extension time in this standardized thermal cycling program is 45 s, although this may need to be increased for larger PCR fragments according to the rule of 60 s of extension per 1,000 base pairs amplified.
17. We use a standard template spreadsheet to easily calculate volumes for a master mixture for PCR (*see* Fig. 2), SAP/Exo pretreatment reactions (*see* Fig. 3), or dilute BDT sequencing reactions (*see* Fig. 4). The total reaction volume for each PCR reaction can be 10–20 μL, and the standard template for the

mastermix can be adjusted accordingly (Fig. 2). We add 7 % overage to our final volumes, mix these reagents in a sterile reservoir, and then easily transfer an appropriate volume to reaction wells on a 96-well plate using a 12-channel pipette. SAP/Exo pretreatment reactions (Fig. 3) are run in a total reaction volume of 9 μL, of which 5 μL is derived from the mastermix. BDT sequencing reactions have a volume of 10 μL. The template for the dilute BDT sequencing reaction option is presented in Fig. 4 (Subheading 3.6), although the template can be adjusted accordingly for the standard BDT sequencing reaction protocol described in Subheading 3.5.

18. The Option 1 sequencing reactions are optimized for manual sequencing reaction cleanup with Sephadex. These sequencing reactions use eight times the amount of BigDye Terminator mix (2 μL) than reactions prepared for the magnetic bead purification under Option 2. The more dilute Option 2 sequencing reactions are only for use with the Beckman Coulter Biomek® FX Laboratory Automated Workstation purification method.

19. Forward sequencing reactions are primed with M13 forward, TGTAAAACGACGGCCAGT, while reverse reactions are primed with M13 reverse, CAGGAAACAGCTATGACC. These primers can be purchased from any vendor, and are prepared as a 3.2 μM stock solution. M13 primers are stored at either −20 °C or −80 °C.

20. Completed sequencing reactions may be removed from the thermal cycler and stored at 4 °C for up to 24 h prior to proceeding with the sequencing reaction purification (according to either Option 1 or Option 2) and loading on the DNA sequencer. If samples will not be loaded on the DNA sequencer within 24 h, the sealed sequencing reaction plate may be stored at −20 °C or −80 °C for up to 1 month. Because sequencing dyes are fluorescent, stored sequencing reactions should ideally be kept in the dark.

21. Relative *g* force will vary depending upon the centrifuge rotor used. For an average of 1,273 × *g*, we centrifuge the Sephadex multiscreen plate at 1,730 × *g* for 5 min using a Beckman Coulter GH 3.8A rotor.

22. In the transfer of the sequencing reaction into the Sephadex matrix, it is imperative that the reaction is pipetted directly into the center of the Sephadex column. If the reaction contents are transferred to the edge of a well on the multiscreen plate, the contents will not be effectively filtered through the Sephadex. When run on the DNA sequencer, ineffectively purified reactions will have significant fluorescent background and poor overall sequence quality.

23. The 96-well reaction plates into which Sephadex or CleanSEQ-purified sequencing reactions are transferred must be compatible for use with a DNA sequencer. We recommend Thermo-Fast 96 Detection Plates (product number AB-1100, ABgene, Surrey, UK) for use with ABI DNA Analyzers.
24. CleanSEQ magnetic beads collect at the bottom of the reagent container during storage. Therefore, the magnetic beads should be resuspended in solution by vigorously shaking the CleanSEQ container prior to its use.
25. All subsequent steps are to be carried out using the Beckman Coulter Biomek FX Laboratory Automated Workstation equipped with V.3 CleanSEQ 96 script software (batch method). The FX Workstation can accommodate four 96-well plates at one time on the robot's stage.
26. During this step, the nucleic acid sequencing products are bound to the magnetic beads which are adhered to the side of each reaction well while the plate is on the magnet. Prior to running the wash script, a reservoir containing 85 % ethanol is placed on the stage of the workstation.
27. Excessive drying of the DNA sequence sample complexed to the magnetic beads can lead to sample degradation. We recommend that the drying step does not exceed the recommended 10 min.
28. We use ABI 3730xl DNA Analyzers as the platform for DNA sequence analysis.

Acknowledgments

This work was supported by the Division of Intramural Research, NHLBI, National Institutes of Health. The views expressed are those of the authors and do not represent endorsement by the National Institutes of Health.

References

1. Lander ES, Linton LM, Birren B et al (2001) Initial sequencing and analysis of the human genome. Nature 409:860–921
2. Venter JC, Adams MD, Myers EW et al (2001) The sequence of the human genome. Science 291:1304–1351
3. Frazer KA, Ballinger DG, Cox DR et al (2007) A second generation human haplotype map of over 3.1 million SNPs. Nature 449:851–861
4. Cohen JC, Kiss RS, Pertsemlidis A, Marcel YL, McPherson R, Hobbs HH (2004) Multiple rare alleles contribute to low plasma levels of HDL cholesterol. Science 305:869–872
5. Cohen JC, Pertsemlidis A, Fahmi S, Esmail S, Vega GL, Grundy SM, Hobbs HH (2006) Multiple rare variants in NPC1L1 associated with reduced sterol absorption and plasma low-density lipoprotein levels. Proc Natl Acad Sci USA 103:1810–1815
6. Kathiresan S, Manning AK, Demissie S, D'Agostino RB, Surti A, Guiducci C, Gianniny L, Burtt NP, Melander O, Orho-Melander M, Arnett DK, Peloso GM, Ordovas JM, Cupples

LA (2007) A genome-wide association study for blood lipid phenotypes in the Framingham Heart Study. BMC Med Genet 8(Suppl 1):S17

7. Manolio TA, Brooks LD, Collins FS (2008) A HapMap harvest of insights into the genetics of common disease. J Clin Invest 118: 1590–1605
8. McPherson R, Pertsemlidis A, Kavaslar N, Stewart A, Roberts R, Cox DR, Hinds DA, Pennacchio LA, Tybjaerg-Hansen A, Folsom AR, Boerwinkle E, Hobbs HH, Cohen JC (2007) A common allele on chromosome 9 associated with coronary heart disease. Science 316:1488–1491
9. Romeo S, Pennacchio LA, Fu Y, Boerwinkle E, Tybjaerg-Hansen A, Hobbs HH, Cohen JC (2007) Population-based resequencing of ANGPTL4 uncovers variations that reduce triglycerides and increase HDL. Nat Genet 39:513–516
10. Ley TJ, Mardis ER, Ding L et al (2008) DNA sequencing of a cytogenetically normal acute myeloid leukaemia genome. Nature 456:66–72
11. Bentley DR, Balasubramanian S, Swerdlow HP et al (2008) Accurate whole human genome sequencing using reversible terminator chemistry. Nature 456:53–59
12. Wang J, Wang W, Li R et al (2008) The diploid genome sequence of an Asian individual. Nature 456:60–65
13. Kumkhaek C, Taylor JG, Zhu J, Hoppe C, Kato GJ, Rodgers GP (2008) Fetal haemoglobin response to hydroxycarbamide treatment and sar1a promoter polymorphisms in sickle cell anaemia. Br J Haematol 141:254–259
14. Taylor JG, Ackah D, Cobb C, Orr N, Percy MJ, Sachdev V, Machado R, Castro O, Kato GJ, Chanock SJ, Gladwin MT (2008) Mutations and polymorphisms in hemoglobin genes and the risk of pulmonary hypertension and death in sickle cell disease. Am J Hematol 83:6–14
15. Sjoblom T, Jones S, Wood LD et al (2006) The consensus coding sequences of human breast and colorectal cancers. Science 314:268–274
16. Pruitt KD, Tatusova T, Maglott DR (2007) NCBI reference sequences (RefSeq): a curated non-redundant sequence database of genomes, transcripts and proteins. Nucleic Acids Res 35:D61–D65
17. Wood LD, Parsons DW, Jones S et al (2007) The genomic landscapes of human breast and colorectal cancers. Science 318:1108–1113

Chapter 7

Introduction to Next-Generation Nucleic Acid Sequencing in Cardiovascular Disease Research

Lena Diaw, Victoria Youngblood, and James G. Taylor VI

Abstract

The identification of new genomic paradigms in lipoprotein and cardiovascular diseases will be accelerated by the application of the recent technological advances in nucleic acid sequencing. Presently, large-scale genomics facilities are equipped to accomplish this objective with a combination of "next-generation" DNA sequencing chemistries, largely focused on assembling massively parallel sequence reads corresponding to complete genes, entire exomes, or whole genomes from populations of individuals. In the future, individual laboratories will also use this emerging technology for focused genomic studies with the use of a combination of next-generation sequencing and automated Sanger sequencing. In particular, next-generation sequencing will play an increasingly important role when applied to chromatin immunoprecipitation, RNA transcriptome analysis, and studies of human genetic variation and mutation in carefully phenotyped healthy and disease populations. In this chapter, a brief overview of recent technological advances in next-generation nucleic acid sequencing is presented, with emphasis on practical application to clinical studies in cardiovascular diseases.

Key words Next-generation sequencing, DNA, Single nucleotide polymorphism, Mutation, Genetic variant, Gene expression

1 Introduction

Cardiovascular diseases (CVDs), affecting the heart and blood vessels, are the number one cause of death in the United States. Collectively, CVDs refer to a class of diseases including their most common subtypes: coronary artery disease, stroke, systemic hypertension, and heart failure [1, 2]. CVDs are complex, multigenic diseases attributable to the co-inheritance of both common and rare variants in susceptibility genes as well as environmental influences including lifestyle [3–5]. Rapid technological advances in genetic research after the completion of the Human Genome Project have facilitated the identification of some novel CVDs genes, as well as progress towards understanding underlying disease mechanisms. These advances hold great promise for cardiovascular medicine as investigators embark upon more in-depth

Lita A. Freeman (ed.), *Lipoproteins and Cardiovascular Disease: Methods and Protocols*, Methods in Molecular Biology, vol. 1027, DOI 10.1007/978-1-60327-369-5_7, © Springer Science+Business Media, LLC 2013

studies of CVD genes using deep genomic sequencing in increasingly larger study populations.

Since DNA sequencing was first introduced in 1977 [6, 7], Sanger's original approach has been modified to become the accepted method for nucleic acid sequencing. This method utilizes chain-terminating dideoxynucleotide (ddNTP) analogs which act as base-specific terminators of DNA polymerization. Fluorescent labeling of each ddNTP with a different color fluorophore that can be detected with laser excitation and a photomultiplier, polymerase chain reaction cycle sequencing, and resolution of sequencing products by capillary electrophoresis were key improvements allowing for individual sequencing reads of 500–1,000 base pairs in a single capillary run on a DNA sequencer [8, 9]. These modifications to automated sequencing contributed to completion of the first draft sequence of the human genome [10, 11], and this sequencing chemistry is presently the gold standard for clinical molecular diagnostics [8]. However, the high labor and financial costs associated with this sequencing technology for large genome-scale projects led to a challenge to develop new technologies capable of sequencing an entire human genome for about $1,000 [12]. As a result of the 1,000-dollar genome challenge, new non-electrophoresis, high-throughput sequencing instruments have been introduced over the past 5 years. With increased availability of massively parallel sequencing, genomics investigators are poised to decipher sequence data from millions of different nucleic acid fragments during a single instrument run [13]. This technology has already been applied to a diverse array of studies with different research objectives for nucleic acid analysis.

This introduction will provide a brief overview of current second-generation sequencing platforms and emerging third-generation sequencing technologies, as well as the potential for using these instruments for investigating different genetic paradigms in CVD.

2 Second-Generation Sequencing

High-throughput, second-generation sequencing refers to amplification of single nucleic acid strands from a prepared fragment library, followed by simultaneous sequencing reactions for millions of individual amplified fragments [14, 15]. Creation of a sequencing fragment library is the initial step common to all of the second-generation platforms, where randomly sheared DNA fragments are ligated to adaptor sequences (Fig. 1). Ligation of adaptors allows for PCR amplification of all DNA fragments of interest prior to sequencing using universal primers complementary to adaptors. Sequencing reactions then proceed with individual amplified DNA fragment templates immobilized at specific locations within a platform matrix (either a bead or a glass slide).

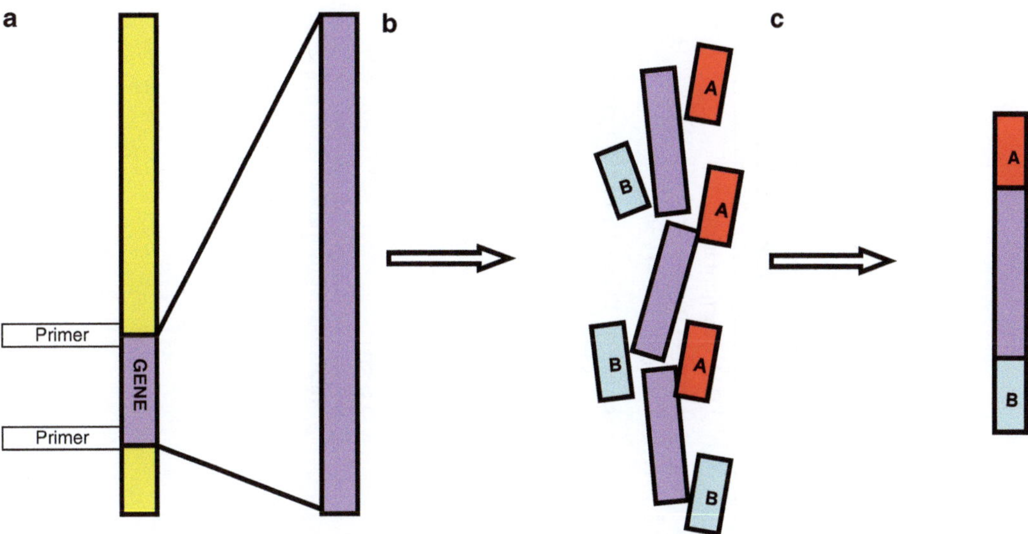

Fig. 1 Fragment library preparation for next-generation sequencing. (**a**) If the investigator wishes to sequence a specific gene or target region (shown in *purple*), then traditional PCR using specific primers (shown) or array hybridization is performed to enhance recovery of this DNA. (**b**) The nucleic acid is randomly fragmented by nebulization and adaptor oligonucleotides are ligated to ends of sheared DNA fragments. (**c**) The library is then purified to recover only those DNA fragments with adaptors attached to each end prior to amplification

Thereafter, the addition of nucleotides during sequencing reactions is monitored using fluorescence or another light-based detection method and recorded. Short sequence reads derived from the library fragments are then assembled into a consensus sequence for further analysis.

To date, all of the second-generation sequencing instruments have demonstrated their ability to generate high-throughput DNA sequence, including the recent completion of several entire human genomes [16–23]. Three platforms are widely available, including the Roche 454 FLX, the Illumina/Solexa Genome Analyzer II, and the Applied Biosystems SOLiD™ 3.0. Collectively these instruments share a common concept, based on the use of a dense array of different DNA templates, repeated enzymatic sequencing reactions, and serial imaging to detect light signals corresponding to specific nucleotides [24, 25]. Despite their similarity in concept, each of the second-generation nucleic acid sequencers has unique differences in sequencing chemistry, immobilization arrays, and technological limitations (Table 1). A brief overview of these sequencing platforms is presented below, highlighting their key features.

2.1 Roche 454 Life Sciences FLX Genome Sequencer (http://www.454.com)

The 454 sequencing system, the first next-generation system to be introduced commercially, utilizes emulsion PCR and pyrosequencing [24, 26, 27]. The principle underlying pyrosequencing is that pyrophosphate is released after nucleotide incorporation by DNA

Table 1
Comparison of second-generation sequencing platforms

	Roche 454 GS-FLX	Illumina Genome Analyzer	Applied Biosystems SOLiD
Library preparation	DNA fragmentation Ligation of adaptors both DNA fragment ends	DNA fragmentation Ligation of adaptors both DNA fragment ends	DNA fragmentation Ligation of adaptors both DNA fragment ends
Template amplification	Emulsion PCR of library fragments on agarose beads	"Bridging" PCR of library fragments affixed to a glass flow cell	Emulsion PCR of library fragments on beads
Sequencing reaction	DNA polymerase and pyrosequencing	DNA polymerase and reversible dye terminators	Addition of 8-mers by DNA ligase
Nucleotide detection	Sequential flow of single nucleotides Pyrophosphate activation of luciferase	Reversible terminator ddNTPs Individual ddNTPs with 4 fluorescent labels	Dinucleotide pairs coded by 4 fluorescent labels
Read length (bp)	250–400	30–75	35–50
Source of errors	Insertions and deletions	Nucleotide substitutions	Nucleotide substitutions
Limitations	Homopolymer stretches	Difficult de novo sequence assembly	Difficult de novo sequence assembly and detection of insertion–deletions
Human genomes (reference)	1 [19]	5 [16–18, 20, 22, 23]	1 [21]

polymerase. Each nucleotide added to the sequencing product is detected indirectly though pyrophosphate, which initiates a series of downstream enzymatic reactions to create a firefly luciferase light reaction. As is true for all second-generation sequencing methods, the first step in pyrosequencing is library preparation, where nucleic acid (genomic DNA) is randomly fragmented with nebulization and ligated to two oligonucleotide adaptors (A and B adaptors) (Fig. 1). The second step in pyrosequencing is emulsion PCR. Emulsion PCR clonally amplifies nucleic acid library fragments within individual micelles that function as PCR microreactors. Emulsion offers a convenient method for performing thousands of parallel PCR reactions and then efficiently isolating individual fragment beads for sequencing, all within a much smaller space than with traditional PCR. In preparation for this step, each individual nucleotide strand is captured and immobilized onto a single agarose bead by mixing the purified DNA fragment library with an excess of agarose beads (Fig. 2). Roche 454 agarose beads are coated with oligonucleotide sequences complementary to the A or B library adaptor sequences such that nucleotide fragment capture occurs through a sequence-specific hybridization reaction between an adaptor on a library nucleic acid fragment and a complementary sequence bound to agarose beads. This process is optimized such that only a single library fragment is bound to an individual bead. DNA capture bead–library fragment complexes are then compartmentalized into micelles composed of oil, water, and consumable reagents for approximately 8 h of emulsion PCR during which each library fragment is clonally amplified. By the end of thermal cycling, each individual agarose capture bead is coated with 10^6 or more copies of the original nucleic acid library fragment. At this point, amplified beads are removed from the micelles and subjected to a magnetic bead avidin–biotin purification step. Avidin–biotin purification isolates only double-stranded nucleotide fragments on the beads with adaptor sequences at both ends of these strands. In preparation for sequencing, double-stranded DNA is made single-stranded by treating the capture beads with sodium hydroxide.

Beads containing amplified, single-stranded DNA for sequencing are then arrayed into slots on a picotiter plate, where each agarose bead is immobilized into one of several hundred thousand 44-μm slot wells contained on the plate. Each single bead will now become an individual library fragment sequence read. Pyrosequencing of all beads is then performed simultaneously, with each bead immobilized at a specific location on the picotiter plate for pyrophosphate detection. Packing and reaction beads containing reagents for activation of the luciferase reactions in the presence of pyrophosphate are added to surround the immobilized agarose beads on the plate. Sequencing primers are included for both library adaptors (designated A and B adaptors) to allow for bidirectional nucleic acid sequencing. The sequencing reaction consists of

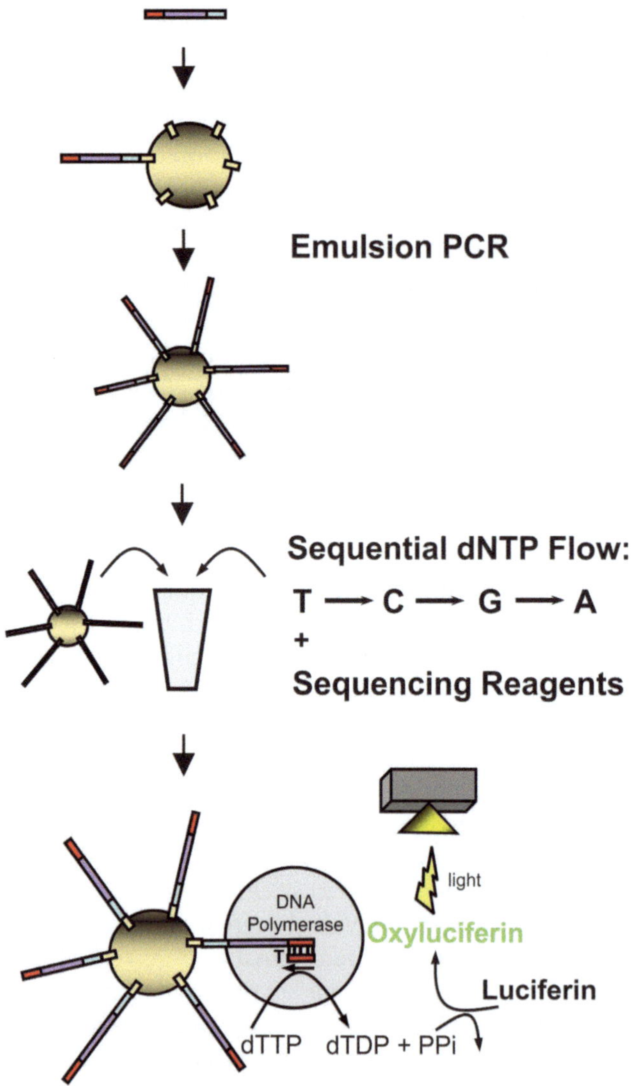

Fig. 2 Emulsion PCR and pyrosequencing with the 454 FLX system. A purified library with each fragment ligated to adaptor sequences is added to agarose beads that are coated with oligonucleotide complements of the adaptor sequences. Individual beads are immersed in a micelle of oil and PCR reagents for amplification by emulsion PCR. Beads with amplified DNA are then arrayed into individual wells on a picotiter plate. Pyrosequencing occurs in the wells with single dNTPs flowed in at a time. Complementary nucleotides are added to the extending DNA strand by the polymerase, and pyrophosphate is liberated by this reaction. The liberation of pyrophosphate into the solution in the well initiates a series of reactions that generate light by luciferase, which is then detected by imaging

successive flows of individual dNTPs, where each is added, one at a time, as a pure nucleotide. Addition of a nucleotide complementary to the template strand generates a luminescent luciferase signal (initiated by the liberation of pyrophosphate) that is recorded by an imaging step after each nucleotide flow (Fig. 2). By-products of the luciferase reaction and any other unincorporated nucleotides are degraded by another enzymatic reaction, eliminating the need for a washing step between dNTP flows. Thus, pyrosequencing is a single-color detection system (luciferase) where specific bases are specified by the specific dNTP flow. If no signal is detected during an individual dNTP flow, then a nucleotide was not incorporated at this position, indicating the absence of this specific base.

The sequencing instrument run is approximately 8 h long, during which 100 nucleotide flows/imaging steps occur. The first four pure dNTP flows occur in the order corresponding to the first four bases (TCGA) on the adaptor fragments after the reaction primer sequence (e.g., the sequencing primer binding site). All subsequent dNTP flows to the picotiter plate occur in this order (TCGA). The sequential addition of T, C, G, and A from the adaptor fragment allows the instrument to calibrate the light signal intensity detected after the addition of a nucleotide (or several nucleotides) to the agarose bead in each slot well on the picotiter plate. Finally, FLX software determines the intensity of the luminescent signal at a specific position during a single flow step to assign a read base to specific positions on the picotiter plate. Furthermore, the FLX software also performs a quality filtering step to remove sequences that lack the TCGA initiation sequence or bead slots on the picotiter plate that generate sequences corresponding to more than one nucleic acid fragment. Moreover, the unique features of this technology are longer sequence read lengths (averaging four nucleotides added per flow × 100 flows = 400-base-pair read average) and again a detectable single-color system. Average read length has improved to nearly half the length achieved with automated sequencing (400 base pairs), and a 400-base-pair read length is more than sufficient to assemble de novo sequence into consensus sequence contigs.

Based on this technology, Roche launched the Genome Sequencer 20 in 2005 as the first commercially available next-generation sequencing platform. This instrument was able to read and assemble the 25 million-base-pair genome from *Mycoplasma genitalium* from a single 4-h instrument run [24]. While the basic technology utilized by emulsion PCR/pyrosequencing is unchanged, the current FLX Genome Sequencer introduced in 2007, along with more recent software upgrades, now boasts longer read lengths, paired-end bidirectional reads (*see* **Note 1**), and improved workflow [28]. This second-generation instrument was also the first to sequence a complete human genome without the use of an intermediate bacterial cloning strategy [19]. 454 has also developed a software package called Newbler for de novo sequence assembly from raw 454 data which can be run through a Java graphical user interface or via the command line [29]. Roche has also recently incorporated Amplicon Variant

Analyzer (AVA) software into the workstation and this should facilitate identification and analysis of single nucleotide polymorphisms (SNPs) and mutations.

There are several advantages to pyrosequencing which investigators may wish to consider prior to initiating a next-generation sequencing experiment. In addition to the long sequence read length, pyrosequencing has a low error rate, producing highly accurate sequence. Furthermore, it is capable of accurately identifying copy number variants (CNVs) or insertion/deletion mutations ranging in size from single to millions of base pairs [19]. A high percentage of generated sequences (95 % in a recent pilot study) are usable for analysis compared to the two other next-generation sequencing platforms [29]. Unfortunately, pyrosequencing has difficulty sequencing through long homopolymer regions that exceed about six bases due to "slippage" (incorporation of different numbers of the dNTP in the homopolymer region from individual templates) in the region of the repeated nucleotide [9].

2.2 Illumina Genome Analyzer II/ Solexa (http://www.Illumina.com)

The first Illumina Genome Analyzer was introduced in 2007, and it represented a significant improvement in high-throughput sequencing compared to automated Sanger sequencing and the 454 FLX sequencer. Unlike the Roche 454 instrument, the Genome Analyzer has several key differences: it is a short-read sequencing platform that uses a reversible nucleotide terminator chemistry, bridging PCR amplification of the sequencing template (to produce colonies or "polonies" of identical DNA molecules), and a nucleotide flow step incorporating all fluorescent nucleotides at once in a four-color system [30, 31]. However, the Solexa is similar to 454 in that the workflow consists of steps for (1) library preparation, (2) library amplification/cluster generation by PCR, and (3) a proprietary DNA sequencing chemistry. The current Illumina Genome Analyzer II instrument is comprised of a flow cell with eight individual sealed glass channels. The glass surfaces of the flow cell devices are covalently bound to a dense lawn of two Illumina-specific oligonucleotide adaptors used for nucleic acid library preparation and their complementary sequences. This eight-channel instrument allows for a single library to be run across all eight channels or, alternatively, a different library in each channel.

As with the other second-generation platforms, Solexa sequencing starts with an initial nucleic acid library preparation step (Figs. 1 and 3). Nucleic acid samples are randomly sheared into fragments, the overhanging single-stranded DNA ends are enzymatically repaired, and overhanging "A" nucleotides are added to the ends of DNA fragments to allow ligation of two different Illumina-specific oligonucleotide adaptors (with complementary overhanging "T" nucleotides). The nucleic acid library DNA is then purified and denatured into single-stranded molecules in a Cluster Station instrument in preparation for the PCR

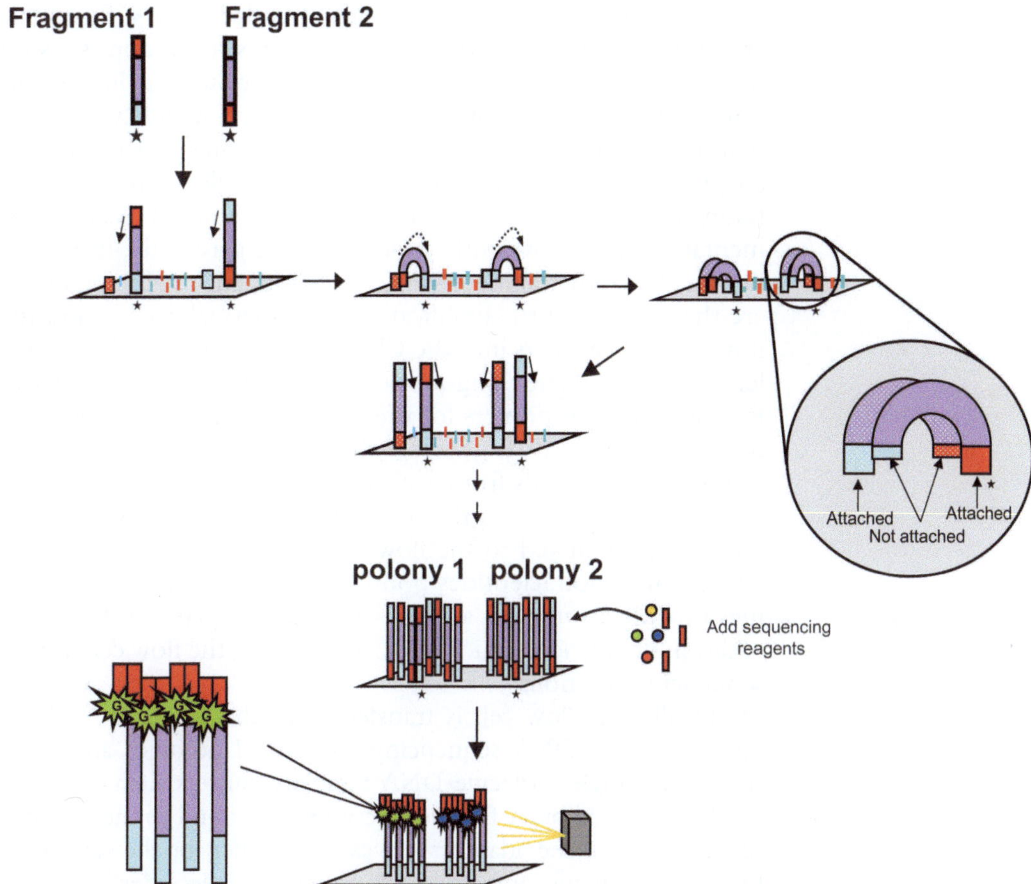

Fig. 3 Bridging PCR and reversible terminator sequencing with the Illumina/Solexa system. First, single-stranded DNA library fragments with adaptor sequences at both ends are randomly bound to the glass flow cell. Two different fragments ligated to adaptors are shown binding to different locations on the glass flow cell. The glass flow cell is also coated with a dense lawn of oligonucleotides complementary to the library adaptor fragments. These oligonucleotides are also covalently bound to the flow cell. The "free" end of each single-stranded DNA library fragment will hybridize, in a reversible manner, to a nearby complementary oligonucleotide affixed to the slide. This oligonucleotide is then able to serve as a primer for bridging PCR in order to generate double-stranded DNA. Amplified strand copies are covalently bound to the flow cell via adaptor primers incorporated during bridge amplification. Denaturation of amplified double-stranded DNA yields two complementary single-stranded DNA fragments covalently bound to the slide, for each single-stranded fragment originally bound. The "free" end of each single-stranded DNA library fragment can again hybridize reversibly to a nearby bound complementary oligonucleotide, which again serves as a primer for another round of bridging PCR. Amplification is repeated until there are 10^6 copies of each starting fragment to form a cluster or polony. Amplified polonies are denatured, leaving single-stranded DNA permanently affixed to the glass flow cell. Single-stranded DNA fragments become the template for sequencing, which proceeds using labeled reversible terminators (ddNTPs), primers complementary to one of the adaptor fragments, and DNA polymerase. Nucleotides are added to the elongating sequencing product one at a time, followed by an imaging step using laser excitation to emit fluorescence from the newly added nucleotide. Sequential nucleotide additions and imaging steps continue to yield sequence reads of 35–50 nucleotides

amplification step. Individual single-stranded library fragments are immobilized at one of the adaptor ends to random sites on the surface of a flow cell within the Cluster Station (Fig. 3). As noted above, surfaces of the cells are coated with a very high-density lawn of oligonucleotides complementary to the library adaptor sequences. The free end of each single-stranded library fragment, ligated to an adaptor sequence, hybridizes to a complementary adaptor covalently attached to the glass slide, forming a "bridged" DNA fragment (Fig. 3). "Bridged" DNA fragments are then subjected to amplification with thermal cycling and the flow of PCR reagents into the Cluster Station (Fig. 3). The covalently bound adaptor oligonucleotides hybridized to the library fragment serve as primers for this "bridging" PCR amplification. Several rounds of PCR generate clusters of many identical copies of both DNA strands from each starting fragment (termed polonies) (Fig. 3). Copies of the original hybridized library fragment are covalently linked to the flow cell, because these copies were primed by oligonucleotides bound to the flow cell. These polonies provide a sufficient amount of DNA template to detect the addition of individual nucleotides by imaging the flow cell during sequencing reactions.

Finally, the flow cell is transferred to the Genome Analyzer instrument for DNA sequencing. Amplified polonies are denatured into single-molecule DNA template clusters such that only single-stranded library fragments covalently bound to the flow cell remain. One of the adaptors serves as a primer for the sequence-by-synthesis incorporation of fluorescent nucleotides (Fig. 4). After the addition of each individual fluorescent dideoxynucleotide, an image of the flow cell is taken to detect the identity of the specific fluorescent nucleotide incorporated at each cluster site. After the image is processed, the fluorophore and the dideoxy terminator group are enzymatically removed from the incorporated nucleotide, so as to allow for subsequent rounds of fluorescent ddNTP base incorporation. Thus, the ddNTPs used for Solexa sequencing are reversible terminators. Repeated rounds of nucleotide incorporation and imaging steps generate up to 75-base-pair sequence reads from each polony fragment (Fig. 4). However, Solexa can accommodate tens of millions of different polonies affixed to specific locations on the flow cell, allowing for millions of individual sequencing reactions from different DNA template fragments to be run simultaneously in parallel. Overall, this platform is able to produce at least 1.5 gigabases of sequence in a single 2- to 3-day instrument run.

The obvious advantage of the Illumina platform is its ability to produce huge sequence data sets over the course of several days. A module (what kind of module?) is now available, and this is likely to be most useful for sequencing complex genetic organisms like mammals. (Note: Paired-end sequencing allows two separated

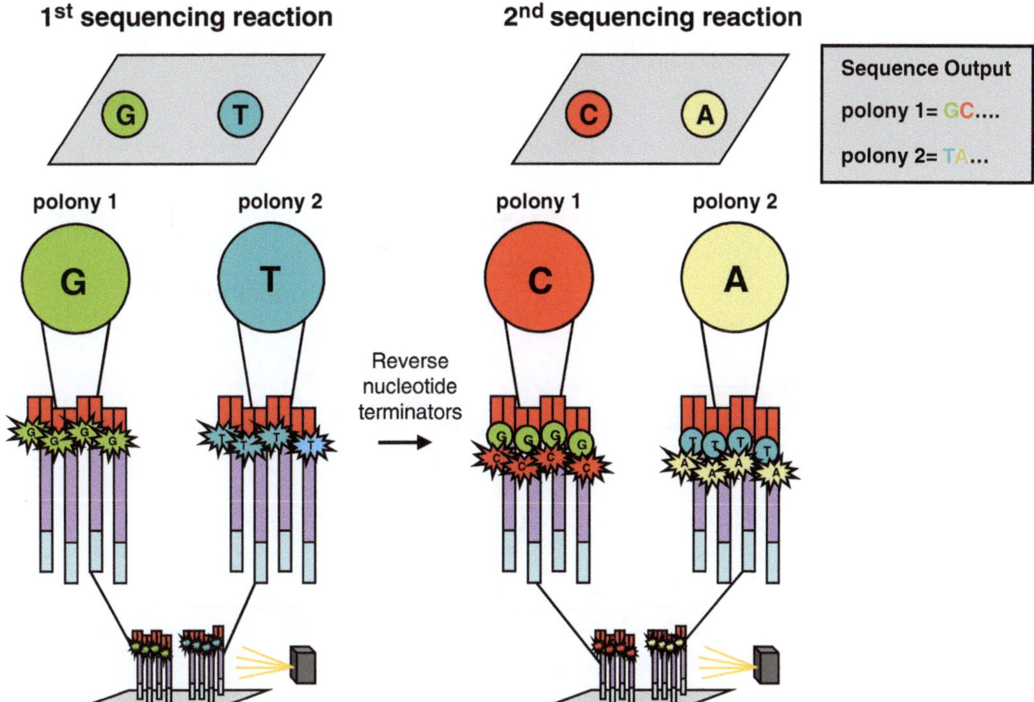

Fig. 4 Reversible terminator sequencing with the Illumina/Solexa Genome Analyzer. Schematic representation of sequencing two polonies where a G fluorescent nucleotide is added to a sequence product for polony 1 and a T fluorescent nucleotide is added to polony 2, where each polony is affixed to a specific location on the flow cell. The flow cell image for sequence reaction 1 is illustrated at the *top left* of the figure. After this image step, the G and T nucleotides have their 3′-OH chain-terminating groups removed so that sequencing reaction 2 can proceed

regions of DNA from the same chromosomal strand to be sequenced from a single library fragment and this is described in **Note 1**.) Similar to the 454 FLX instrument, Solexa has demonstrated the ability to accurately identify insertion/deletion polymorphisms [16]. However, there is concern that this instrument may have a higher rate of false-positive base calls and a relatively low percentage of usable sequence, as up to 57 % of sequence reads have been removed during informatics quality control steps [17, 29]. Of note, this platform has been the most widely used by independent centers to successfully sequence entire human genomes [16–18, 20, 22, 23].

2.3 Applied Biosystems SOLiD™ 3.0 Sequencer (http://solid.appliedbiosystems.com)

The Applied Biosystems SOLiD™ (Supported Oligonucleotide Ligation and Detection) system was the second short-read sequencing platform commercially released during 2007, with the most recent upgrade introduced in Spring, 2009. Conceptually, SOLiD™ is designed to generate a complex matrix of data that facilitates analysis by computers, which is later decoded and aligned into

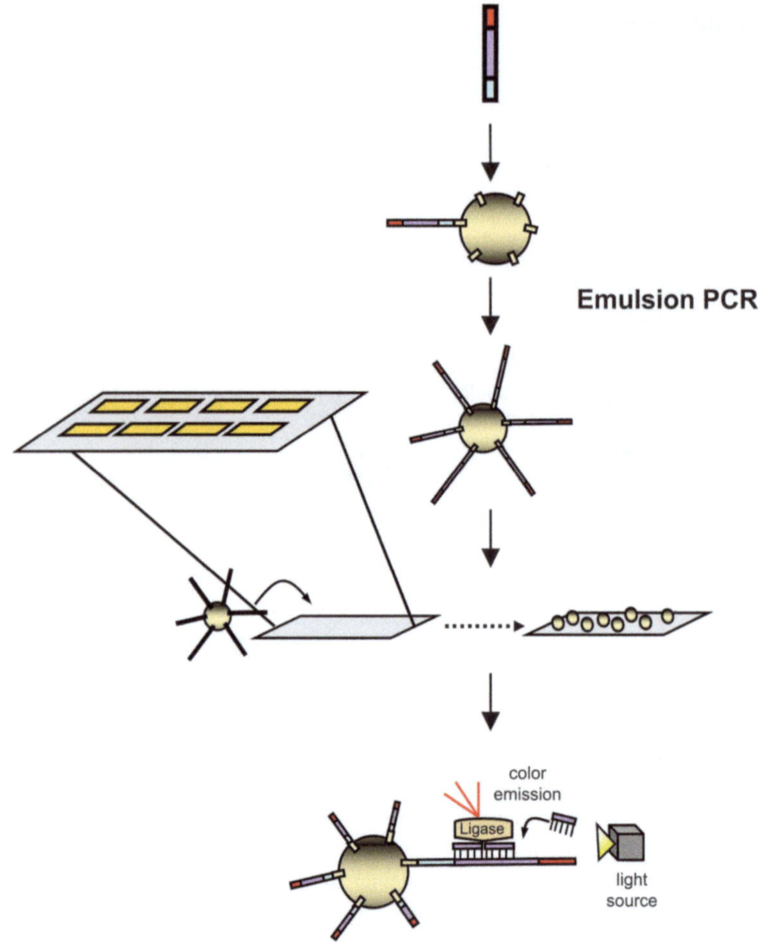

Fig. 5 The emulsion PCR and oligonucleotide ligation sequencing method used by the SOLiD system. Library fragments are mixed with beads so that each unique DNA fragment is attached to a different bead. Beads then undergo emulsion PCR to amplify copies of each library fragment. Millions of these beads are next attached to specific locations on a glass slide flow cell. Sequencing proceeds with the addition of 8-mer oligonucleotides complementary to the DNA template by a DNA ligase. One of four fluorescent dyes are attached to each 8-mer, and these only code for two of the nucleotides (a dinucleotide pair) on the 8-mer (Fig. 7). These four fluorescent dyes encode for 16 possible two-base combinations, and this allows for the identity of a specific dinucleotide addition to be determined (Fig. 7). The flow cell is imaged after each 8-mer addition by ligation

standard sequence format for investigators. The sequencing methodology is unique as it detects specific dinucleotides with a DNA ligase chemistry instead of interrogating single bases [25] (Fig. 5). The SOLiD™ 2.0 platform was released during 2008, and it was capable of sequencing tens of millions of polony beads with an output of ten gigabases of sequence from a single instrument run over nearly 5 days. The recently upgraded SOLiD™ 3.0 further provided needed improvements in workflow, sequencer run time, and software for data analysis. Despite the underlying complexity

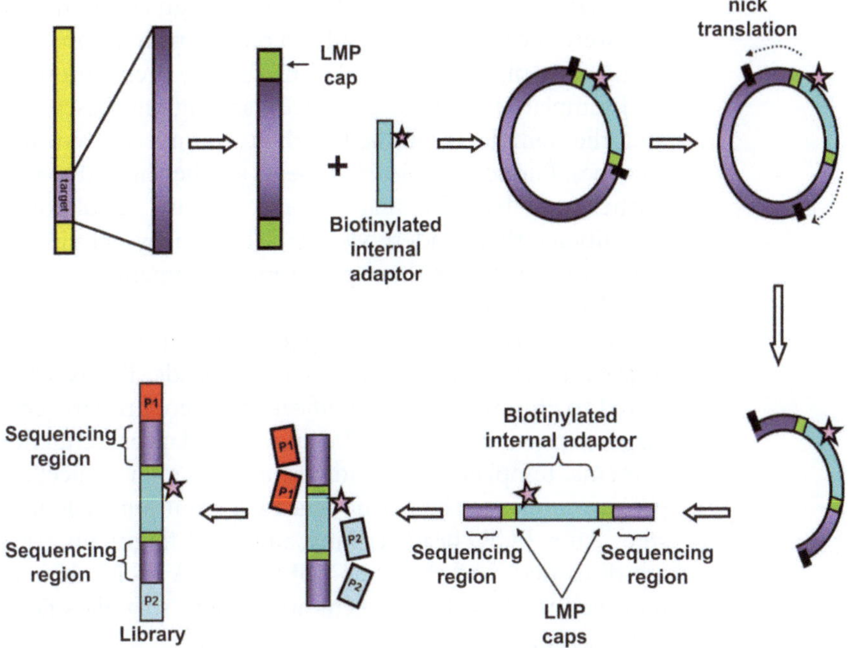

Fig. 6 Mate-paired library preparation for the SOLiD DNA sequencing system. Genomic DNA is fragmented and ligated to CAP adaptors. A second biotinylated internal adaptor is then ligated to one of the CAP adaptors, and this internal adaptor allows for the fragment to be circularized. DNA nicks present in the CAP adaptors are next extended into the target DNA sequence by nick translation. The DNA beyond these extended nicks is digested, and a new fragment consisting of two target DNA sequences separated by the internal adaptor is generated. Note that these two target regions for sequencing on this new library fragment only represent the ends of the original sheared piece of DNA (the intervening sequence was removed by digestion). The new fragment is then ligated to library adaptors. This method helps to map two short-read segments (35–50 bases) back to a specific genomic location because both segments are derived from the same chromosomal strand

of this platform, it still shares the same three fundamental workflow steps common to second-generation sequencing, including nucleic acid library preparation, amplification (emulsion PCR), and sequence generation with a proprietary sequencing chemistry (Fig. 5).

The initial step in SOLiD™ sequencing is library preparation, and the reader is referred to the Applied Biosystems library preparation guides for specific information on options for different types of sequencing libraries (http://solid.appliedbiosystems.com). While the exact details of library preparation are beyond the scope of this overview, humans have sufficient genetic complexity that sequencing human DNA requires either mate-paired libraries or a combination of fragment and mate-paired libraries [21]. Figure 6 presents the details of preparing a mate-paired library (also *see* **Note 2**). Mate-paired libraries improved the proportion of short SOLiD sequence reads successfully aligned to unique locations on the human genome consensus sequence, compared to the small proportion (34 %) of simple human fragment library sequences that were successfully mapped in a recent study comparing different sequencing platforms [29]. During library preparation, the nucleic acid template is first sheared, overhanging end bases are repaired, and the sample is purified with a column to remove reaction enzymes. Purified template fragments are then ligated to two adaptors (either a 41-base P1 adaptor or a 28-base P2 oligonucleotide). The library then undergoes size selection by gel electrophoresis, where only 150–200-base-pair library fragments are recovered for actual sequencing.

The second step is template amplification, where library fragments are first attached to 1-μm beads. Beads are then subjected to emulsion PCR amplification where primers (complementary to portions of the P1 and P2 adaptor sequences), PCR reagents, template, and beads are immersed in a micelle of oil in a process similar to the Roche 454 FLX emulsion PCR amplification step. After PCR, beads lacking amplified fragments (or polonies) are discarded, and the 3′ end of the DNA molecule is chemically modified to allow for covalent attachment to a glass flow cell slide in the sequencer.

Amplified products are transferred and immobilized on a glass flow cell surface in the sequencer for the final sequencing step. The SOLiD™ 3.0 instrument presently accommodates two flow cell slides, and each individual slide can accommodate 1, 4, or 8 different libraries (Fig. 7). Thus, 16 different libraries (or samples) can be run simultaneously, and this number can be further increased to as many as 256 different libraries by using a library barcoding system. During a single SOLiD™ run, the instrument performs five sequential sequencing runs, where each reaction series is primed with one of five different sequencing primers (complementary to sequences within the P1 and P2 adaptors). Each of the five sequencing runs consists of 7–10 sequential rounds of hybridization and ligation of fluorescently labeled oligonucleotides that are eight base pairs in length (8-mers) (Fig. 7). The 8-mers consist of a pair of nucleotides coded by one of the four fluorophores, three nonspecific bases, and the three 5′ bases covalently linked to the fluorophore dye that are removed with the dye from the 8-mer at the end of the ligation cycle (*see* Fig. 7a, f). After each ligation reaction, the flow cell is imaged prior to the removal of the three 5′ bases and fluorophore. Thus, individual ligation reactions add a net five nucleotides to the elongating sequencing product at a time, and the fluorescent imaging step allows a computer to determine the identity of two of these five nucleotides (Fig. 8). The 7–10 ligations performed during a round of sequencing, where each ligation reaction adds a net of five bases, means that an average SOLiD™ sequence read length is 35–50 bases (5 bases added per ligation × 7–10 ligations = 35–50 base read). Construction of the 50-base sequence contig requires the complete data set for all five primer sequencing runs and informatic analysis with the SOLiD™ software suite or an alternative software (Fig. 8).

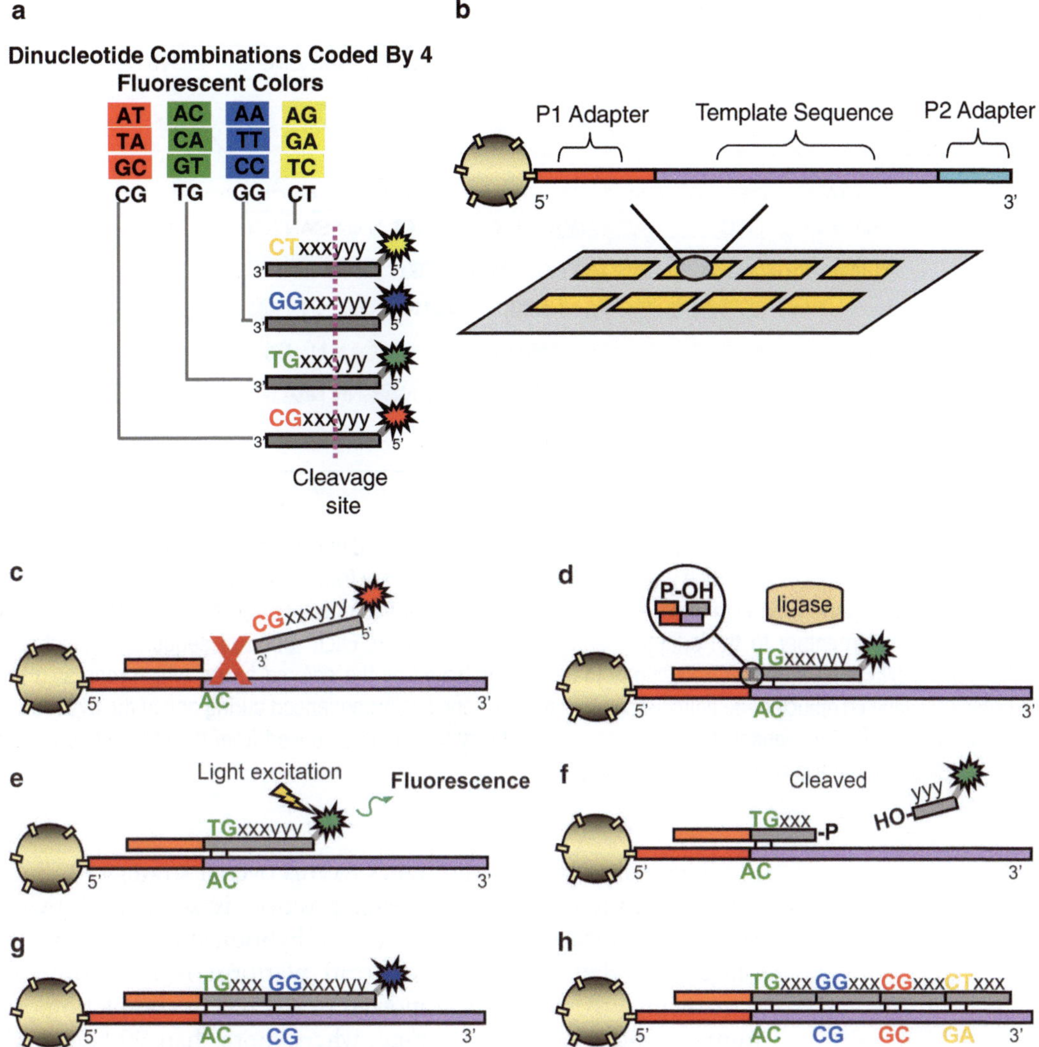

Fig. 7 Overview of the SOLiD oligonucleotide ligation sequencing chemistry. (**a**) Specific dinucleotide combinations coding for the four different fluorescent dyes on 8-mer SOLiD oligonucleotides. The cleavage site for removal of three nucleotides (the "y" nucleotides) and the fluorescent dye is indicated by the *dotted line*. (**b**) A SOLiD flow cell slide illustrating sites for placement of up to eight different libraries on an individual flow cell. The location of a single library nucleic acid fragment–bead complex on the flow cell is also indicated. (**c**) A complementary 8-mer oligonucleotide must bind adjacent to the sequencing primer on the library template. Here, an 8-mer lacking the nucleotide sequence complementary to template sequence is unable to undergo hybridization. (**d**) A fluorescently labeled 8-mer oligonucleotide undergoes sequence-specific hybridization to the template DNA and ligation to the sequencing primer. (**e**) The flow cell is imaged after the ligation of each 8-mer oligonucleotide to the growing sequencing product. (**f**) After the flow cell imaging step, the 3 "y" nucleotides and the fluorescent dye are removed from the newly ligated 8-mer. (**g**) During the next round of ligation sequencing, a new fluorescently labeled and complementary 8-mer is added to the growing sequencing product. (**h**) At the completion of a round of sequencing with a specific sequencing primer, 7–10 ligation reactions have been performed. This figure illustrates four different ligation reactions added to the sequencing product. Bases that are not specifically coded by the fluorescent dyes are indicated by "x"

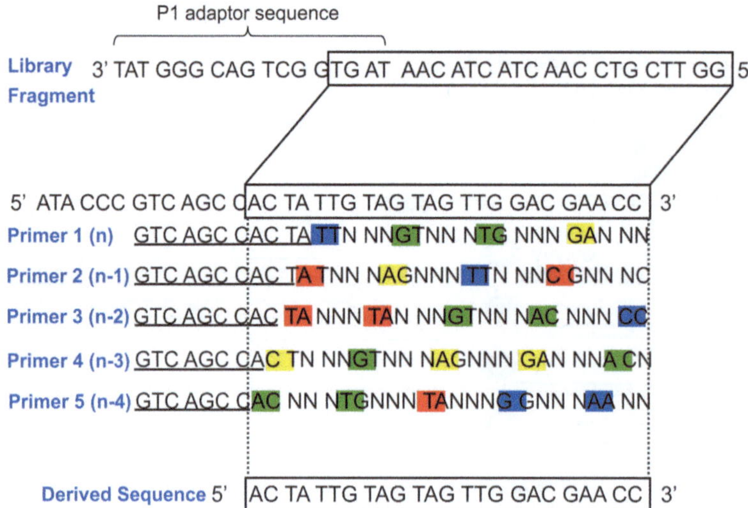

Fig. 8 Consensus sequence generation using fluorescently labeled dinucleotides with SOLiD sequencing illustrated in an example of sequence analysis for a hypothetical library fragment. In this example, five rounds of sequencing are performed with each reaction primed by one of five different primers. The specific primer sequences complementary to the adaptor sequence are *underlined*. Each round of sequencing illustrates either four or five different ligation reactions to 8-mers indicated by the *colored boxes*, which represent the fluorescently labeled dinucleotide pairs. Bases that are not specifically sequenced during one of the five rounds are indicated by N's. The consensus sequence for the library fragment is derived from fluorescently encoded dinucleotides from all five rounds of sequencing

There are obvious advantages to this overall strategy of dinucleotide sequencing, in that each position is sequenced twice. When combined with the specificity of hybridization of a dinucleotide pair, the SOLiD™ short-read platform is less prone to sequencing errors. This is supported by the recent completion of a human genome at 17X coverage, where more than 99 % of SNP sites determined by sequencing had the same genotype calls as HapMap genotypes at the same sites [21]. The false-positive call rate was similarly low at 0.76 %, and these values are comparable to those obtained by the Illumina Genome Analyzer [16, 21]. When sequencing human DNA with mate-paired libraries, it is also possible to generate highly accurate haplotypes [21]. In addition, the platform has the potential for future scalability, possibly with a higher density of library fragments sequenced on the flow cell. However, there are some limitations to this technology. Preliminary comparison of genomic sequence from the same individual suggested that ligation sequencing identified only about half of the insertion/deletion polymorphisms identified with Solexa [16, 21]. At present, accurate identification of these types of polymorphisms might persuade some investigators to utilize a different second-generation platform. In addition, a high proportion of sequence reads are not able to be mapped to the human genome due to their short read length and other factors [21, 29]. Finally, SOLiD™ has not performed well in sequencing repetitive elements in the human genome [29]. Overall, this is a conceptually elegant method of sequencing, and readers can anticipate the release of further SOLiD™ upgrades during late 2009.

3 Applications for Second-Generation Sequencing

The increased capacity and throughput of these new sequencers present a new challenge to apply these technologies in ways that are designed to test biological and medical hypotheses, not only with genome sequencing but also with the analysis of nucleic acids in other research arenas. Examples of different types of nucleic acid analyses include both sequencing and quantification of genomic DNA (including SNPs and CNVs or copy number variants); small amounts of genomic DNA from fossils, rare samples, or even single cells; messenger RNA transcriptome analysis including alternatively spliced RNAs, serial analysis of gene expression, and microRNAs; and protein–DNA interactions.

The most direct application of second-generation sequencing is to elucidate the genetic basis of human diseases, as this may help to identify new approaches for targeted or personalized therapies [32]. The early success of this approach has recently been demonstrated by completely sequencing the tumors from two patients with acute myeloid leukemia (AML), where either 10–12 somatic coding mutations were identified in gene exons when compared to germline DNA sequence from these same patients [17, 20]. The significance of whole-genome tumor sequencing is that it provides an unbiased means for identifying new cancer-initiating mutations that, with additional functional studies, might be candidates for targeted therapies [32]. However, a more cost-effective alternative could be whole-exome sequencing, where the template DNA is enriched for exons with an oligonucleotide hybridization array step prior to library construction. This has been demonstrated in a proof-of-principle study where *MYH3* mutations in individuals known to have Freeman–Sheldon syndrome were successfully identified using exome capture arrays and Solexa sequencing [33]. These technologies are also being applied to a comprehensive study of the spectrum human genetic mutation and variation across the genomes of 1,000 individuals (www.1000genomes.org). Finally, the small amount of input template required for second-generation sequencing makes this a useful method for analyzing ancient DNA samples, like the Neanderthal mitochondrial genome, or screening small amounts of DNA extracted from archived newborn screen blood spots for mutations [34, 35].

Second-generation sequencing of other nucleic acids beyond genomic DNA is also being used to examine fundamental biological processes. The original concept of serial analysis of gene expression (SAGE) was to analyze the pattern of messenger RNA transcripts over time by cloning and sequencing these species [36]. With next-generation sequencing, serial analysis of mRNA can now performed without the intermediate cloning step, and RNA sequencing has also been used to catalogue both human mRNA expression patterns and alternative splicing [37, 38]. Not surprisingly, this appears to be a more sensitive method for gene expression profiling than hybridization microarrays [38]. Finally, the role of regulatory transcription factors and histones bound to nuclear DNA during gene transcription is being examined on a genomic scale by combining the techniques of chromatin immunoprecipitation and next-generation sequencing [39, 40].

4 Future Technologies

Several new "third-generation" technologies are currently under development, and these hold tremendous promise for improving both scale and accuracy compared to second-generation sequencing chemistries. They also will help to lower the cost of sequencing individual genomes, and potentially fulfill the ultimate goal of sequencing an entire human genome for about $1,000 [12]. Conceptually, these technologies seek to eliminate a major source of sequencing errors experienced with second-generation sequencing: those introduced by the PCR amplification step. Instead, third-generation technologies directly sequence single DNA molecules with real-time detection of individual nucleotide additions.

4.1 Helicos Biosciences (http://www.helicosbio.com)

The Helicos Biosciences sequencer was the first single-molecule sequencing platform capable of direct DNA or RNA measurement in parallel through its True Single Molecule Sequencing (tSMS) technology [41]. This instrument performs sequencing by synthesis reactions simultaneously on large numbers of single nucleic acid molecules immobilized on a flow cell surface. The key feature to this system is that no clonal amplification of template nucleic acids is required. Instead of enhancing fluorescent sequence detection through sequencing template amplification, a highly sensitive fluorescence detection system is used. Briefly, template DNA is sheared, polyadenylated and labeled, and then hybridized to poly(dT) oligonucleotides immobilized on a flow cell surface. The location of each labeled template DNA strand is imaged with a CCD camera and recorded. Millions of DNA strands are immobilized on the surface and imaged in this step. For sequencing, the template labels are cleaved and washed away and fluorescently labeled nucleotides are introduced, one at a time. If the added nucleotide is complementary to a particular template it will be incorporated into the fragmented DNA by a DNA polymerase, resulting in a labeled fragment that will emit a fluorescent signal. If the label is not complementary to the template, no label will be present at that particular fragment location and no signal is emitted... After acquisition of images by a CCD camera, the label is chemically cleaved and washed away before proceeding with further sequencing. Sequencing information at each template location is captured with each new base addition. The average read length is 32 base pairs (range 25–70) with greater than 99.99 % accurate base calls, and this platform can generate up to 28 gigabases from a single run. It is likely to become a powerful tool for genomic analysis, as recent proof-of-principle studies have demonstrated its ability to sequence both simple and complex organisms, including one viral and one human genome, respectively [42, 43]. Thus, the Helicos Biosciences sequencer is the fifth sequencing methodology used to completely sequence a human genome. However, the utility of the present platform is partially limited by short read lengths, because short reads from any of the next-generation sequencing platforms significantly limit usable sequence that can be mapped back to a specific genomic location [29, 43]. The major source of sequencing errors is insertion–deletions, which occur more than 4 % of the time, whereas nucleotide substitution errors, which are the most common errors observed with second-generation sequencers, occur at a lower rate [29, 43].

4.2 Pacific Biosciences (http://pacificbiosciences.com)

Single Molecule Real Time (SMRT) DNA sequencing technology was originally developed at Cornell University, and it is now being commercialized by Pacific Biosciences [44, 45]. This sequencing chemistry provides direct, real-time observation of DNA synthesized by a bacteriophage DNA polymerase attached to the surface of a nanostructure unit called a zero-mode waveguide (ZMW). During DNA synthesis in the ZMW, high concentrations of fluorescently labeled phospholinked nucleotides are added to the reaction chamber. A photon detector aimed at an aperture in the ZMW is able to record an increase in fluorescent output with the incorporation of a single labeled nucleotide into an elongating DNA molecule at the bottom of the ZMW. Phosphodiester bond formation catalyzed by the DNA polymerase also removes the fluorophore from the newly incorporated nucleotide because the dye is bound to pyrophosphate via a linker. The detected fluorescent signal is lost when the now free dye–pyrophosphate molecule rapidly diffuses out of the ZMW [44, 46]. The nanostructure ZMWs can be arrayed to allow for sequencing of many different DNA molecules in parallel. This ability to scale ZMW arrays combined with long reads of synthesized DNA and short sequencer run times holds the potential for even more rapid sequencing in the future [45]. Published reports indicate that sequence read lengths of 3,000 base pairs can be generated over 5 min from a circular DNA template (ten bases per second) [44, 46]. The major sources of sequencing errors appear to be deletions, insertions, and occasional base mismatches. Despite these limitations, the SMRT platform has great potential for improving both the accuracy and throughput of nucleic acid sequencing.

4.3 Oxford Nanopore Technologies (http://nanoporetech.com)

Oxford Nanopore Technologies is developing a label-free, single-molecule sequencing system called BASE technology, which is based upon research by Hagan Bayley's team [47–49]. This also has the potential for extremely high-throughput DNA sequencing by passing an individual DNA molecule through a nanopore embedded in a lipid bilayer on a silicon chip. When an electrical current is applied to the topside of the chip, the nanopore is the only point at which current can pass across to the silicon chip. However, the nanopore is attached to a processive enzyme that cleaves individual nucleotides and fires them into the nanopore. Cleaved nucleotides travel down through the nanopore and transiently bind to a cyclodextran molecule. When this occurs, the electrical current is blocked to a different degree by each of the four different nucleotides, and it is possible to determine the identity based upon subtle differences in electrical signal. This technology, still under commercial development, also holds great promise for long direct-sequencing reads.

5 Conclusion: Clinical Applications in Cardiovascular Diseases

Identification of the genetic basis of monogenic and complex CVDs is of considerable importance in successfully translating fundamental findings towards improved therapies in cardiovascular medicine. New genetic knowledge will not only provide better assessments of individual CVD risks, but it will also help to further an in-depth understanding of CVD pathophysiology and speed the rational design of new therapeutic agents. Second-generation nucleic acid sequencing represents an essential next step to identify functional mutations underlying disease susceptibility loci which have been localized by genome-wide association studies, and it is a major technological advance that will speed the process of discovery. However, continued development of the single-molecule sequencing chemistries may render second-generation sequencing obsolete in the future. In the meantime, clinical investigators can anticipate increasing utilization of second-generation sequencing in both basic and clinical CVD studies.

6 Notes

1. Paired-end sequencing means that genomic DNA is sheared into small, several hundred base-pair fragments. Each fragment is then ligated to adaptors as in Fig. 1. The prepared library fragment is then sequenced from both ends to produce a longer sequence contig. Paired-end sequencing facilitates mapping a longer sequence from complex organisms like humans back onto the consensus sequence for the human genome. Paired-end libraries are presently used with the 454 FLX and Solexa sequencers. Paired-end libraries also allow for the detection of structural variation in DNA, including insertions and deletions.

2. Mate-paired libraries, like paired-end sequencing, allow two separated regions of DNA from the same chromosomal strand to be sequenced from a single library fragment. The steps involved in the generation of a mate-paired library fragment are illustrated in Fig. 6. The key difference is that the two fragments from the same chromosomal strand may be separated by a much greater distance that the mate-paired fragments.

Acknowledgments

This work was supported by the Division of Intramural Research, NHLBI, National Institutes of Health. The views expressed are those of the authors and do not represent an endorsement by the National Institutes of Health.

References

1. Nabel EG (2003) Cardiovascular disease. N Engl J Med 349:60–72
2. NHLBI, Department of Health and Human Services (2008) NHLBI fact book 2008; Chapter 4. Disease statistics. Bethesda: DHHS
3. Cohen JC, Kiss RS, Pertsemlidis A, Marcel YL, McPherson R, Hobbs HH (2004) Multiple rare alleles contribute to low plasma levels of HDL cholesterol. Science 305:869–872
4. McPherson R, Pertsemlidis A, Kavaslar N, Stewart A, Roberts R, Cox DR, Hinds DA, Pennacchio LA, Tybjaerg-Hansen A, Folsom AR et al (2007) A common allele on chromosome 9 associated with coronary heart disease. Science 316:1488–1491
5. Romeo S, Pennacchio LA, Fu Y, Boerwinkle E, Tybjaerg-Hansen A, Hobbs HH, Cohen JC (2007) Population-based resequencing of ANGPTL4 uncovers variations that reduce triglycerides and increase HDL. Nat Genet 39:513–516
6. Maxam AM, Gilbert W (1977) A new method for sequencing DNA. Proc Natl Acad Sci USA 74:560–564
7. Sanger F, Nicklen S, Coulson AR (1977) DNA sequencing with chain-terminating inhibitors. Proc Natl Acad Sci USA 74:5463–5467
8. Mitchelson KR (2003) The use of capillary electrophoresis for DNA polymorphism analysis. Mol Biotechnol 24:41–68
9. Youngblood V, Taylor JG (in press) Sequencing PCR amplified DNA in cardiovascular and lipoprotein research. Methods Mol Biol
10. Lander ES, Linton LM, Birren B, Nusbaum C, Zody MC, Baldwin J, Devon K, Dewar K, Doyle M, FitzHugh W et al (2001) Initial sequencing and analysis of the human genome. Nature 409:860–921
11. Venter JC, Adams MD, Myers EW, Li PW, Mural RJ, Sutton GG, Smith HO, Yandell M, Evans CA, Holt RA et al (2001) The sequence of the human genome. Science 291:1304–1351

12. Mardis ER (2006) Anticipating the 1,000 dollar genome. Genome Biol 7:112
13. Rogers YH, Venter JC (2005) Genomics: massively parallel sequencing. Nature 437:326–327
14. Ansorge WJ (2009) Next-generation DNA sequencing techniques. N Biotechnol 25:195–203
15. Mardis ER (2008) Next-generation DNA sequencing methods. Annu Rev Genomics Hum Genet 9:387–402
16. Bentley DR, Balasubramanian S, Swerdlow HP, Smith GP, Milton J, Brown CG, Hall KP, Evers DJ, Barnes CL, Bignell HR et al (2008) Accurate whole human genome sequencing using reversible terminator chemistry. Nature 456:53–59
17. Ley TJ, Mardis ER, Ding L, Fulton B, McLellan MD, Chen K, Dooling D, Dunford-Shore BH, McGrath S, Hickenbotham M et al (2008) DNA sequencing of a cytogenetically normal acute myeloid leukaemia genome. Nature 456:66–72
18. Wang J, Wang W, Li R, Li Y, Tian G, Goodman L, Fan W, Zhang J, Li J, Zhang J et al (2008) The diploid genome sequence of an Asian individual. Nature 456:60–65
19. Wheeler DA, Srinivasan M, Egholm M, Shen Y, Chen L, McGuire A, He W, Chen YJ, Makhijani V, Roth GT et al (2008) The complete genome of an individual by massively parallel DNA sequencing. Nature 452: 872–876
20. Mardis ER, Ding L, Dooling DJ, Larson DE, McLellan MD, Chen K, Koboldt DC, Fulton RS, Delehaunty KD, McGrath SD et al (2009) Recurring mutations found by sequencing an acute myeloid leukemia genome. N Engl J Med 361:1058–1066
21. McKernan KJ, Peckham HE, Costa GL, McLaughlin SF, Fu Y, Tsung EF, Clouser CR, Duncan C, Ichikawa JK, Lee CC et al (2009) Sequence and structural variation in a human genome uncovered by short-read, massively parallel ligation sequencing using two-base encoding. Genome Res 19:1527–1541
22. Ahn SM, Kim TH, Lee S, Kim D, Ghang H, Kim DS, Kim BC, Kim SY, Kim WY, Kim C et al (2009) The first Korean genome sequence and analysis: full genome sequencing for a socio-ethnic group. Genome Res 19:1622–1629
23. Kim JI, Ju YS, Park H, Kim S, Lee S, Yi JH, Mudge J, Miller NA, Hong D, Bell CJ et al (2009) A highly annotated whole-genome sequence of a Korean individual. Nature 460:1011–1015
24. Margulies M, Egholm M, Altman WE, Attiya S, Bader JS, Bemben LA, Berka J, Braverman MS, Chen YJ, Chen Z et al (2005) Genome sequencing in microfabricated high-density picolitre reactors. Nature 437:376–380
25. Shendure J, Porreca GJ, Reppas NB, Lin X, McCutcheon JP, Rosenbaum AM, Wang MD, Zhang K, Mitra RD, Church GM (2005) Accurate multiplex polony sequencing of an evolved bacterial genome. Science 309: 1728–1732
26. Ronaghi M, Karamohamed S, Pettersson B, Uhlen M, Nyren P (1996) Real-time DNA sequencing using detection of pyrophosphate release. Anal Biochem 242:84–89
27. Dressman D, Yan H, Traverso G, Kinzler KW, Vogelstein B (2003) Transforming single DNA molecules into fluorescent magnetic particles for detection and enumeration of genetic variations. Proc Natl Acad Sci USA 100:8817–8822
28. Droege M, Hill B (2008) The Genome Sequencer FLX System—longer reads, more applications, straight forward bioinformatics and more complete data sets. J Biotechnol 136:3–10
29. Harismendy O, Ng PC, Strausberg RL, Wang X, Stockwell TB, Beeson KY, Schork NJ, Murray SS, Topol EJ, Levy S et al (2009) Evaluation of next generation sequencing platforms for population targeted sequencing studies. Genome Biol 10:R32
30. Adessi C, Matton G, Ayala G, Turcatti G, Mermod JJ, Mayer P, Kawashima E (2000) Solid phase DNA amplification: characterisation of primer attachment and amplification mechanisms. Nucleic Acids Res 28:E87
31. Fedurco M, Romieu A, Williams S, Lawrence I, Turcatti G (2006) BTA, a novel reagent for DNA attachment on glass and efficient generation of solid-phase amplified DNA colonies. Nucleic Acids Res 34:e22
32. Taylor JG, Cheuk AT, Tsang PS, Chung JY, Song YK, Desai K, Yu Y, Chen QR, Shah K, Youngblood V et al (2009) Identification of FGFR4-activating mutations in human rhabdomyosarcomas that promote metastasis in xenotransplanted models. J Clin Invest 119:3395–3407
33. Ng SB, Turner EH, Robertson PD, Flygare SD, Bigham AW, Lee C, Shaffer T, Wong M, Bhattacharjee A, Eichler EE et al (2009) Targeted capture and massively parallel sequencing of 12 human exomes. Nature 461:272–276
34. Druley TE, Vallania FL, Wegner DJ, Varley KE, Knowles OL, Bonds JA, Robison SW, Doniger SW, Hamvas A, Cole FS et al (2009) Quantification of rare allelic variants from pooled genomic DNA. Nat Methods 6:263–265

35. Green RE, Malaspinas AS, Krause J, Briggs AW, Johnson PL, Uhler C, Meyer M, Good JM, Maricic T, Stenzel U et al (2008) A complete Neandertal mitochondrial genome sequence determined by high-throughput sequencing. Cell 134:416–426
36. Velculescu VE, Zhang L, Vogelstein B, Kinzler KW (1995) Serial analysis of gene expression. Science 270:484–487
37. Mortazavi A, Williams BA, McCue K, Schaeffer L, Wold B (2008) Mapping and quantifying mammalian transcriptomes by RNA-Seq. Nat Methods 5:621–628
38. Sultan M, Schulz MH, Richard H, Magen A, Klingenhoff A, Scherf M, Seifert M, Borodina T, Soldatov A, Parkhomchuk D et al (2008) A global view of gene activity and alternative splicing by deep sequencing of the human transcriptome. Science 321:956–960
39. Barski A, Cuddapah S, Cui K, Roh TY, Schones DE, Wang Z, Wei G, Chepelev I, Zhao K (2007) High-resolution profiling of histone methylations in the human genome. Cell 129:823–837
40. Johnson DS, Mortazavi A, Myers RM, Wold B (2007) Genome-wide mapping of in vivo protein-DNA interactions. Science 316:1497–1502
41. Braslavsky I, Hebert B, Kartalov E, Quake SR (2003) Sequence information can be obtained from single DNA molecules. Proc Natl Acad Sci USA 100:3960–3964
42. Harris TD, Buzby PR, Babcock H, Beer E, Bowers J, Braslavsky I, Causey M, Colonell J, Dimeo J, Efcavitch JW et al (2008) Single-molecule DNA sequencing of a viral genome. Science 320:106–109
43. Pushkarev D, Neff NF, Quake SR (2009) Single-molecule sequencing of an individual human genome. Nat Biotechnol 27:847–852
44. Eid J, Fehr A, Gray J, Luong K, Lyle J, Otto G, Peluso P, Rank D, Baybayan P, Bettman B et al (2009) Real-time DNA sequencing from single polymerase molecules. Science 323:133–138
45. Levene MJ, Korlach J, Turner SW, Foquet M, Craighead HG, Webb WW (2003) Zero-mode waveguides for single-molecule analysis at high concentrations. Science 299:682–686
46. Korlach J, Bibillo A, Wegener J, Peluso P, Pham TT, Park I, Clark S, Otto GA, Turner SW (2008) Long, processive enzymatic DNA synthesis using 100 % dye-labeled terminal phosphate-linked nucleotides. Nucleosides Nucleotides Nucleic Acids 27:1072–1083
47. Astier Y, Braha O, Bayley H (2006) Toward single molecule DNA sequencing: direct identification of ribonucleoside and deoxyribonucleoside 5′-monophosphates by using an engineered protein nanopore equipped with a molecular adapter. J Am Chem Soc 128:1705–1710
48. Clarke J, Wu HC, Jayasinghe L, Patel A, Reid S, Bayley H (2009) Continuous base identification for single-molecule nanopore DNA sequencing. Nat Nanotechnol 4:265–270
49. Stoddart D, Heron AJ, Mikhailova E, Maglia G, Bayley H (2009) Single-nucleotide discrimination in immobilized DNA oligonucleotides with a biological nanopore. Proc Natl Acad Sci USA 106:7702–7707

Part III

Transgenic, Knockout, and Knockdown Methodologies

Chapter 8

Strategies for Designing Transgenic DNA Constructs

Chengyu Liu

Abstract

Generation and characterization of transgenic mice are important elements of biomedical research. In recent years, transgenic technology has become more versatile and sophisticated, mainly because of the incorporation of recombinase-mediated conditional expression and targeted insertion, site-specific endonuclease-mediated genome editing, siRNA-mediated gene knockdown, various inducible gene expression systems, and fluorescent protein marking and tracking techniques. Site-specific recombinases (such as PhiC31) and engineered endonucleases (such as ZFN and Talen) have significantly enhanced our ability to target transgenes into specific genomic loci, but currently a great majority of transgenic mouse lines are continuingly being created using the conventional random insertion method. A major challenge for using this conventional method is that the genomic environment at the integration site has a substantial influence on the expression of the transgene. Although our understanding of such chromosomal position effects and our means to combat them are still primitive, adhering to some general guidelines can significantly increase the odds of successful transgene expression. This chapter first discusses the major problems associated with transgene expression, and then describes some of the principles for using plasmid and bacterial artificial chromosomes (BACs) for generating transgenic constructs. Finally, the strategies for conducting each of the major types of transgenic research are discussed, including gene overexpression, promoter characterization, cell-lineage tracing, mutant complementation, expression of double or multiple transgenes, siRNA knockdown, and conditional and inducible systems.

Key words Transgene, DNA construct, Position effects, BAC, Plasmid, Overexpression, Insulator, siRNA, Recombinase, Cre–loxP, FLP-FRT, Conditional, Inducible, Animal model

1 Introduction

Since its development over 30 years ago [1], the pronuclear microinjection method for making transgenic mice has been widely used to address a variety of biological questions, including lipoprotein and atherosclerosis research (see reviews 2–5). Transgenics enables the in vivo modulation of gene activity in a spatial- and temporal-specific manner, greatly enhancing our ability to analyze the functions of genes involved in lipoprotein biology as well as to build animal models for atherosclerosis and other disorders of lipoprotein metabolism. Transgenic models also complement well with

Lita A. Freeman (ed.), *Lipoproteins and Cardiovascular Disease: Methods and Protocols*, Methods in Molecular Biology, vol. 1027, DOI 10.1007/978-1-60327-369-5_8, © Springer Science+Business Media, LLC 2013

the increasing number of available knockout/knockin and other mutant mouse lines generated by both ethyl nitrosourea (ENU) and gene-trapping mutagenesis. These loss-of-function and gain-of-function approaches synergistically increase our ability to understand complex physiological processes and mechanisms of disease.

Pronuclear microinjection is an efficient and facile method for delivering foreign genes into the mouse genome, although extensive training is required to become proficient at it. Recent advancements in site-specific recombinase systems, such as Cre-loxP, FLP-FRT and Dre-rox systems [6], and site-specific endonucleases, such as zinc finger nucleases (ZFNs) and transcription activator-like effector nucleases (Talens) [7], have begun to launch the era of site-specific or targeted transgenics [8]. However, as of today, the great majority of transgenic projects are still carried out using the conventional random integration method. In spite of three decades of experimentation and optimization, our ability to predict and control the expression of randomly inserted transgenes is still far from satisfactory. Nevertheless, some important knowledge and general guidelines for controlling transgene expression have accumulated over the past three decades. Following these guidelines can significantly improve, although not guarantee, the odds of proper transgene expression. This chapter is mainly aimed at providing beginners with a broad perspective on the applications of the conventional transgenic technology. It is impossible to discuss each topic in much detail, and therefore many important innovations and contributions are not mentioned here.

2 Major Challenges for Controlling Transgene Expression

Microinjected transgenes integrate into the genome at random positions. One important lesson learned from the past three decades is that the genomic environment surrounding the site of integration can exert a profound influence on expression of the transgene. In the cell nucleus, chromatin interacts extensively with the nuclear matrix and other proteins to form looped and topologically constrained domains. Distinct boundary elements or insulators embedded in the genome separate it into regions where gene expression is often facilitated or repressed. Consequently, it is not surprising that the promoter activity of the inserted transgene is subject to the influence of the local genomic environment as well as distant transcriptional enhancers or repressors.

These chromosomal position effects are typically represented in the following forms. First, it is quite common that some of the transgenic lines do not express the transgene at all or express it at a very low level. These silenced transgenic lines are often eliminated during the initial period of transgenic mouse characterization. Second, altered transgene expression refers to transgenic lines with spatial or temporal expression patterns that do not match those expected based on the promoters used to drive the transgenes. This is often a result of the influence of a strong promoter/enhancer/repressor near the integration site. Third, mosaic transgene expression is another often-encountered problem, i.e., the transgene is expressed only in a portion of cells of the same tissue. Fourth, progressive extinction also happens to some transgenes. In this case, the transgene gives correct and robust expression when the transgenic line is first established, but its expression decreases or even completely stops after multiple generations of breeding. Chromosomal epigenetic modifications such as DNA methylation and histone acetylation probably play important roles in diminishing the expression of these transgenes. Most researchers carefully characterize their transgenic lines when they are first created, but then do not reexamine transgene expression thereafter. Many scientists who receive mice from another laboratory tend to trust the published initial characterization results. Therefore, failure to periodically check transgene expression can potentially lead to erroneous scientific conclusions.

Some of the patterns of uncontrolled transgene expression cannot be explained by chromosomal position effect alone. For example, genes of mammalian origin usually express better than genes of nonmammalian origin [9]. Prokaryote-derived cloning vector sequences often strongly suppress transgene expression if they are not removed from the transgene. Genomic DNA usually expresses better than cDNA [10, 11]. When many copies of transgenes are inserted as tandem repeats at the same location, they are more likely to be silenced [12, 13]. While introns may facilitate the export of mRNA out of the nucleus, which at least partially explains why genomic DNA is better than cDNA for achieving gene expression, the other observed phenomena still remain to be explained, although chromosomal epigenetic modifications are believed to be the main suspect.

The importance of ensuring correct DNA construct design for transgenic projects can be puzzling to beginners, because their experience is often grounded in transfecting and expressing foreign genes in cultured mammalian cells. After all, shouldn't cultured cells and cells in live animals have similar, if not exactly the same, genomic organization and transcriptional machinery? To help clear the confusion, it should be pointed out that the majority of in vitro transfection experiments involve transient transfection. The DNA vectors do not even integrate into the genome, and therefore they are not subjected to complicated genomic environmental influences. In fact, for generation of stable cell lines, the same problems of chromosomal position effects and epigenetic modifications do exist, as reviewed by Kwaks and Otte [14]. However, because hundreds of thousands or millions of cells are normally used in each cell transfection experiment, it is usually not very difficult to find some cells with the expected gene expression. In the case of transgenic animals, creating and maintaining each transgenic line is labor-intensive and costly. Most scientists do not have the luxury of selecting several perfect lines from many available transgenic lines. Furthermore, each transgenic animal contains hundreds of different cell types, whereas each cell transfection experiment normally uses only one type of cell. Therefore, unexpected patterns of tissue-specific gene expression are far more likely to show up in transgenic mice than in a particular cell line. Lastly, useful transgenic models need to be maintained for a very long time, for multiple generations. Moreover, authentic transgene expression must also be properly maintained through the changes in gene activities and overall genomic modifications associated with the different stages of embryonic and postnatal development. In contrast, the changes in gene expression and epigenetic modification in stably transfected cell lines are rather limited.

3 Commonly Used Transgenic Cloning Vectors

One of the key advantages of the pronuclear microinjection method over the newer lentiviral method is that there is essentially no limit on the size of DNA constructs that can be microinjected, whereas the lentiviral vector can only accept inserts of 10 Kb or smaller [15]. Currently, the most commonly used vectors for cloning transgenes are plasmids and bacterial artificial chromosomes (BACs).

3.1 Plasmid Vectors

Plasmids are convenient and efficient cloning vectors for carrying out a variety of recombinant DNA procedures. Generating a typical transgenic construct involves assembling three basic DNA elements: (1) a promoter and/or enhancer which confers the desired spatial and temporal pattern of transgene expression; (2) the gene to be transcribed, which may or may not encode a protein; and (3) a transcription termination or polyadenylation signal sequence to stop transcription and enable 3′ end processing. An optional fourth DNA element is a genomic boundary element for reducing position effects. The promoter and transcribed region will be discussed in more detail in Subheading 4.1. Commonly used polyadenylation sequences are derived from SV40, the growth hormone (GH) gene, and the β-globin gene, and are quite effective at terminating transgene expression.

One controversial area is the inclusion of genomic boundary elements for combating chromosomal position effects. Plasmid-based transgenes are relatively small, and hence they are prone to influences of the neighboring chromatin environment. Over the years, several genomic DNA elements have been employed to reduce chromosomal position effects. The locus control region (LCR) initially derived from the β-globin gene can improve transgene expression by positively stimulating transcription, presumably by imposing a chromatin configuration that favors gene expression [16]. Another class of elements is the so-called matrix- or scaffold-attachment regions (MARs/SARs). They appear to mediate their effects by acting as a barrier to both positive and negative signals from the surrounding environment [17]. Enhancer-blocking (EB) insulators and barrier insulators [18] have also been tested for improving transgene expression. Many positive results have been reported with these elements, but negative results are also frequently found (see reviews 19, 20). Therefore, use of boundary elements in transgenic constructs is not yet widespread. Further characterization of these genomic boundary elements should increase their use not only for controlling transgene expression but also in gene therapy.

Prokaryote-derived cloning vector sequences often interfere with adjacent promoter activity when inserted into a mammalian genome, so it is necessary to remove the cloning vector from the transgene before microinjection. Therefore, when designing the DNA construct, it is important to leave suitable restriction enzyme digestion sites at both ends of the transgene so that the transgenic DNA fragment (promoter + transgene + polyadenylation signal) can be separated from the plasmid backbone. At the time of transgene integration, it is possible that a few base pairs at the ends of the DNA fragment may be deleted [21]. This usually will not affect the expression of the transgene. However, if the ends of the released transgenic fragment are critical to the function of the transgene, then restriction enzymes that leave at least several base pairs of polylinker sequence at the ends should be used.

Although plasmid vectors do not have an absolute limit on the size of DNA that can be cloned, both the transformation efficiency and plasmid stability decrease dramatically when the DNA fragments become too big (>25 kb). Cosmids are modified plasmids which contain the cos sequences from bacteriophage lambda [22]. Cosmids can be handled and propagated in *Escherichia coli* similar to plasmid vectors, but they can accept DNA fragments up to 47 kb.

After finishing the transgenic DNA construct, it is recommended that it be tested by transfecting it into appropriate mammalian cell lines. Correct expression in cell lines cannot guarantee the same in transgenic mice, but negative results are usually indicative of problems with the constructs.

3.2 Bacterial Artificial Chromosomes (BACs) and Other Large Pieces of DNA

One of the by-products of the genome sequencing projects was the creation of a series of overlapping BAC clones that essentially covers the entire human and mouse genomes. For the great majority of mouse and human genes, BAC clones are readily available from repositories such as the BACPAC Resource Center at Oakland Children's Hospital (http://bacpac.chori.org/). These fully sequenced BAC clones are usually 100–300 kb long, and can be used directly for generating transgenic mouse lines. The main advantage of BACs is that they are more likely to contain all the genomic regulatory elements required for mimicking the endogenous gene expression pattern. Also, large pieces of DNA can better shield the transgene from unwanted position effects. When overlapping BAC clones containing the same gene are available, it is best to choose the clone that contains not only the entire transcribed region of the gene but also as much 5′ and 3′ flanking regions as possible. When more than 50 kb of genomic region flanks both the 5′ and 3′ ends of the gene, the probability of faithful and reproducible transgene expression is as high as 85 % [23]. However, one potential problem with using large BACs is that larger clones are more likely to include neighboring genes. Whether these extra full-length or partial genes pose any complications to the study will depend on the purpose of the project.

The recombineering method [24–26] has enabled the precise modification of BAC DNA molecules. Therefore, BAC clones are not only good for overexpressing native genes, but they also can be used to drive reporter gene or modified gene expression. The open reading frame of a reporter gene can be inserted into the 5′ UTR or the coding region to replace the native gene, or be inserted into the 3′ UTR for expression in addition to the native gene. BACs can be used for driving expression of Cre and FLP recombinase genes in desired tissues, but as will be discussed in Subheading 4.6, the commonly used BAC vectors already contain loxP sites, which can be problematic when crossbred to floxed mouse lines.

Unmodified mouse BAC clones are exactly the same as the endogenous genomic DNA except for the linked cloning vector. Some knowledge of vector DNA sequence is needed for designing probes or primers with which to genotype the transgenic mice. The mRNA and protein expressed from the transgene are also indistinguishable from their endogenous counterparts, which makes it difficult to assess the level of transgene expression although quantitative real-time RT-PCR can be used to determine relative expression of a transcript in a transgenic mouse compared to a control, nontransgenic mouse, as described in Chapter 2 of this volume. BAC clones from a different species such as human or rat can be used, which are often different enough to allow differentiation of the transgene from the endogenous gene. Alternatively, molecular tags such as FLAG tag or GFP can be added to the transgene.

Removal of the BAC vector sequence from the transgene is generally not required for microinjection, because (1) the influence of the prokaryote-derived vector sequence is often diminished by the large eukaryotic DNA fragment in the BAC and (2) releasing and separating the large BAC insert from its cloning vector can be technically challenging. It has been reported that linear and circular BAC DNAs were equally efficient in resulting in transgenic mice

[27], but in our hands linear BACs appear to be more efficient for producing positive founders (unpublished observation). Shearing is a potential problem during preparation, storage, and microinjection of BAC as well as other large DNA molecules. BAC DNA should be stored in TE buffer or microinjection buffer supplied with salt and polyamines to make the long DNA strands more compact and hence less prone to shearing. Wide-bore pipette tips should be used for transferring BAC DNA solutions. When microinjecting linearized BACs, slightly larger needle openings and lower injection pressures are preferable for minimizing shearing.

Besides BACs, other large DNA fragments have also been successfully used to generate transgenic mice. For example, the bacteriophage P1 cloning system can accept DNA inserts of 100 kb, and its modified form, PAC (P1 artificial chromosome) vector, can typically carry 100–250 kb of DNA. For a small portion of mammalian genes, even these large vectors are not large enough to contain the entire gene or the flanking regions that are important for the desired expression pattern. Yeast artificial chromosomes (YACs) can accommodate a couple of megabases of DNA [28], and human artificial chromosomes (HACs) can carry more than 10 Mb [29]. Both vector types have been successfully used to generate transgenic mice. Gene-targeting methods are also increasingly being used to ensure that a transgene assumes the expression pattern of the endogenous gene, which is the ultimate method for recapitulating in vivo gene expression.

4 Main Applications of Transgenic Technology

Transgenic methods have been used to address a wide range of biomedical questions. A good transgenic construct must suit the purpose of the study. Below we discuss the strategies for each of the major applications of transgenic technology.

4.1 Gene Overexpression

An obvious application of transgenic methods is to increase the level of gene expression and then observe what effect this has on phenotype. If the goal is to modestly increase the expression of an endogenous protein in its native environment, then BAC DNA is a good choice, particularly for genes whose promoters have not been extensively characterized. This approach is not ideal for producing severe phenotypes because both the level and tissue specificity of gene expression are not dramatically changed. Nevertheless, it is very useful for studying subtle physiological differences, and for building animal models for disorders resulting from overexpression of a gene, such as Down syndrome.

Thus far, most overexpression constructs are assembled in plasmid vectors using heterologous promoters. Both genomic DNA and cDNA (minigenes) can be used, but genomic DNA is preferable because introns contained in the genomic DNA can facilitate export of the mRNA from the nucleus to the cytoplasm [10]. Unfortunately, the genomic DNAs of many mammalian genes are too large to be conveniently manipulated in plasmid vectors. Consequently, full-length cDNAs are often used to make transgenic constructs, and many of them are expressed well. However, many experienced transgenic researchers prefer to incorporate a heterologous intron, such as introns from β-globin, SV40, or adenovirus, into their constructs to increase the odds of successful expression. For instance, the pCI and pSI vectors marketed by Promega contain a chimeric intron (β-globin donor site and immunoglobulin acceptor site) at the 5′ end of the cloned gene. Alternatively, a chimeric gene can be created by creating an in-frame fusion between the cDNA and genomic DNA of the same gene [30].

Besides full-length cDNA, mouse and human full-length ORFs (open reading frames) are also commercially available. These clones contain the entire coding region, but not the 5′ and 3′ untranslated regions. It is not recommended to use ORFs to directly make transgenic animals, because it has been well documented that the 5′ and 3′ UTRs can play important roles in regulating mRNA stability, intracellular localization, and efficiency of protein translation. However, it is perfectly acceptable to insert these ORFs into other genes, such as the UTRs of BACs, for expression as transgenes.

When choosing a promoter, it is recommended to consider promoters that have been shown to be able to drive heterologous gene expression. A promoter that can drive its native gene expression may not necessarily correctly drive heterologous gene expression, because some regulatory genomic elements may lie downstream of the transcription initiation site. It is noteworthy that some of the most commonly used promoters for transfecting cultured cells are not the best choice for directing ubiquitous gene expression in transgenic animals. Such promoters are small in size and very convenient, but they are subject to heavy position effects. For example, CMV is considered a ubiquitous promoter for transfecting a wide variety of cultured mammalian cells. However, in transgenic mice its effects are not as ubiquitous [31]. For ubiquitous transgene expression, good results have been achieved using the ROSA26 and the CAGGS (chicken β-actin promoter with CMV early enhancer) promoters. For tissue-specific expression, proven promoters for most major tissues or cell types can be found in the literature. For instance, the Tie2 promoter has been shown to be able to direct transgene expression specifically in the vascular endothelium [32]. Tissue-specific promoters for liver, intestine, and macrophage, which are of particular interest to lipoprotein researchers, are discussed in Chapter 12 of this volume. If an untested promoter needs to be used, it is advisable to use as big a piece of genomic DNA as practically possible, such as an entire BAC clone.

Mutated genes can also be expressed in transgenic animals, which may result in knockdown or dominant negative phenotypes, depending on the dominance or recessiveness of the mutation.

An extreme form of gene overexpression is to use transgenic animals as bioreactors for producing recombinant proteins (see reviews 33–35). Although the main purpose of this approach is to use large farm animals to produce biopharmaceutical products, transgenic mice have also been used to test DNA constructs and to conduct biomedical research. For example, the mouse ZP3 protein has been successfully produced in mouse milk by expressing it under the control of the goat β-casein promoter [36].

4.2 Promoter Characterization and Cell Linage Markers

Transgenic animals are often generated for characterizing the temporal and spatial patterns of gene expression governed by promoter elements. Knowing the exact cell types and developmental stages in which a particular gene is normally expressed can shed light on its physiological functions. Generating and characterizing transgenic mice is the most efficient method for examining in vivo gene expression patterns. A good starting point is to place several kilobases of promoter region in front of a suitable reporter gene. Depending on whether the reporter gene expression is able to match the pattern of endogenous gene expression, the length of the promoter can be subsequently increased or decreased to identify important genomic regulatory elements. For most genes a few kilobases or even several hundred base pairs of DNA upstream of the transcription initiation sites are enough to confer tissue-specific gene expression. However, for some other genes, dozens of kilobases of promoter region or even entire BACs are not sufficient for recapitulating the in vivo expression pattern of the gene. In the latter cases, targeting a reporter gene into the native genomic locus has increasingly been used for recapitulating the expression pattern of the endogenous gene.

Various reporter genes have been successfully used for characterizing promoter activity [37]. The firefly luciferase and chloramphenicol O-acetyltransferase (CAT) reporters are excellent for quantitative measurement of gene activity in various organs at various time points of development, while beta-galactosidase, alkaline phosphatase, or fluorescent proteins can provide a direct visualization of promoter activity at the tissue and cell levels. Fluorescent proteins are particularly good for providing a live and dynamic view of gene activity, because no tissue fixation and disruption are required. Concerns over GFP toxicity have been raised [38, 39], but a large number of healthy GFP-expressing transgenic lines have been successfully generated.

It should be pointed out that the presence or absence of reporter proteins may not precisely represent the true state of the native protein, because mRNA and protein stabilities as well as the regulation of mRNA translation may be different between the native and reporter genes.

The promoter–reporter systems described above are not only useful for analyzing promoter activity but also very useful as cell-lineage markers. For example, there are many types of neurons in the brain which synthesize different neuropeptides or neurotransmitters. It is very difficult to distinguish them morphologically. The GENSAT (Gene Expression Nervous System Atlas) project is aimed at labeling various neurons with fluorescent protein using neuron-specific promoters [40]. Besides marking specific cell types, GFP can also be fused to various subcellular localization signals as well as to individual proteins to track individual cellular organelles or proteins in animal tissues. It is recommended to verify cellular localization of endogenous proteins using antibody-based methods, if possible.

4.3 Gene Knockdown by RNA Interference

RNA interference (RNAi) technology has been widely used in cultured cells to knock down gene expression, but its use in transgenic animals is still not widespread. This is partly because gene knockout technology has been firmly adopted as the gold standard for developing loss-of-function animal models. Another reason is that the uncontrolled transgene expression described in Subheading 2 seems also to affect RNA polymerase III promoters, causing large variations in shRNA expression among different tissues [41]. Nevertheless, it has been shown that the shRNA method can work well in transgenic animals [42, 43]. Both the H1 and U6 promoters can direct RNA polymerase III to transcribe enough shRNA to knock down endogenous gene expression by as much as 80–90 %. RNA polymerase II-dependent promoters can also be used to drive the expression of double-strand RNA, which opened the door for utilizing many well-characterized RNA polymerase II promoters for directing tissue-specific siRNA knockdown [44–46].

Much more work is needed before siRNA transgenesis can challenge the knockout method. However, it should be noted that the siRNA method is a much quicker (one generation vs. three generations of mouse breeding) and easier (small transgenic construct vs. large multifragmented gene-targeting construct plus tedious ES cell work) than the gene knockout approach. One of the main problems of the siRNA approach is that it can only reduce, but never completely abolish, endogenous gene expression. However, in the cases when complete knockout results in lethality, the knockdown approach can actually be advantageous. As will be discussed in Subheadings 4.6 and 4.7, the conditional and inducible systems developed for RNA polymerase II promoters also appear to work for the U6 and H1 promoters, which significantly enhance the flexibility and usefulness of the siRNA knockdown approach [47].

4.4 Complementation and Mutation Mapping

For spontaneously occurring or chemically (such as ENU) or physically (such as irradiation) induced mutant mouse lines, extensive genetic mapping is required to identify the mutated gene(s).

Transgenic animals can be used to confirm or nullify the mapping results. Large pieces of DNA, such as BAC or YAC clones, from the suspected genomic regions can be used to create transgenic mice for rescuing the mutant phenotype if the phenotype is recessive. If the mutant phenotype is dominant, the mutated form of the gene can be introduced into wild-type mice to recreate the mutant phenotype. Some users of transgenic core facilities request that their transgenic lines be made directly on the mutant background, but it is usually easier and faster to create the transgenic lines in standard wild-type mice and then cross them into the mutant background. Even if transgenic mice are created using the mutant mice as egg donors, it is likely that the mutated gene and the newly inserted transgene will segregate in subsequent generations. The BAC and YAC clones used in the complementation studies may contain multiple genes. When interpreting experimental results, the possibility that these extra genes may be contributing to the phenotype needs to be considered.

Transgenic mice can also be used to rescue knockout mouse lines, and more interestingly tissue-specific transgenic lines can be used to restore gene expression to only some of the knockout tissues. This strategy is particularly useful for studies of gene function in a specific tissue, when floxed conditional knockout mouse lines are not yet available.

4.5 Double or Multiple Transgenes

Sometimes it is necessary to co-express two or more transgenes in the same mouse line. Of course, this can be achieved by crossbreeding two or multiple transgenic lines together. The disadvantage is that extensive breeding and genotyping are required; furthermore, the obtained genotypes are not stable and hence the transgenes may segregate in subsequent generations. Several alternative methods exist for incorporating multiple transgenes into the same transgenic line. First, two or more separate transgenes can be mixed and co-microinjected. Because multiple copies of transgenes are often cointegrated at the same genomic location, there is a good chance that both transgenes are integrated in the same locus in the same transgenic line. Such transgenic lines can be stably maintained without tedious crossbreeding and genotyping. When mixing DNA constructs for microinjection, the DNA concentration for each construct should be proportionally reduced so that the total DNA concentration remains in the normal 1–2 μg/ml range. Otherwise the elevated DNA concentration may reduce the survivability of the injected embryos. Second, two transgenes, each containing its own promoter and polyadenylation sequence, can be cloned into the same transgenic vector. Third, an internal ribosomal entry site (IRES) can be used to create bi- or polycistronic mRNA, which enables the production of two or more proteins from the same mRNA. However, the efficiency of the polycistronic system still varies greatly, depending on the sources of the IRESs, the genes to be expressed, and the distance between the two genes relative to the IRES sequence, as if the cistrons interfere with each other. The rules for governing IRES efficiency are not clear, but it is generally recommended to place the more important gene in front of, and the less important gene (such as the reporter or marker gene) after, the IRES. Fourth, 2A peptides have been increasingly used as alternatives to the IRES in expressing multiple proteins from one DNA construct. The 2A sequences were first identified in picornaviruses, which encode relative short peptides (~20 amino acids) that can cause ribosomal skipping and consequently result in multiple polypetides being made from the same mRNA. They function efficiently in a wide variety of eucaryotes, ranging from yeast to human cells, including in transgenic mice [48].

4.6 Site-Specific Recombinase and Conditional Transgenic Lines

Cre and FLP transgenic mouse lines are extremely useful tools for carrying out conditional knockout and knockin studies. Cre [49] and FLP [50] are site-specific recombinases that can specifically recognize the loxP and FRT DNA sequences, respectively. FLPe is the most used variant of FLP recombinases thus far, but the codon-optimized FLPo has been shown to be more efficient at mediating homologous recombination between FRT sites [51]. When two copies of loxP or FRT sites exist in the same mouse line, the homologous recombination events mediated by these recombinases can result in deletion, inversion, and translocation, depending on the orientation and location of the two sites [52].

Cre or FLP transgenic lines can be generated following standard transgenic methods using a wide variety of ubiquitous or tissue-specific promoters. Because BAC clones are excellent for blunting chromosomal position effects and maintaining authentic gene expression patterns, they are candidates for driving recombinase gene expression. This can be done by inserting Cre into the 5′ UTR of the BAC clones. However, it must be pointed out that the commonly used BAC vectors (RPCI-23 library from the C57BL/6 J mouse strain and RPCI-22 from the 129/SvEv mouse strain) already possess loxP sites. These sites can potentially cause serious troubles when crossed into the floxed conditional knockout mouse lines. Therefore, it is strongly recommended to eliminate the loxP sites embedded in the BAC vector before using it to drive Cre gene expression.

The expression pattern of Cre and FLP transgenic lines must be systematically characterized by crossing them to commonly used recombinase reporter mouse lines, including R26R [53], Z/AP [54], and Z/EG [55] for Cre and R26:FRAP for FLPe [56]. Generally, these reporter lines contain a floxed (for Cre) or flirted (for FLP) translation STOP cassette between a ubiquitous promoter and a reporter gene. In the absence of the recombinase, the reporter protein is not expressed. However, in tissues where recombinase is expressed, the reporter gene product is produced because the STOP cassette has been deleted by the recombinase.

It is important to note that the pattern of tissue-specific promoter activity revealed by ordinary reporter gene assay, as discussed in Subheading 4.2, can only report the status of promoter activity at the moment when the tissues were harvested. However, the patterns revealed by the recombinase reporter mice are more like fate mapping. Recombinase-mediated DNA excision is irreversible, and therefore even brief expression of the recombinase can permanently delete the DNA flanked by the loxP or FRT sites not only in that particular cell but also in all of its progeny. This characteristic makes it more difficult to conduct tissue-specific knockout experiments in certain cell types, because many seemingly tissue-specific promoters in adult mice also transiently express in some other cell types during embryonic development. That is why inducible recombinase transgenic lines are often needed to achieve tissue-specific knockout (*see* Subheading 4.7).

All the adverse consequences of chromosomal position effects and epigenetic modification discussed in Subheading 2 also apply to Cre and FLP transgenic lines. Because the recombinase lines usually need to be crossbred with other conditional knockout mouse lines to conduct complicated studies, the importance of thoroughly and rigorously examining the pattern of Cre or FLP expression cannot be overemphasized [57]. Sometimes the Cre protein can be transported from one cell to another, or even from the uterine lining of the mother to an early embryo. Sometimes, males and females from the same transgenic line behave differently in terms of Cre expression [58]. The genetic background of the mouse can also apparently influence Cre expression [59]. To avoid potential problems, some researchers even characterize each individual mouse that will be used for certain conditional mutagenesis experiments [60].

Of note, germline-specific Cre transgenic lines, ZP3-Cre for oocytes and prm-Cre for spermatids, can be used to specifically knock out genes in reproductive cells. This feature can be used to bypass the sterility or lethality of heterozygous mice to generate homozygous knockout embryos [61].

Besides being used as a tool for achieving conditional gene knockouts, the Cre and loxP-STOP cassette can also be used to generate conditional transgenic mice. The loxP-STOP cassette can be inserted between the gene of interest and its promoter. The transgene can be turned on in tissues where Cre recombinase is expressed. This approach is more complicated than making conventional tissue-specific transgenic lines, but it is a useful strategy when constitutive expression of your gene causes lethality or sterility, which makes it impossible to maintain the transgenic lines. Furthermore, the same STOP cassette-containing transgenic line can be crossed to various Cre lines for achieving different expression patterns.

A similar Cre–loxP strategy has also been shown to function in RNA polymerase III-transcribed transgenes [62, 63]. LoxP-flanked stuffer DNA sequences (similar to the STOP cassette) can be inserted between the Pol III promoter and the shRNA sequence to block gene expression. Upon excision of the stuffer sequence by Cre recombinase in specific tissues, the shRNA will be activated to exert its knockdown effects only in those tissues where Cre is expressed. Alternatively, the entire shRNA transgene can also be flanked by two loxP sites. In this case, the shRNA knockdown is absent in the tissues where Cre is expressed, and is present in other tissues. These conditional systems allow tissue-specific gene knockdown using the ubiquitous H1 and U6 promoters.

All of above strategies rely on the recombinases' capability to delete DNA sequences. As in all reversible chemical reactions, the catalysts/enzymes should also be able to facilitate the reverse reaction to an equal degree, i.e., to insert DNA sequence into a specific site. However, Cre and FLP are not efficient for inserting DNA, mainly because the inserted DNA is quickly deleted by the same enzyme due to the reversible nature of the reaction. For achieving site-specific insertion, another site-specific recombinase, ϕC31, is utilized. It catalyzes homologous recombination between two nonidentical sites, attB and attP [64]. When one DNA molecule containing one site is inserted into another molecule containing the other site, recombination results in two chimeric sites which cannot be recognized by the ϕC31 recombinase, and therefore are not subsequently deleted. This feature has made it possible for targeted insertion of transgenic constructs with the attB site into specific genomic loci, where the attP docking site had been previously inserted through gene-targeting process [8].

It should be noted that new site-specific recombinase-mediated conditional systems [65, 66] as well as new variants of existing recombinases [67] or recognitions sites are continually being developed. These variations in recombinases and recognition sites enable the insertion of nonidentical targeting sites into the genome, which can dramatically facilitate the insertion of transgenes in a site-specific and direction-specific manner, through recombination-mediated cassette exchange (RMCE). In this respect, it should also be mentioned that the ZFN, Talen, as well as the recently emerged RNA-guided CRISPR genome-editing methods [68] have substantially increased our ability to insert transgenes into specific genomic loci without using any site-specific recombinases and embryonic stem (ES) cells. These latest technological developments are beginning to launch the era of site-specific transgenics, but they are clearly out of the scope of this chapter.

Site-specific recombinases can also result in DNA inversion when the two recognition sites are in reverse orientation. Livet et al. [69] took advantage of this feature and generated transgenic lines that can stochastically recombine the expression of several fluorescent proteins to label brain neurons with many different colors, which greatly facilitated the identification of individual neurons in a complicated neural network.

4.7 Inducible Transgene Expression

Over the years, various inducible gene expression systems have been developed and applied in transgenic research, including metallothionein- [70], interferon- [71], ecdysone- [72], and cytochrome P-450-inducible [73] systems, but the most advanced systems are the tetracycline-inducible systems (see review 74). These systems require the interbreeding of two transgenic mouse lines, one carrying the gene of interest under the control of a tetracycline-inducible promoter (fusion of CMV minimal promoter and seven copies of the TetO sequence termed the tetracycline response element or TRE), and the other carrying the transcriptional activator (tTA for the Tet-off system or rtTA for the Tet-on system). In the absence of tTA or rtTA, the CMV minimal promoter is not enough to drive expression of the transgene. However, when tTA or rtTA binds to the TetO sequences of the tetracycline-inducible promoter, RNA polymerase can be recruited to the CMV minimum promoter to begin transcription. Both tTA and rtTA are created by fusion of a TetR protein and the herpes simplex virus VP16 protein, but they behave oppositely in terms of their requirements for tetracycline to activate the tetracycline-inducible promoter. tTA requires the absence of tetracycline for binding to TRE, whereas rtTA requires the presence of tetracycline to be able to bind to the promoter. Therefore, when a tetracycline analog such as doxycycline is administered through food, water, or IP injection, rtTA can turn on the transgene whereas tTA can turn it off.

Both the Tet-on and Tet-off systems have been used to obtain inducible transgene expression. There are dozens of ubiquitous or tissue-specific tTA and rtTA transgenic lines available at The Jackson Laboratory. One advantage of the rtTA over the tTA system is that induction of gene expression can be achieved quickly. This is because the added doxycycline can enter the tissues in a matter of minutes or hours, whereas clearing it out of the system usually takes days. It is worth mentioning that for brain tissues, it takes a longer period of time for the added doxycycline to take effect because of the blood–brain barrier. However, the disadvantage of the rtTA system is that basal level (leaky) expression appears to be higher [75, 76].

After generating the inducible transgenic lines, it is essential to carefully characterize them before conducting any experiments. A convenient approach for examining expression pattern and inducibility is to include a reporter gene in the same transgenic construct. This can be done by using a bidirectional inducible promoter, which contains two oppositely oriented CMV minimum promoters adjacent to the shared TRE sequences [77].

The conditional system described in the previous section and the inducible systems described here can be combined to create inducible conditional systems. Schonig et al. [78] reported that a tetracycline-inducible Cre transgene line can be successfully generated, but some other studies [79] have had trouble generating efficient tetracycline-inducible Cre lines. Because the tetracycline-inducible system itself requires two mouse lines, when they are further crossed to the conditional knockout/knockin mouse line, breeding and genotyping become very complicated. Therefore, the tamoxifen-inducible system is often adopted for achieving inducible recombinase transgene expression. In this case, the Cre protein is fused to a modified estrogen receptor (ER) protein. Depending on the presence or absence of estrogen analogs, particularly tamoxifen or 4-hydroxy (OH) tamoxifen, ER can drag the recombinase in and out of cell nuclei [80]. In the absence of tamoxifen, the Cre–ER fusion proteins stay in the cytoplasm. Upon tamoxifen administration, tamoxifen binds to ER to cause a conformational change in the receptor and result in transport of the Cre–ER fusion protein into the nucleus. Because Cre needs to be located in the cell nucleus to perform its function of mediating homologous recombination between two loxP sites, its activity can be effectively turned on and off by tamoxifen. Several versions of Cre–ER fusion proteins are currently available, and CreERT2 appears to be the most sensitive variant to date. Hunter et al. [81] created the FLPeERT2 fusion gene, which can be used for inducibly removing sequences flanked by FRTs. Besides ER, the progesterone receptor (PR) can also be fused to Cre recombinase to create a Cre–PR fusion protein, which can be induced by the synthetic steroid RU486 [82].

Acknowledgments

This work was supported by the Intramural Research Program of the National Heart, Lung, and Blood Institute.

References

1. Gordon JW, Scangos GA, Plotkin DJ, Barbosa JA, Ruddle FH (1980) Genetic transformation of mouse embryos by microinjection of purified DNA. Proc Natl Acad Sci USA 77: 7380–7384
2. Heeneman S, Lutgens E, Schapira KB, Daemen MJ, Biessen EA (2008) Control of atherosclerotic plaque vulnerability: insights from transgenic mice. Front Biosci 13:6289–6313
3. de Winther MP, Hofker MH (2002) New mouse models for lipoprotein metabolism and atherosclerosis. Curr Opin Lipidol 13:191–197
4. Bock HH, Herz J, May P (2007) Conditional animal models for the study of lipid metabolism and lipid disorders. Handb Exp Pharmacol 178:407–439
5. Zadelaar S, Kleemann R, Verschuren L, de Vries-Van der Weij J, van der Hoorn J, Princen HM, Kooistra T (2007) Mouse models for atherosclerosis and pharmaceutical modifiers. Arterioscler Thromb Vasc Biol 27:1706–1721
6. Rossant J, Nutter LM, Gertsenstein M (2011) Engineering the embryo. Proc Natl Acad Sci USA 108(19):7659– 7660

7. Xiao A, Wu Y, Yang Z, Hu Y, Wang W, Zhang Y, Kong L, Gao G, Zhu Z, Lin S, Zhang B (2013) EENdb: a database and knowledge base of ZFNs and TALENs for endonuclease engineering. Nucleic Acids Res 41:D415–422
8. Tasic B, Hippenmeyer S, Wang C, Gamboa M, Zong H, Chen-Tsai Y, Luo L (2011) Site-specific integrase-mediated transgenesis in mice via pronuclear injection. Proc Natl Acad Sci USA 108(19):7902– 7907
9. Montoliu L, Chavez S, Vidal M (2000) Variegation associated with lacZ in transgenic animals: a warning note. Transgenic Res 9:237–239
10. Palmiter RD, Sandgren EP, Avarbock MR, Allen DD, Brinster RL (1991) Heterologous introns can enhance expression of transgenes in mice. Proc Natl Acad Sci U S A 88: 478–482
11. Qiu L, Wang H, Xia X, Zhou H, Xu Z (2008) A construct with fluorescent indicators for conditional expression of miRNA. BMC Biotechnol 8:77
12. Garrick D, Fiering S, Martin DI, Whitelaw E (1998) Repeat-induced gene silencing in mammals. Nat Genet 18:56–59
13. Lau S, Jardine K, McBurney MW (1999) DNA methylation pattern of a tandemly repeated LacZ transgene indicates that most copies are silent. Dev Dyn 215:126–138
14. Kwaks TH, Otte AP (2006) Employing epigenetics to augment the expression of therapeutic proteins in mammalian cells. Trends Biotechnol 24:137–142
15. Lois C (2006) Generation of transgenic animals using lentiviral vectors. In: Pease S, Lois C (eds) Mammalian and avian transgenesis—new approaches. Springer, Heidelberg, Germany, pp 1–22
16. Li Q, Harju S, Peterson KR (1999) Locus control regions: coming of age at a decade plus. Trends Genet 15:403–408
17. Harraghy N, Gaussin A, Mermod N (2008) Sustained transgene expression using MAR elements. Curr Gene Ther 8:353–366
18. Gaszner M, Felsenfeld G (2006) Insulators: exploiting transcriptional and epigenetic mechanisms. Nat Rev Genet 7:703–713
19. Giraldo P, Rival-Gervier S, Houdebine LM, Montoliu L (2003) The potential benefits of insulators on heterologous constructs in transgenic animals. Transgenic Res 12:751–755
20. Recillas-Targa F, Valadez-Graham V, Farrell CM (2004) Prospects and implications of using chromatin insulators in gene therapy and transgenesis. Bioessays 26:796–807
21. Pawlik KM, Sun CW, Higgins NP, Townes TM (1995) End joining of genomic DNA and transgene DNA in fertilized mouse eggs. Gene 165:173–181
22. Sambrook J, Fritsch EF, Maniatis T (1989) Molecular cloning—a laboratory manual, 2nd edn. Cold Spring Harbor Laboratory Press, Cold Spring Harbor, NY
23. Gong S, Zheng C, Doughty ML, Losos K, Didkovsky N, Schambra UB, Nowak NJ, Joyner A, Leblanc G, Hatten ME, Heintz N (2003) A gene expression atlas of the central nervous system based on bacterial artificial chromosomes. Nature 425:917–925
24. Yang XW, Model P, Heintz N (1997) Homologous recombination based modification in Escherichia coli and germline transmission in transgenic mice of a bacterial artificial chromosome. Nat Biotechnol 15:859–865
25. Copeland NG, Jenkins NA, Court DL (2001) Recombineering: a powerful new tool for mouse functional genomics. Nat Rev Genet 2:769–779
26. Heintz N (2001) BAC to the future: the use of bac transgenic mice for neuroscience research. Nat Rev Neurosci 2:861–870
27. Van Keuren ML, Gavrilina GB, Filipiak WE, Zeidler MG, Saunders TL (2009) Generating transgenic mice from bacterial artificial chromosomes: transgenesis efficiency, integration and expression outcomes. Transgenic Res 18:769–785
28. Schedl A, Larin Z, Montoliu L, Thies E, Kelsey G, Lehrach H, Schutz G (1993) A method for the generation of YAC transgenic mice by pronuclear microinjection. Nucleic Acids Res 21:4783–4787
29. Henning KA, Novotny EA, Compton ST, Guan XY, Liu PP, Ashlock MA (1999) Human artificial chromosomes generated by modification of a yeast artificial chromosome containing both human alpha satellite and single-copy DNA sequences. Proc Natl Acad Sci U S A 96:592–597
30. Barash I, Faerman A, Richenstein M, Kari R, Damary GM, Shani M, Bissell MJ (1999) In vivo and in vitro expression of human serum albumin genomic sequences in mammary epithelial cells with beta-lactoglobulin and whey acidic protein promoters. Mol Reprod Dev 52:241–252
31. Zhan Y, Brady JL, Johnston AM, Lew AM (2000) Predominant transgene expression in exocrine pancreas directed by the CMV promoter. DNA Cell Biol 19:639–645
32. Schlaeger TM, Bartunkova S, Lawitts JA, Teichmann G, Risau W, Deutsch U, Sato TN (1997) Uniform vascular-endothelial-cell-

specific gene expression in both embryonic and adult transgenic mice. Proc Natl Acad Sci U S A 94:3058–3063

33. Echelard Y, Meade H (2002) Toward a new cash cow. Nat Biotechnol 20:881–882
34. Keefer CL (2004) Production of bioproducts through the use of transgenic animal models. Anim Reprod Sci 82–83:5–12
35. Niemann H, Kues WA (2007) Transgenic farm animals: an update. Reprod Fertil Dev 19:762–770
36. Litscher ES, Liu C, Echelard Y, Wassarman PM (1999) Zona pellucida glycoprotein mZP3 produced in milk of transgenic mice is active as a sperm receptor, but can be lethal to newborns. Transgenic Res 8:361–369
37. Saunders TL (2003) Reporter molecules in genetically engineered mice. Methods Mol Biol 209:125–143
38. Huang WY, Aramburu J, Douglas PS, Izumo S (2000) Transgenic expression of green fluorescence protein can cause dilated cardiomyopathy. Nat Med 6:482–483
39. Liu HS, Jan MS, Chou CK, Chen PH, Ke NJ (1999) Is green fluorescent protein toxic to the living cells? Biochem Biophys Res Commun 260:712–717
40. Heintz N (2004) Gene expression nervous system atlas (GENSAT). Nat Neurosci 7:483
41. Peng S, York JP, Zhang P (2006) A transgenic approach for RNA interference-based genetic screening in mice. Proc Natl Acad Sci U S A 103:2252–2256
42. Kuhn R, Streif S, Wurst W (2007) RNA interference in mice. Handb Exp Pharmacol 178:149–176
43. Premsrirut PK, Dow LE, Kim SY, Camiolo M, Malone CD, Miething C, Scuoppo C, Zuber J, Dickins RA, Kogan SC, Shroyer KR, Sordella R, Hannon GJ, Lowe SW (2011) A rapid and scalable system for studying gene function in mice using conditional RNA interference. Cell 145(1):145–158
44. Shinagawa T, Ishii S (2003) Generation of Ski-knockdown mice by expressing a long double-strand RNA from an RNA polymerase II promoter. Genes Dev 17:1340–1345
45. Stein P, Svoboda P, Schultz RM (2003) Transgenic RNAi in mouse oocytes: a simple and fast approach to study gene function. Dev Biol 256:187–193
46. Fedoriw AM, Stein P, Svoboda P, Schultz RM, Bartolomei MS (2004) Transgenic RNAi reveals essential function for CTCF in H19 gene imprinting. Science 303:238–240
47. Szulc J, Wiznerowicz M, Sauvain MO, Trono D, Aebischer P (2006) A versatile tool for conditional gene expression and knockdown. Nat Methods 3:109–116
48. Trichas G, Begbie J, Srinivas S (2008) Use of the viral 2A peptide for bicistronic expression in transgenic mice. BMC Biol 6:40
49. Gu H, Zou YR, Rajewsky K (1993) Independent control of immunoglobulin switch recombination at individual switch regions evidenced through Cre-loxP-mediated gene targeting. Cell 73:1155–1164
50. Dymecki SM (1996) Flp recombinase promotes site-specific DNA recombination in embryonic stem cells and transgenic mice. Proc Natl Acad Sci U S A 93:6191–6196
51. Raymond CS, Soriano P (2007) High-efficiency FLP and PhiC31 site-specific recombination in mammalian cells. PLoS One 2:e162
52. Nagy A (2000) Cre recombinase: the universal reagent for genome tailoring. Genesis 26:99–109
53. Soriano P (1999) Generalized lacZ expression with the ROSA26 Cre reporter strain. Nat Genet 21:70–71
54. Lobe CG, Koop KE, Kreppner W, Lomeli H, Gertsenstein M, Nagy A (1999) Z/AP, a double reporter for cre-mediated recombination. Dev Biol 208:281–292
55. Novak A, Guo C, Yang W, Nagy A, Lobe CG (2000) Z/EG, a double reporter mouse line that expresses enhanced green fluorescent protein upon Cre-mediated excision. Genesis 28:147–155
56. Awatramani R, Soriano P, Mai JJ, Dymecki S (2001) An Flp indicator mouse expressing alkaline phosphatase from the ROSA26 locus. Nat Genet 29:257–259
57. Wamhoff BR, Sinha S, Owens, GK (2007) Conditional mouse models to study developmental and pathophysiological gene function in muscle, Handb Exp Pharmacol 178: 441–468
58. Mishina Y, Hanks MC, Miura S, Tallquist MD, Behringer RR (2002) Generation of Bmpr/Alk3 conditional knockout mice. Genesis 32:69–72
59. Hebert JM, McConnell SK (2000) Targeting of cre to the Foxg1 (BF-1) locus mediates loxP recombination in the telencephalon and other developing head structures. Dev Biol 222:296–306
60. Murali D, Yoshikawa S, Corrigan RR, Plas DJ, Crair MC, Oliver G, Lyons KM, Mishina Y, Furuta Y (2005) Distinct developmental programs require different levels of Bmp signaling during mouse retinal development. Development 132:913–923

61. Akiyama H, Chaboissier MC, Martin JF, Schedl A, de Crombrugghe B (2002) The transcription factor Sox9 has essential roles in successive steps of the chondrocyte differentiation pathway and is required for expression of Sox5 and Sox6. Genes Dev 16:2813–2828
62. Kasim V, Miyagishi M, Taira K (2003) Control of siRNA expression utilizing Cre-loxP recombination system. Nucleic Acids Res Suppl 3:255–256
63. Fritsch L, Martinez LA, Sekhri R, Naguibneva I, Gerard M, Vandromme M, Schaeffer L, Harel-Bellan A (2004) Conditional gene knock-down by CRE-dependent short interfering RNAs. EMBO Rep 5:178–182
64. Belteki G, Gertsenstein M, Ow DW, Nagy A (2003) Site-specific cassette exchange and germline transmission with mouse ES cells expressing phiC31 integrase. Nat Biotechnol 21:321–324
65. Anastassiadis K, Fu J, Patsch C, Hu S, Weidlich S, Duerschke K, Buchholz F, Edenhofer F, Stewart AF (2009) Dre recombinase, like Cre, is a highly efficient site-specific recombinase in E. coli, mammalian cells and mice. Dis Model Mech 2(9–10):508–515
66. Nern A, Pfeiffer BD, Svoboda K, Rubin GM (2011) Multiple new site-specific recombinases for use in manipulating animal genomes. Proc Natl Acad Sci USA 108(34):14198– 14203
67. Suzuki E, Nakayama M (2011)VCre/VloxP and SCre/SloxP: new site-specific recombination systems for genome engineering. Nucleic Acids Res 39(8):e49
68. Chang N, Sun C, Gao L, Zhu D, Xu X, Zhu X, Xiong J, Xi JJ (2013) Genome editing with RNA-guided Cas9 nuclease in Zebrafish embryos. Cell Res 23(4):465–472
69. Livet J, Weissman TA, Kang H, Draft RW, Lu J, Bennis RA, Sanes JR, Lichtman JW (2007) Transgenic strategies for combinatorial expression of fluorescent proteins in the nervous system. Nature 450:56–62
70. Palmiter RD, Norstedt G, Gelinas RE, Hammer RE, Brinster RL (1983) Metallothionein-human GH fusion genes stimulate growth of mice. Science 222:809–814
71. Kuhn R, Schwenk F, Aguet M, Rajewsky K (1995) Inducible gene targeting in mice. Science 269:1427–1429
72. Albanese C, Reutens AT, Bouzahzah B, Fu M, D'Amico M, Link T, Nicholson R, Depinho RA, Pestell RG (2000) Sustained mammary gland-directed, ponasterone A-inducible expression in transgenic mice. FASEB J 14:877–884
73. Jones SN, Jones PG, Ibarguen H, Caskey CT, Craigen WJ (1991) Induction of the Cyp1a-1 dioxin-responsive enhancer in transgenic mice. Nucleic Acids Res 19:6547–6551
74. Sprengel R, Hasan MT (2007) Tetracycline-controlled genetic switches. Handb Exp Pharmacol 49–72
75. Keyvani K, Baur I, Paulus W (1999) Tetracycline-controlled expression but not toxicity of an attenuated diphtheria toxin mutant. Life Sci 64:1719–1724
76. Imhof MO, Chatellard P, Mermod N (2000) A regulatory network for the efficient control of transgene expression. J Gene Med 2:107–116
77. Baron U, Freundlieb S, Gossen M, Bujard H (1995) Co-regulation of two gene activities by tetracycline via a bidirectional promoter. Nucleic Acids Res 23:3605–3606
78. Schonig K, Schwenk F, Rajewsky K, Bujard H (2002) Stringent doxycycline dependent control of CRE recombinase in vivo. Nucleic Acids Res 30:e134
79. Leneuve P, Colnot S, Hamard G, Francis F, Niwa-Kawakita M, Giovannini M, Holzenberger M (2003) Cre-mediated germline mosaicism: a new transgenic mouse for the selective removal of residual markers from tri-lox conditional alleles. Nucleic Acids Res 31:e21
80. Feil R, Brocard J, Mascrez B, LeMeur M, Metzger D, Chambon P (1996) Ligand-activated site-specific recombination in mice. Proc Natl Acad Sci U S A 93:10887–10890
81. Hunter NL, Awatramani RB, Farley FW, Dymecki SM (2005) Ligand-activated Flpe for temporally regulated gene modifications. Genesis 41:99–109
82. Wunderlich FT, Wildner H, Rajewsky K, Edenhofer F (2001) New variants of inducible Cre recombinase: a novel mutant of Cre-PR fusion protein exhibits enhanced sensitivity and an expanded range of inducibility. Nucleic Acids Res 29:E47

Chapter 9

Purification of Plasmid and BAC Transgenic DNA Constructs

Chengyu Liu, Yubin Du, Wen Xie, and Changyun Gui

Abstract

Pronuclear microinjection is the most used method for generating transgenic mice. The quality of DNA to be microinjected is a key determinant of the success rate of this method. DNA purity is a critical factor because trace amounts of many substances, when microinjected into the pronucleus of the fertilized egg, can kill or prevent the further development of the embryo. Avoiding all contaminants is not a trivial issue, because most transgenic fragments need to be purified from agarose gels. Small particles and viscous materials in the DNA solution can also dramatically reduce the efficiency of microinjection because they tend to clog the injection needles. DNA shearing or breakage during purification and microinjection is also a potential problem, particularly when linearized bacterial artificial chromosomes (BAC) DNAs are used. The overall quantity and the final DNA concentration are also important considerations, because egg pronuclei are very sensitive to the amount of foreign DNA. In this chapter, we first discuss the general guidelines and cautions for preparing microinjection-quality DNA, and then describe in detail two protocols, one for gel purification of transgenic fragments from plasmid vectors and the other for isolating high-quality BAC DNA from bacteria.

Key words Transgenic, BAC, Plasmid, DNA purification, Microinjection

1 Introduction

Transgenic and knockout mouse models have been extensively used in lipoprotein and atherosclerosis research [1–3], just as in many other fields of biomedical research. A great majority of these transgenic lines were generated using the so-called pronuclear microinjection method. Presently, the most commonly used vectors for cloning transgenes are plasmids and bacterial artificial chromosomes (BACs). Plasmids are versatile and convenient for precise DNA recombination, and most biomedical scientists are already familiar with the use of plasmids. The main disadvantage of plasmids is that they can accept only relatively small DNA fragments. Although there is not a specific size limit for plasmid vectors, cloning efficiency decreases dramatically when the DNA insert is larger than 20–30 kb. Small transgenes are more prone to

Lita A. Freeman (ed.), *Lipoproteins and Cardiovascular Disease: Methods and Protocols*, Methods in Molecular Biology, vol. 1027, DOI 10.1007/978-1-60327-369-5_9, © Springer Science+Business Media, LLC 2013

chromosomal position effects, which often lead to uncontrolled transgene expression.

In recent years, BACs have gained in popularity in making transgenic mice for the following reasons. First, the Genome Sequencing Project has made it very easy to obtain BAC clones for almost any mouse or human gene from repositories such as the BACPAC Resource Center at Oakland Children's Hospital (http://bacpac.chori.org/). Second, the large DNA fragment contained in the BAC (100–300 kb) is more resistant to chromosomal position effects and is more likely to contain all the cis genomic elements required for conferring authentic gene expression patterns. Third, the recently developed recombineering method [4–6] has enabled precise modification of BAC DNA molecules.

No matter which type of vector is used, the purity and integrity of the transgene are critical factors in determining the efficiency of microinjection and transgenic mouse production. Both BACs and transgenic inserts from plasmid vectors can be purified using essentially the same methods as those used in conventional molecular biology experiments, but the requirements for purity, integrity, and quantity are much higher for transgenic applications. Extreme caution must be exercised throughout the entire purification procedures, so that the highest quality of DNA is obtained to maximize the efficiency of the laborious and expensive micromanipulation and surgical procedures involved in transgenic mouse production.

2 Materials

2.1 Commercial Kits

1. QIAquick Gel Extraction Spin Column Kit (Qiagen Inc., Valencia, CA, Cat: 29706), including the following components:
   (a) QIAquick spin columns.
   (b) Buffer QG.
   (c) Buffer PE. Before use, add the appropriate volume of ethanol to the PE buffer as instructed on the bottle label.
   (d) Collection tubes (2 ml).
2. Qiagen Large-Construct Kit (Qiagen, Cat: 12462), including the following components:
   (a) QIAGEN-tip 500.
   (b) Buffer P1. Before use, add one vial of supplied RNase A to each bottle of Buffer P1 to give a final concentration of 100 μg/ml. Store Buffer P1 containing RNase at 4 °C.
   (c) Buffer P2. Under low temperature storage conditions, SDS may precipitate. If this occurs, heat the P2 buffer in a 37 °C water bath to redissolve the SDS.

(d) Buffer P3. Prechill Buffer P3 on ice before using and store at 4 °C.

(e) Buffer QBT.

(f) Buffer QC.

(g) Buffer QF. Prewarm Buffer QF to 65 °C before eluting DNA to increase DNA recovery.

(h) ATP-dependent exonuclease. Completely resuspend each vial of exonuclease (80 μg) in 225 μl of supplied exonuclease solvent before use.

(i) Buffer EX.

(j) Buffer QS.

(k) Folded paper filters.

(l) QIAGEN-tip 500 holders.

2.2 Equipment and Supplies

1. Beckman Allegra 64R centrifuge (Beckman Coulter, Inc., Fullerton, CA), Beckman Conical C0650 rotor, and Beckman F0650 rotor.
2. Beckman Microfuge 18 microcentrifuge.
3. PowerPac 1000 electrophoresis power supply (Bio-Rad Laboratories, Hercules, CA).
4. Gel boxes for electrophoresis.
5. UV spectrophotometer (Bio-Rad, SmartSpec 3000) with 100 μl cuvette.
6. Gel documentation system (Bio-Rad, GelDoc 2000).
7. Water bath.
8. Shaker incubator for bacteria culture.
9. 2-l flask.
10. 50-ml disposable conical tubes.
11. 15-ml disposable conical tubes.
12. 50-ml round bottom superspeed centrifuge tubes (Sorvall, Newtown, CT, Cat: 03530).
13. 5-ml Falcon culture tube (Becton Dickinson, Franklin Lakes, NJ, Cat: 5-2058-0).
14. 1.5-ml microcentrifuge tubes.

2.3 Chemicals

1. Hydro PicoPure-2 UV Plus distilled water (Hydro, Durham, NC).
2. UltraPure agarose (Invitrogen, Carlsbad, CA, Cat: 15510-027).
3. 10× TAE (400 mM Tris acetate, 2 mM EDTA) (Advanced Biotechnology Inc., Columbia, MD).

4. DNA gel loading buffer (10× BlueJuice, Invitrogen, Cat: 10816015).
5. Ethidium bromide (10 mg/ml solution, Invitrogen, Cat: 15585011).
6. 3 M sodium acetate (pH 5.2) (Quality Biologicals Inc., Gaithersburg, MD, Cat: 351-035-060).
7. Isopropanol.
8. TE buffer (10 mM Tris–HCl, pH 7.4, 1 mM EDTA; Quality Biologicals Inc., Cat: 351-011-131).
9. Phenol:Chloroform:Isoamyl Alcohol (25:24:1 v/v), saturated with 0.1 M Tris–HCl buffer (pH 8.0) (Invitrogen, Cat: 15593031). Store at 2–8 °C.
10. Chloroform (Sigma, St. Louis, MO, Cat: C2432).
11. Ethanol.
12. Microinjection buffer (5 mM Tris–HCl, pH 7.4, 0.2 mM EDTA).
13. LB bacterial culture medium (10 g tryptone, 5 g yeast extract, 10 g NaCl per liter).
14. Spermidine (Sigma, Cat: S2501).
15. Spermine (Sigma, Cat: S1141).
16. ATP (Sigma, Cat: A3377). Dissolve 2.75 g ATP (dehydrated disodium salt) in 50 ml distilled water. Adjust the pH to 7.5 and store at −20 °C.
17. BAC microinjection buffer (5 mM Tris–HCl, pH 7.4; 0.2 mM EDTA, 100 mM NaCl, 30 μM spermine, and 70 μM spermidine).

3 Methods

3.1 General Guidelines and Cautions for Purifying Microinjection-Quality DNA

3.1.1 DNA Integrity

DNA samples stored at 4 °C for extended periods of time or those that have been frozen and thawed multiple times may contain extensive nicks and breakages. Therefore, it is important to use freshly prepared plasmids or BACs, or ones that have been stored in alcohol in precipitated form. For BACs, DNA shearing is a potential problem. Avoid vigorous shaking and repetitive pipetting. Wide-bore pipette tips should be used. When commercial kits are used for BAC preparation, make sure the kits are specifically designed for processing large DNA molecules. Inclusion of salt and polyamines in the solution may help make the long DNA strands more compact and hence reduce shearing. When microinjecting linearized BACs, slightly larger needle openings and lower injection pressures are preferable for minimizing shearing when the DNA passes through the needle tip.

3.1.2 DNA Purity

When microinjected into the pronuclei of fertilized eggs, trace amount of many substances, including agarose, phenol, chloroform, alcohol, and detergents, can kill the embryos or inhibit their further development. Always use the highest grade of water and chemical reagents, and avoid carrying any residual unintended materials into the final DNA solution. Prewash glass beads and columns with elution buffer before loading the DNA for binding. Perform extra extraction and washing steps and use a longer drying time to evaporate residual alcohol.

3.1.3 Particles

The tip of a microinjection needle is usually ~1 μm in diameter, and hence it is fairly easy to become clogged by tiny particles such as dust, agarose particles, powder from gloves, and microorganisms. Special attention must be paid to avoid including any particles in the final DNA solution. All solutions used for dissolving DNA should be filtered through a 0.22 μm filter. All test tubes and pipette tips should be rinsed with filtered water. Avoid wearing powered gloves during the last several steps of transgenic DNA purification. Immediately before microinjection, the DNA solution should be centrifuged at a high speed to pellet any particles.

3.1.4 DNA Quantity and Concentration

Although microinjection needs only a couple of micrograms of DNA, this amount is still at least one order of magnitude more than the amount required for most molecular biological experiments, such as DNA cloning. For BAC clones, DNA quantity is usually not a problem, because no gel purification is involved. However, for transgenes cloned in plasmids, purifying a substantial amount of the transgenic insert is often a challenge for beginners, especially when the extra precautions have to be taken to avoid any contamination. It is recommended to start with 20–60 μg of plasmid DNA, depending on the size of the transgenic insert relative to the cloning vector. When separating this large amount of DNA on an agarose gel, it is important to avoid overloading by using a sufficiently wide loading well in a large-size gel.

The optimal DNA concentration for pronuclear microinjection is 1–2 μg/ml, but a higher DNA concentration is preferable for storage. For measuring DNA concentration, in addition to UV absorbance, an aliquot of the final DNA solution should be run on an agarose gel in parallel with a known concentration of DNA marker to confirm the concentration and integrity of the transgene DNA.

3.2 Isolation of the Transgenic Fragment from the Plasmid Vector

Plasmid preparation (Maxi-prep) for a transgenic project is the same as a plasmid preparation for most molecular biology experiments. For most plasmid extraction kits, simply following the vendor-supplied protocols should suffice. Some researchers suggest using an endotoxin-free kit to prepare plasmid DNA, but this is probably not necessary because the many subsequent steps, particularly the electrophoresis and gel extraction procedures, make the endotoxin status of the initial DNA samples irrelevant. The plasmid backbone needs to be removed from the transgenic insert before microinjection, because the prokaryote-derived cloning vector sequence often interferes with transgene expression after integrating into the mouse genome.

Sucrose gradient centrifugation is an old-fashioned and very reliable method for separating the transgene from the cloning vector [7]. It is still used by some transgenic laboratories for preparing microinjection-quality DNA, but most laboratories currently use agarose gel electrophoresis to separate the DNA pieces. Most of the methods used in routine molecular biology experiments can be adapted for recovering transgenes from agarose gel slices, including electroelution, absorption to glass beads, DNA affinity column chromatography, and low-melting-point agarose gel electrophoresis followed by GELase (Epicentre Biotechnologies, Madison, WI) digestion. A convenient method using the QIAquick Gel Extraction Spin Column Kit is described in detail below. This protocol is adapted from the QIAquick® Spin Handbook (Qiagen, March 2008).

3.2.1 Procedure

1. Digest 20–60 μg of freshly prepared plasmid DNA using appropriate restriction enzymes. The exact amount of plasmid varies depending on the size of the insert relative to the vector backbone. Ideally, 5–10 μg of transgene insert should be released from the vector. We usually digest DNA for 1–3 h only. Extended incubation at 37 °C can increase DNA nicking. Completion of digestion can be checked by running an aliquot of the digestion mixture (0.2 μg DNA) on an agarose gel (*see* **Note 1**).
2. Load the digested DNA sample into a wide slot (formed by taping multiple teeth of the comb together) in an agarose gel. Run the gel at a relatively low voltage of 2–4 V/cm. Low voltages can increase the gel's resolution.
3. Stain the gel for 15 min with 0.5 μg/ml ethidium bromide in distilled water. Under long-wavelength UV light, cut out the gel slice containing the transgenic insert (*see* **Note 2**).
4. Weigh the gel slice in a 15-ml conical tube. Add 3 volumes of Buffer QG from the QIAquick Gel Extraction Spin Column Kit to 1 volume of gel (3 ml per gram of gel). For concentrated agarose gels (>1.5 %), double the volume of Buffer QG (6 ml per gram of gel).
5. Incubate the tube in a 50 °C water bath and mix by inverting the tube every 2 min. When the gel is completely solubilized, the color of the solution should be yellow. If the color of the solution is orange or violet, slowly add drops of 3 M sodium acetate (pH 5.2) into the tube until the color turns to yellow. The yellow color indicates that the pH of the solution is below 7.5, which is required for efficient DNA binding in the following steps.

6. If the DNA fragment is >4 kb or <0.5 kb, add 1 gel volume of isopropanol to the sample and mix.
7. Determine the number of columns needed, assuming that each QIAquick spin column can process 400 mg of agarose gel. Place each spin column into a manufacturer-provided 2-ml collection tube and put the assembly into a microcentrifuge.
8. Prewash the columns by adding 0.8 ml of TE (pH 7.4) into each QIAquick column, and centrifuge at 13,000 × *g* in a microcentrifuge for 1 min. This prewash step will wash out any contaminants that cannot be washed out by the sample loading buffer (QG) and washing buffer (PE), because TE's ionic strength is much less than that of these buffers.
9. Discard the flow-through and place the washed QIAquick columns back into the same collection tubes. Load 0.8 ml of DNA mixture into each QIAquick column, and centrifuge for 1 min at 13,000 × *g*. Discard the flow-through and place the QIAquick columns back into the same collection tube. More DNA mixture can be loaded into the same spin column until the loading capacity of 400 mg gel per column is reached.
10. Add 0.8 ml of Buffer QG to each QIAquick column and centrifuge at 13,000 × *g* for 1 min to wash out residual agarose.
11. Add 0.8 ml of Buffer PE to each QIAquick column and leave the wash buffer in the column for 2–3 min, and then centrifuge at 13,000 × *g* for 1 min. Discard the flow-through and repeat the PE wash step.
12. Centrifuge the QIAquick column for an additional 1 min at 13,000 × *g* to remove residual PE buffer.
13. Place the QIAquick column into a clean 1.5-ml microcentrifuge tube, and add 50 μl of TE buffer (pH 7.4) to the center of the QIAquick membrane. Wait 2 min and then centrifuge the column for 1 min. Add another 30 μl of TE buffer to the center of the column membrane. Wait for another 2 min and then centrifuge to recover the second wash in the same microcentrifuge tube.
14. Combine the eluted DNA solutions from all QIAquick spin columns. Add a tenth volume of 3 M sodium acetate (pH 5.2).
15. Extract the combined DNA solution twice with an equal volume of equilibrated phenol/chloroform/isoamyl alcohol, and then once with chloroform alone.
16. After the final extraction, carefully transfer the aqueous phase into a rinsed microcentrifuge tube using a rinsed pipette tip. Both the tube and tip are rinsed with filtered (0.22 μm) TE buffer to eliminate any particles. In all subsequent steps, tips and tubes should be rinsed with filtered TE. All buffers or

water used to dissolve or dilute DNA should be filtered through 0.22 μm filters.

17. Precipitate the DNA by adding 2 volumes of ethanol, and then store at −20 °C for at least 30 min.
18. Recover the DNA by centrifuging at 13,000 × *g* for 10 min at room temperature. At this stage, the DNA pellet should be a small, slightly yellowish (*see* **Note 3**) translucent mass at the bottom of the tube.
19. Discard the supernatant. Wash the DNA pellet with 1 ml of 70 % ethanol. Centrifuge at 13,000 × *g* for 3 min. Discard the supernatant, and then briefly spin all the liquid to the bottom of the tube. Carefully remove as much liquid as possible using a pipette tip.
20. Air-dry the DNA pellet for 20 min, and then redissolve in 50 μl of TE buffer (pH 7.4).
21. Remove a small aliquot for DNA quantification by UV spectrophotometry. Ideally, the OD_{260}/OD_{280} should be 1.8.
22. Confirm the DNA concentration, purity, and integrity by running another small aliquot of the DNA on an agarose gel in parallel with a known concentration marker.
23. Store the DNA solution at 4 °C. Before microinjection, remove an aliquot and dilute to 1.5 μg/ml using microinjection buffer (5 mM Tris–HCl, pH 7.4, 0.2 mM EDTA). Immediately before loading the needles, the diluted DNA should be centrifuged at 13,000 × *g* for 5 min to pellet any particles.

3.3 Purification of BAC DNA for Microinjection

One controversy about microinjecting BAC DNA is whether the BAC should be linearized before microinjection. Early studies on plasmid DNA have clearly demonstrated that linear DNA molecules are more efficient for producing positive founders. Saunders and colleagues reported that linear and circular BAC DNAs are equally efficient in giving rise to transgenic mice [8], but in our hands, linear BAC appears to be more efficient (unpublished observation). We normally do not eliminate the BAC vector sequence from the transgene to be microinjected because of the difficulty of separating and purifying large BAC DNA fragments from gels and the fact that the influence of the prokaryote-derived vector sequence is often diminished by the enormous piece of mammalian DNA contained in the BAC.

BACs can be extracted from bacteria using the same alkaline lysis method that is often used for plasmid extraction. However, precautions to avoid DNA shearing should be practiced throughout the purification and microinjection procedures. Unlike plasmid-derived transgenes, BACs will not be further gel-purified. Therefore, greater caution is required for minimizing inclusion of impurities and particles in the DNA solution when extracting BACs from bacteria.

After supercoiled BACs are extracted by alkaline lysis, an additional purification step is required. Cesium chloride gradient centrifugation followed by dialysis is a reliable method for purifying a large amount of supercoiled DNA [9]. However, for researchers who do not have the necessary ultracentrifuge and rotor or who prefer columns and kits, Sepharose 4B-CL chromatography [10] and the Nucleobond column method (see University of Michigan Transgenic Animal Model Core website http://www.med.umich.edu/tamc/BACDNA.html for details) are also excellent. Below we describe a protocol using QIAGEN's Large-Construct Kit for BAC DNA purification. The protocol is adapted from the QIAGEN® Large-Construct Kit Handbook (Qiagen, May 2002).

3.3.1 Procedure

1. Inoculate a single *E. coli* colony containing the desired BAC into a tube containing 2 ml LB medium with the appropriate antibiotic selection. Incubate with vigorous shaking (~300 rpm) at 37 °C for ~8 h.
2. Transfer the entire 2-ml starter culture into a 2-l flask containing 500 ml LB medium with antibiotic selection. Grow with vigorous shaking (~300 rpm) at 37 °C for another 12–16 h.
3. Divide the culture into six disposable 50-ml conical tubes. Centrifuge the tubes in a Beckman Conical C0650 rotor (*see* **Note 4**) at 6,000 × *g* for 10 min. Decant the supernatant and drain out as much medium as possible by inverting the tube on paper towels for several minutes.
4. Add 10 ml of P1 buffer (without RNase A) from the Qiagen Large-Construct Kit into each tube. Vortex to resuspend the bacteria. Combine the bacterial suspensions from three tubes into one tube, so that only two tubes are left, each containing 30 ml bacterial suspension in P1 buffer. This step washes out the compounds present in the culture medium, which maybe toxic to embryos when microinjected.
5. Centrifuge the two tubes at 6,000 × *g* for 10 min. Discard the supernatant and completely resuspend the bacterial pellet in 10 ml Buffer P1 containing RNase A.
6. Slowly add 10 ml Buffer P2, and mix gently by inverting the tubes 5–10 times (*see* **Note 5**). Incubate the tubes at room temperature for 5 min. The lysate should appear viscous and free of clumps of bacteria.
7. Add 10 ml ice-cold Buffer P3 and mix gently by inverting the tube repeatedly until a fluffy white material forms and the lysate is no longer viscous.
8. Transfer the entire 30 ml contents from each conical tube into a 50-ml Sorvall superspeed centrifuge tube (*see* **Note 6**).
9. Centrifuge at 20,000 × *g* for 30 min at 4 °C. Filter the supernatant through a pre-wetted folded filter (supplied in the kit) to

eliminate the white material floating at the top or the solid precipitate at the bottom of the tube.

10. Collect the filtered lysate into a new Sorvall centrifuge tube. Add 0.6 volumes of room-temperature isopropanol. Mix and centrifuge immediately at 20,000 × *g* for 15 min at 4 °C.
11. Discard the supernatant, wash the DNA pellet with 10 ml room-temperature 70 % ethanol, and centrifuge at 20,000 × *g* for 5 min.
12. Discard the supernatant. Briefly centrifuge the tube to spin down all liquid. Aspirate as much liquid as possible without disturbing the DNA pellet.
13. Allow the DNA pellet to air-dry for 20 min.
14. Redissolve DNA pellet in 5 ml Buffer EX. Extended incubation at 4 °C and gentle shaking help dissolve the DNA, but avoid repeated pipetting and vortexing.
15. Combine the DNA solutions from the two tubes. Add 200 μl ATP-dependent exonuclease and 300 μl ATP solution. Incubate at 37 °C for about 1 h to digest away *E. coli* genomic DNA and nicked BAC DNA.
16. Add 10 ml Buffer QBT to wet the QIAGEN-tip 500. Then add 15 ml of TE (pH 7.4) buffer to prewash the column (*see* **Note 7**).
17. Add 5 ml Buffer QBT to re-equilibrate the QIAGEN-tip 500.
18. Dilute the DNA sample from **step 15** with 10 ml Buffer QS, and then apply the whole sample to the equilibrated QIAGEN-tip 500.
19. Wash the QIAGEN-tip twice, each with 30 ml Buffer QC.
20. Elute the BAC DNA with 15 ml pre-warmed (65 °C) Buffer QF into a new 50-ml tube.
21. Add 10.5 ml of room-temperature isopropanol to precipitate the DNA and then centrifuge at 20,000 × *g* for 30 min at 4 °C.
22. Discard the supernatant. Briefly centrifuge the tube to spin down any liquid. Aspirate as much liquid as possible without disturbing the DNA pellet.
23. Allow the DNA pellet to air-dry for 20 min.
24. Redissolve the DNA in 500 μl of TE buffer. If the DNA does not readily dissolve, the tube can be left at 4 °C overnight.
25. Add a tenth volume of 3 M sodium acetate (pH 5.2) to the solution containing the dissolved DNA. Extract twice with equilibrated phenol/chloroform/isoamyl alcohol, and once with chloroform alone. Remember to use wide-bore pipette tips during this and all subsequent steps when transferring BAC DNA solution.

26. After the final extraction, carefully transfer the aqueous phase into a rinsed microcentrifuge tube using a rinsed pipette tip. Both the tube and tip are rinsed with filtered (0.22 μm) TE buffer to eliminate any particles. In all subsequent steps, tips and tubes should be rinsed with filtered TE, and the buffers or water used to dissolve or dilute the DNA should also be filtered through 0.22 μm filters.
27. Precipitate the DNA by adding 2 volumes of ethanol. Recover the DNA by centrifuging for 10 min at 13,000 × *g* in a microcentrifuge.
28. Discard the supernatant. Wash the DNA pellet with 1 ml of 70 % ethanol. Centrifuge at 13,000 × *g* for 3 min. Discard the supernatant, and then briefly spin all the liquid to the bottom of the tube. Carefully remove as much liquid as possible using a pipette tip.
29. Air-dry the DNA pellet for 15 min, and then redissolve the DNA pellet in 100 μl of TE buffer (pH 7.4). After the BAC DNA is completely redissolved, NaCl, spermine, and spermidine should be added to a final concentration of 100 mM, 30 μM, and 70 μM, respectively, to compact the large DNA molecule and hence preserve its integrity.
30. Remove a small aliquot for DNA quantification by UV spectrophotometry. Ideally, the OD_{260}/OD_{280} should be 1.8.
31. Confirm the DNA concentration, purity, and integrity by running another small aliquot of the DNA on an agarose gel in parallel with a known concentration of DNA marker. Pulsed-field gel electrophoresis should be used for analyzing intact BAC DNA molecules (*see* **Note 8**).
32. The supercoiled BAC DNA can be microinjected directly after diluting to 1.5 μg/ml using BAC microinjection buffer (5 mM Tris–HCl, pH 7.4; 0.2 mM EDTA, 100 mM NaCl, 30 μM spermine, and 70 μM spermidine). However, if appropriate restriction enzymes exist for linearization, we recommend linearizing the BAC before microinjection.
33. For linearizing BAC DNA, 10 μg of supercoiled DNA can be digested with the chosen restriction enzyme (usually NotI) for 1–3 h. Avoid an extended period of digestion, which may increase DNA nicking.
34. After digestion is completed, the DNA needs to be purified by phenol/chloroform/isoamyl alcohol extraction as described in **steps 25–31**.
35. The linearized BAC DNA can be stored in 10 mM Tris (pH 7.4), 1 mM EDTA, 100 mM NaCl, 30 μM spermine, and 70 μM spermidine at 4 °C. Before each microinjection, take out an aliquot and dilute to 1.5 μg/ml using BAC microinjection buffer (5 mM Tris–HCl, pH 7.4; 0.2 mM EDTA, 100 mM

NaCl, 30 μM spermine, 70 μM spermidine). Immediately before loading the needles, the diluted DNA should be centrifuged at 13,000 × g for 5 min to pellet any particles.

4 Notes

1. If the vector backbone and the transgenic insert are similar in size, additional restriction enzymes should be used to cut the vector into smaller pieces.
2. To minimize the size of the gel slice, trim off the part that contains little DNA.
3. If the pellet is white, it is likely that the DNA is contaminated with other substances; repeating the column binding and phenol/chloroform extraction (**steps 7–17**) may improve DNA purity.
4. The Beckman Conical C0650 rotor nicely fits standard 50-ml conical tubes.
5. Do not vortex or shake vigorously as this may lead to shearing of DNA.
6. 50-ml Sorvall superspeed centrifuge tubes fit nicely into a Beckman F0650 rotor and tolerate much higher centrifugal forces than conical tubes.
7. This step helps to wash out any contaminants that cannot be washed out by the sample loading (QS) and washing (GC) buffers, because TE has a much lower ionic strength than these buffers.
8. Because regular agarose gel electrophoresis has very limited resolution for separating large DNA molecules (>20 Kb), even partially degraded BAC DNA will run as a sharp band at the top of the gel. Therefore, it cannot be used to judge the integrity of BAC DNA preparations. If pulsed-field gel electrophoresis is not available in the laboratory, the BAC DNA can be digested with a common restriction enzyme that recognizes six-base-pair cutting sites. The smaller DNA bands generated from this digestion can be analyzed using conventional agarose gel electrophoresis. The sharpness of these bands can indicate the integrity of the BAC, and the intensity of these bands in comparison with known concentrations of DNA marker can confirm the DNA concentration.

Acknowledgments

This work was supported by the Intramural Research Program of the National Heart, Lung, and Blood Institute.

References

1. Bock HH, Herz J, May P (2007) Conditional animal models for the study of lipid metabolism and lipid disorders. Handb Exp Pharmacol 178:407–439
2. Zadelaar S, Kleemann R, Verschuren L, de Vries-Van der Weij J, van der Hoorn J, Princen HM, Kooistra T (2007) Mouse models for atherosclerosis and pharmaceutical modifiers. Arterioscler Thromb Vasc Biol 27: 1706–1721
3. de Winther MP, Hofker MH (2002) New mouse models for lipoprotein metabolism and atherosclerosis. Curr Opin Lipidol 13: 191–197
4. Yang XW, Model P, Heintz N (1997) Homologous recombination based modification in *Escherichia coli* and germline transmission in transgenic mice of a bacterial artificial chromosome. Nat Biotechnol 15: 859–865
5. Copeland NG, Jenkins NA, Court DL (2001) Recombineering: a powerful new tool for mouse functional genomics. Nat Rev Genet 2:769–779
6. Heintz N (2001) BAC to the future: the use of bac transgenic mice for neuroscience research. Nat Rev Neurosci 2:861–870
7. Maroulakou IG, Muise-Helmericks RC (2001) Guide to techniques in creating transgenic mouse models using SV40 T antigen. Methods Mol Biol 165:269–286
8. Van Keuren ML, Gavrilina GB, Filipiak WE, Zeidler MG, Saunders TL (2009) Generating transgenic mice from bacterial artificial chromosomes: transgenesis efficiency, integration and expression outcomes. Transgenic Res 18:769–785
9. Chandler KJ, Chandler RL, Broeckelmann EM, Hou Y, Southard-Smith EM, Mortlock DP (2007) Relevance of BAC transgene copy number in mice: transgene copy number variation across multiple transgenic lines and correlations with transgene integrity and expression. Mamm Genome 18:693–708
10. Gong S, Yang XW (2005) Modification of bacterial artificial chromosomes (BACs) and preparation of intact BAC DNA for generation of transgenic mice. Curr Protoc Neurosci Chapter 5, Unit 5 21

References

Chapter 10

Pronuclear Microinjection and Oviduct Transfer Procedures for Transgenic Mouse Production

Chengyu Liu, Wen Xie, Changyun Gui, and Yubin Du

Abstract

Transgenic mouse technology is a powerful method for studying gene function and creating animal models of human diseases. Currently, the most widely used method for generating transgenic mice is the pronuclear microinjection method. In this method, a transgenic DNA construct is physically microinjected into the pronucleus of a fertilized egg. The injected embryos are subsequently transferred into the oviducts of pseudopregnant surrogate mothers. A portion of the mice born to these surrogate mothers will harbor the injected foreign gene in their genomes. These procedures are technically challenging for most biomedical researchers. Inappropriate experimental procedures or suboptimal equipment setup can substantially reduce the efficiency of transgenic mouse production. In this chapter, we describe in detail our microinjection setup as well as our standard microinjection and oviduct transfer procedures.

Key words Microinjection, Pronuclear, Transgenic, Embryo transfer

1 Introduction

The first germline-transmissible transgenic mouse line was produced by infecting mouse embryos with Moloney leukemia virus [1]. Several years later, the significantly more efficient pronuclear microinjection method was developed [2–6]. The essence of this method is to physically inject exogenous DNA into the pronuclei of fertilized eggs by using pulled glass needles. Like most, if not all, cell types, the eggs are capable of integrating foreign DNA molecules into their genomes at random positions. The microinjected eggs are then implanted into the oviducts of pseudopregnant foster mothers where they can develop into viable individuals. The successful development of the pronuclear microinjection method is a perfect example of the integration of two areas of scientific research: culturing and manipulation of mammalian embryos, including microinjection

Lita A. Freeman (ed.), *Lipoproteins and Cardiovascular Disease: Methods and Protocols*, Methods in Molecular Biology, vol. 1027, DOI 10.1007/978-1-60327-369-5_10,

technology [7], and recombinant DNA technology, which emerged in the 1970s and made it possible to isolate and recombine specific genes for generating transgenic constructs.

During the past three decades, several alternative methods have also been developed for producing transgenic animals, including sperm-mediated transgenesis, embryonic stem (ES) cell and chimeric mouse approaches (similar to generation of knockout mice), somatic cell nuclear transfer (animal cloning), and retroviral transduction. In recent years, the lentiviral transduction method has increasingly been used for transgenic animal production [8]. Lentivirus is highly efficient at transducing early embryos. Compared with the pronuclear injection method, it not only increases the efficiency of transgenesis, but it is also technically less challenging, although the viral particles may still need to be microinjected into the perivitelline space, the empty space between the egg plasma membrane and the zona pellucida. However, the lentiviral method also has disadvantages. For example, the lentiviral vector can only accept transgenes that are 10 kb or smaller [9], whereas there is essentially no limit on the size of the DNA fragment that can be microinjected. Lentiviral procedures also usually need to be carried out at a higher biosafety level, due to the virus's potential to infect human and other animals. Cloning transgenes into the lentivector and producing high-titer viral particles require more work and expertise than making DNA constructs for microinjection. The high transduction efficiency of lentivirus often leads to multiple copies of the transgene being inserted into multiple genomic loci. This is advantageous if a large number of founder lines are preferred, because these multiple insertion sites can be segregated into multiple transgenic lines by breeding. However, if only a few transgenic lines are needed, the extra breeding and genotyping required for separating these multiple insertion sites are often burdensome. Thus, for generating transgenic mice using standard strains, the pronuclear microinjection method has remained the workhorse in most transgenic laboratories and core facilities.

2 Materials

2.1 Mice

1. Ten B6CBAF1 (The Jackson Laboratory, Bar Harbor, ME) stud males, 2–12 months old, individually caged.
2. Twenty to fifty B6CBAF1 (The Jackson Laboratory) female mice, 1–3 months old, up to five mice per cage.
3. Twenty C57BL/6 (The Jackson Laboratory) stud males, 2–12 months old, individually caged.
4. Fifty to one hundred C57BL/6 (The Jackson Laboratory) females, 1–4 months old, up to five mice per cage.

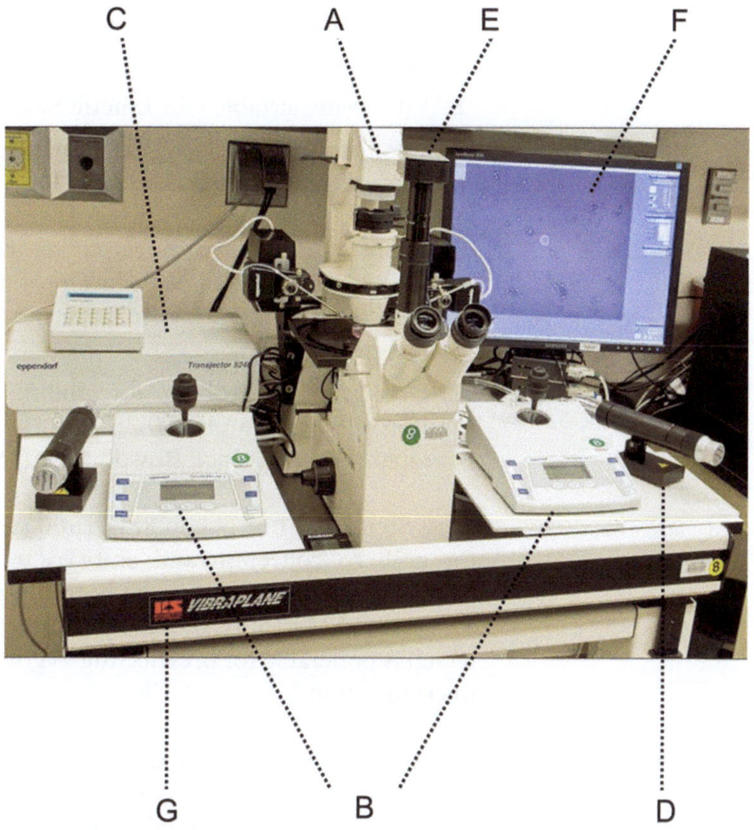

Fig. 1 Micromanipulation system. The microinjection setup consists of the following components: (A) Zeiss Axiovert S100 inverted microscope with ×5, ×10, ×20, and ×40 objective lenses. (B) A pair of Eppendorf TransferMan NK-2 micromanipulators for moving the microinjection needle and holding pipette. (C) An Eppendorf Transjector (Model 5246) for delivering controlled air pressure to the microinjection needle. (D) A mineral oil-filled Eppendorf CellTram Vario for controlling the pressure of the holding pipette. (E) An optional digital video camera. (F) Video monitor. (G) An air table for reducing vibration

5. Twenty vasectomized Swiss Webster (Taconic, Hudson, NY) male mice, 2–12 months old, individually caged. The vasectomy procedure is performed by Taconic for a reasonable fee.
6. Fifty to one hundred Swiss Webster (Taconic) females, 1.5–4 months old, up to five mice per cage.

   See **Note 1** for mouse strain selection and **Note 2** for animal room environmental effects.

2.2 Equipment

2.2.1 Microinjection Setup

As shown in Fig. 1, the microinjection system consists of the following components:

1. Zeiss Axiovert S100 inverted microscope with 5×, 10×, 20×, and 40× objectives, for pronuclear microinjection.
2. Two Eppendorf TransferMan NK-2 micromanipulators.

3. Eppendorf Transjector (Model 5246).
4. Eppendorf CellTram Vario.
5. Vibraplane airtable (RS Kinetic Systems, Lockport, IL).
6. (Optional): A digital video camera and computer system for demonstration and recording.

2.2.2 Pipette Puller and Injection Needles

1. A Sutter P-97 horizontal pipette puller (Sutter Instrument Co., Novato, CA). For pulling pronuclear microinjection needles, our settings are as follows: heat = 590, pull = 65, vel = 50, and time = 100.
2. Glass capillary tubing for pulling microinjection needles: Borosil, 1.0 mm OD × 0.75 mm ID with omega dot fiber for rapid fill (FHC, Inc., Bowdoinham, ME, Cat: 30-30-0).

2.2.3 Stereo Microscope

Zeiss Stemi SV11 or Nikon SMZ1500 stereo microscopes are used for embryo collection and embryo transfer procedures. These microscopes are equipped with both top and bottom light sources. The Nikon SMZ1500 has higher magnification power (15× zoom), which is preferable for preselecting zygotes with clear pronuclei for microinjection.

2.2.4 CO_2 Incubator

Sanyo Scientific (Model MCO-17AI) CO_2 incubators are used for embryo culture. They are normally set at 37 °C, 6 % CO_2, and humidified by a pan containing distilled water.

2.2.5 Sterilizers

1. A Harvey SterileMax table-top autoclave machine (Thermo Scientific, Waltham, MA) for sterilizing surgical instruments.
2. A Germinator 500 hot beads sterilizer (CellPoint Scientific, Inc., Rockville, MD) is used to sterilize instruments between animals when more than one animal is used during a session.

2.3 Tools and Supplies

1. Microinjection chamber (*see* **Note 3**): As shown in Fig. 2, the microinjection chamber is made by attaching a standard histological glass slide to a 2-mm-thick custom-made aluminum frame using rubber cement.
2. Embryo transfer pipette: As shown in Fig. 3, the embryo transfer pipette is assembled by connecting the following components: mouthpiece and rubber aspirator tubing (Sigma, St. Louis, MO, Cat: A5177-5EA), long-tipped glass Pasteur pipette (PGC Scientifics, Gaithersburg, MD, Cat: 71–5215), 25-mm disposable syringe filter (Millipore, Carrigtwohill, Ireland, Cat: SLGS0250S), 1-ml disposable syringe (PGC Scientifics, Cat: 79-4205-03), and a segment of flexible plastic tubing (inner diameter: ¼ in., Nalgene Labware, Vernon Hills, IL, Cat: 8007-0060).
3. Hair clipper (Wahl Model 8980 Pet Trimmer, Sterling, IL).
4. Wound clip applier (Roboz Surgical Instrument Company, Rockville, MD, Cat: RS-9260).

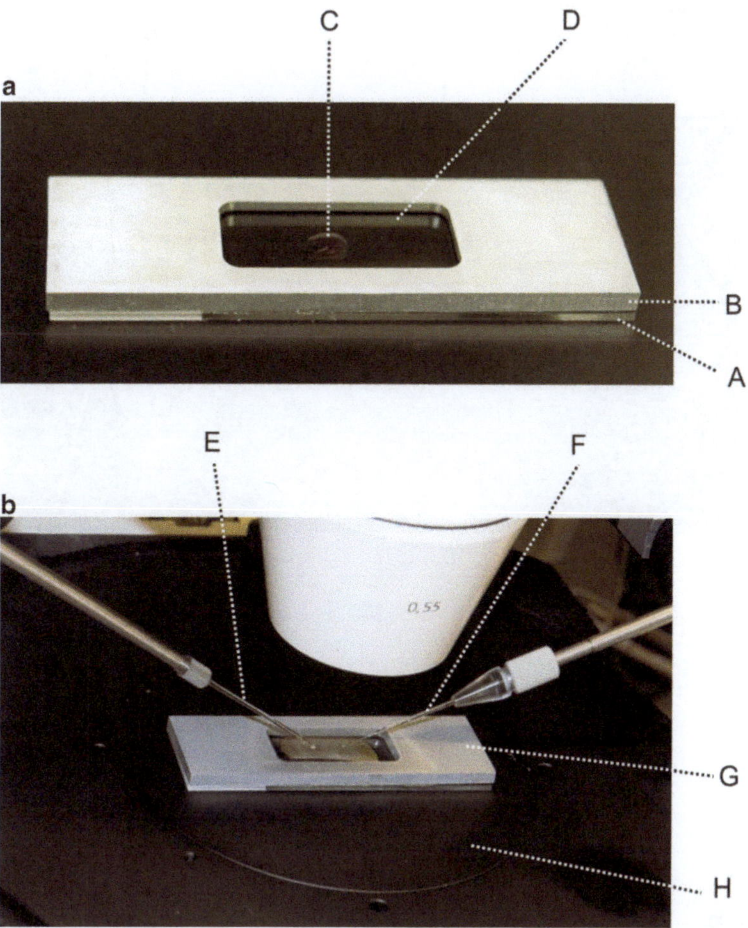

Fig. 2 Microinjection chamber. (**a**) The injection chamber is made by attaching a standard histological glass slide (A) to a 2-mm-thick custom-made aluminum frame (B) using rubber cement. The rectangular hole in the center of the aluminum frame is about 10 × 30 mm. Before microinjection, 4 μl of M2 medium (C) is added to the center of the chamber and then quickly covered with mineral oil (D). (**b**) Setting up the injection chamber on the microscope stage. (E) Holding pipette; (F) microinjection needle; (G) injection chamber; (H) microscope stage

5. Suturing needle holder (Roboz, Cat: RS-7910).
6. Holding pipette for microinjection: Eppendorf VacuTip (Cat: 5175 108.000).
7. 5-0 Vicryl suture (Ethicon, Inc., Somerville, NJ).
8. 29 Gauge × ½ in. 0.3-ml insulin syringe, 28 Ga × ½ in. 1.0-ml insulin syringe, and 26 Ga hypodermic needles (Becton Dickinson, Franklin Lakes, NJ).
9. Thermojet infrared heating lamp (Bel-Art Products, Pequannock, NJ).
10. T/Pump heating water blanket (GayMar Industries, Orchard Park, NY).

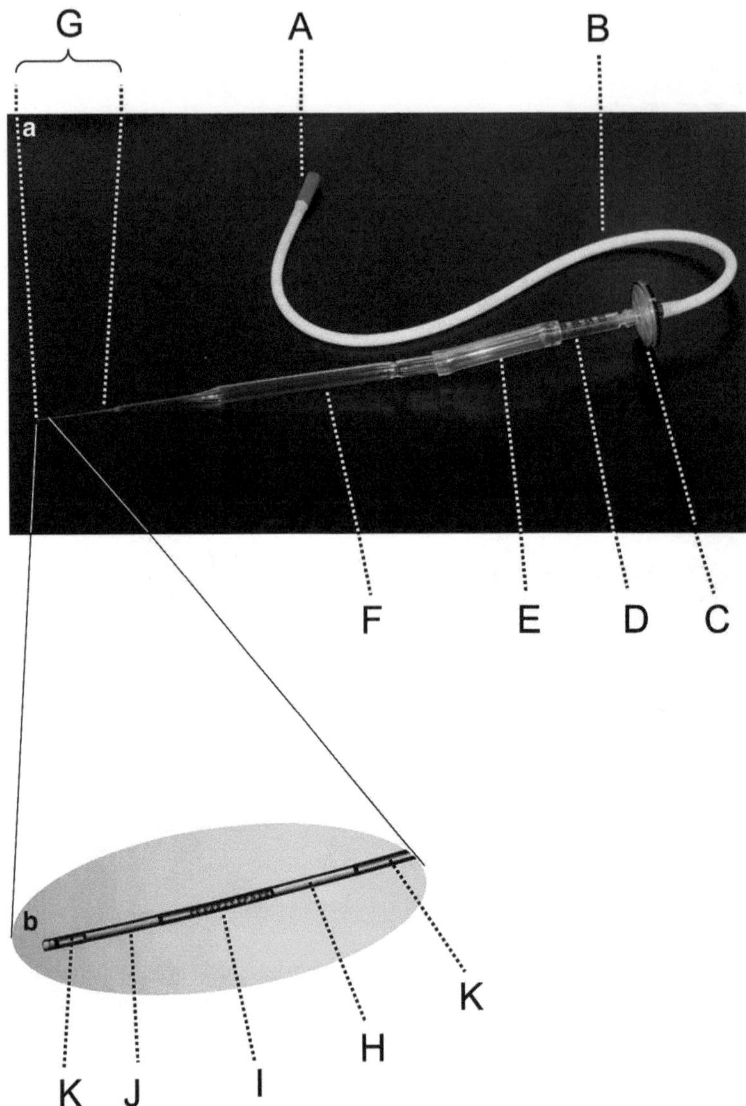

Fig. 3 Embryo transfer pipette. (**a**) The mouth-controlled embryo transfer pipette is constructed by sequential connection of the following components: (A) plastic mouthpiece; (B) flexible rubber tubing; (C) disk filter to prevent transfer of liquid between the culture dish and the mouth; (D) segment of a 1-ml disposable syringe; (E) segment of flexible plastic tubing; (F) long-tipped Pasteur pipette with a drawn-out tip of 90–150 μm in diameter (G). (**b**) Loading embryos for oviduct transfer. A small air bubble (H) is first drawn into the tip of the Pasteur pipette; then, embryos with minimal M2 medium (I) are drawn in; next, a second small air bubble (J) is drawn in. On either side of the two air bubbles, the pipette is filled with M2 medium (K)

11. Betadine (Povidone-Iodine Swabsticks from Medline Industries, Mundelein, IL).
12. Micro dissecting scissors (Roboz, Cat: RS-5882).
13. Operating scissors (Roboz, Cat: RS-6752).
14. Iris forceps (Roboz, Cat: RS-5130).

15. Micro dissecting tweezers #5 (Roboz, Cat: RS-4905).
16. Dieffenbach micro clamp (Roboz, Cat: RS-7422).
17. Surgical instruments sterilization tray (Roboz, Cat: RS-9902).
18. Microloader pipette tips (Eppendorf, Cat: 022351656).
19. Alcohol swab (The Kendall Co., Mansfield, MA).
20. Sterile surgical gloves (Ansell Healthcare, Inc., Dothan).

2.4 Media and Chemicals

1. M2 culture medium (Millipore, Temecula, CA; Cat: MR-015-D).
2. M16 culture medium (Millipore, Cat: MR-010-D).
3. Avertin solution (*see* **Note 4**): Dissolve 5.0 g of 2,2,2-tribromo-ethanol (Aldrich, Milwaukee, WI) in 5 ml tert-amyl alcohol (Aldrich). It may be necessary to warm the solution to 50 °C to achieve complete dissolution. Add 195 ml of isotonic saline (0.9 % NaCl solution) to make a 2.5 % solution. Filter the final solution using a sterile disposable 0.22 μm filter bottle. Aliquot the solution into 5-ml tubes and store the tubes at −20 °C in the dark.
4. Pregnant mare's serum gonadotropin (PMS, Sigma, Cat: G4877).
5. Human chorionic gonadotropin (hCG, Sigma, Cat: C1063).
6. Hyaluronidase (Sigma, Cat: H3506-100 mg). Concentrated solution (10×) is prepared by dissolving the entire bottle (100 mg) in 20 ml M2 medium. Filter through a 0.22-μm filter, aliquot into microcentrifuge tubes, and store at −20 °C in the dark.
7. Embryo culture-tested mineral oil (Sigma, Cat: M8410).
8. 0.25 % Bupivacaine (Hospira, Lake Forest, IL).

3 Methods

3.1 Zygote Collection

1. Three days prior to the scheduled microinjection, inject (i.p.) each egg donor mouse with 5 units of PMS. Skip this step if superovulation is not performed, such as when using C57BL/6 mice as egg donors.
2. Two days (44–48 h) later, inject (i.p.) each of the PMS-injected mice with 5 units of hCG. Immediately after the injection, set up each female mouse with an individually caged stud male for mating. When C56BL/6 mice are used, skip the PMS and hCG injections. Instead, select females which are in estrus and directly set up matings with stud males. Estrus females can be selected from a pool of breeding age animals. A swollen and reddish genital area is a good indication that the female is in estrus. Because a mouse's estrous cycle is 4–5 days, in theory, about 20 % of the females should be in estrus at any given date.

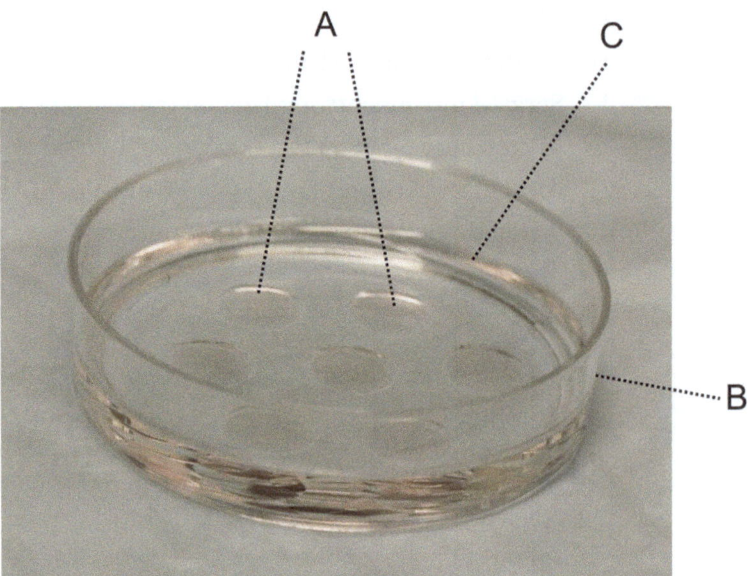

Fig. 4 Drop culture dishes. M2 or M16 drop culture dishes are prepared by quickly adding six or seven 45-µl medium drops (A) into a 60-mm tissue culture dish (B). Then, embryo-tested mineral oil (C) is immediately poured into the dish, so that the medium drops are completely covered with mineral oil (*see* **Note 8**)

3. The next morning, check the females for vaginal plugs. Euthanize plugged females with CO_2 and then cut open the abdominal cavity with a pair of sharp scissors. Dissect out both oviducts from each female using scissors and a pair of microdissecting tweezers. To avoid damaging the oviducts, the ovary as well as a small piece of uterine horn can also be dissected out.
4. Place the dissected tissues in a 35-mm culture dish containing M2 medium. Under a stereo microscope, clean up the oviducts by removing excess fat, the ovary, and the uterine horn. Place the cleaned oviducts into the M2 drop culture dish (*see* Fig. 4 for setting up drop culture dishes). Three or four oviducts can be placed in the same M2 drop.
5. Under the stereo microscope, the ampulla, which is the swollen and translucent segment of the oviducts, usually can be found a few millimeters from the infundibulum, which is the upper end (next to the ovary) of the oviduct. A cumulus mass containing several eggs surrounded by follicular cells is usually visible in the ampulla. Using a 29-gauge insulin needle, tear open the ampulla to allow the cumulus mass to extrude spontaneously into the media. Discard the oviduct tissue. After releasing the cumulus masses from all oviducts, add 5 µl of 10× hyaluronidase solution to each M2 drop. Gently shake the dish and then incubate at room temperature for a few minutes to allow the hyaluronidase to digest the sticky materials, and remove the follicle cells from the eggs. Pick up the eggs using

an embryo transfer pipette and wash them through three M2 drops to remove residual hyaluronidase. The eggs can be stored in an M2 drop in a 37 °C CO_2 incubator for several hours. Normally we isolate eggs in the morning and perform microinjection in the afternoon.

3.2 Pronuclear Microinjection

1. Centrifuge the transgenic DNA solution (1.5 μg/ml) for 5 min at 14,000 × *g* in a microcentrifuge to pellet any possible particles that may clog the injection needle. Use a 1–10 μl pipetman and an Eppendorf microloader tip to deliver 2–3 μl of transgenic DNA solution into the barrel of a microinjection needle through the open (back) end (*see* **Note 5**). The filament inside the needle will facilitate transporting the DNA solution to the needle tip through capillary action.
2. Add 4 μl of M2 medium into the center of the injection chamber. Immediately cover the M2 drop with mineral oil. Place the injection chamber onto the microscope stage, and lower the tip of both the holding pipette and injection needle into the M2 drop (Fig. 2b). Under the microscope, break open the tip of the microinjection needle by hitting the needle tip against the holding pipette.
3. Transfer 20–30 zygotes into the M2 drop in the injection chamber. Moving the tip of the holding pipette to the vicinity of an egg, turn the CellTram Vario counterclockwise until the egg is firmly grasped by the holding pipette (*see* **Note 6**). Focus the microscope on one of the two pronuclei. Use the NK2 micromanipulator to align the microinjection needle next to the pronucleus. It is important to position the needle tip in the same focal plane as the pronucleus to be microinjected. With a quick and smooth forward motion, insert the needle into the pronucleus. Sometimes the needle has to pass through the entire nucleus in order to penetrate the plasma membrane and nuclear envelope. When the needle tip is inside the pronucleus, push the foot peddle of the Transjector to inject the DNA solution. It is very important to adjust and optimize the microscope, so that the expansion of the pronucleus is clearly visible when the DNA solution is injected into it (Fig. 5). Since the needle tip opening varies from needle to needle, it is necessary to adjust the pressure setting of the Transjector for each needle. We normally set the constant pressure (Po), which is continuously applied to prevent M2 medium from being drawn into the needle by capillary action, in the range of 20–60 PSI, and set the injection pressure (Pi), which is applied only when the foot peddle is pressed, in the range of 50–200 PSI. Settings above or below these ranges usually indicate that the needle opening is too small or too large, respectively. After injection, the egg is transferred to a different area using the holding

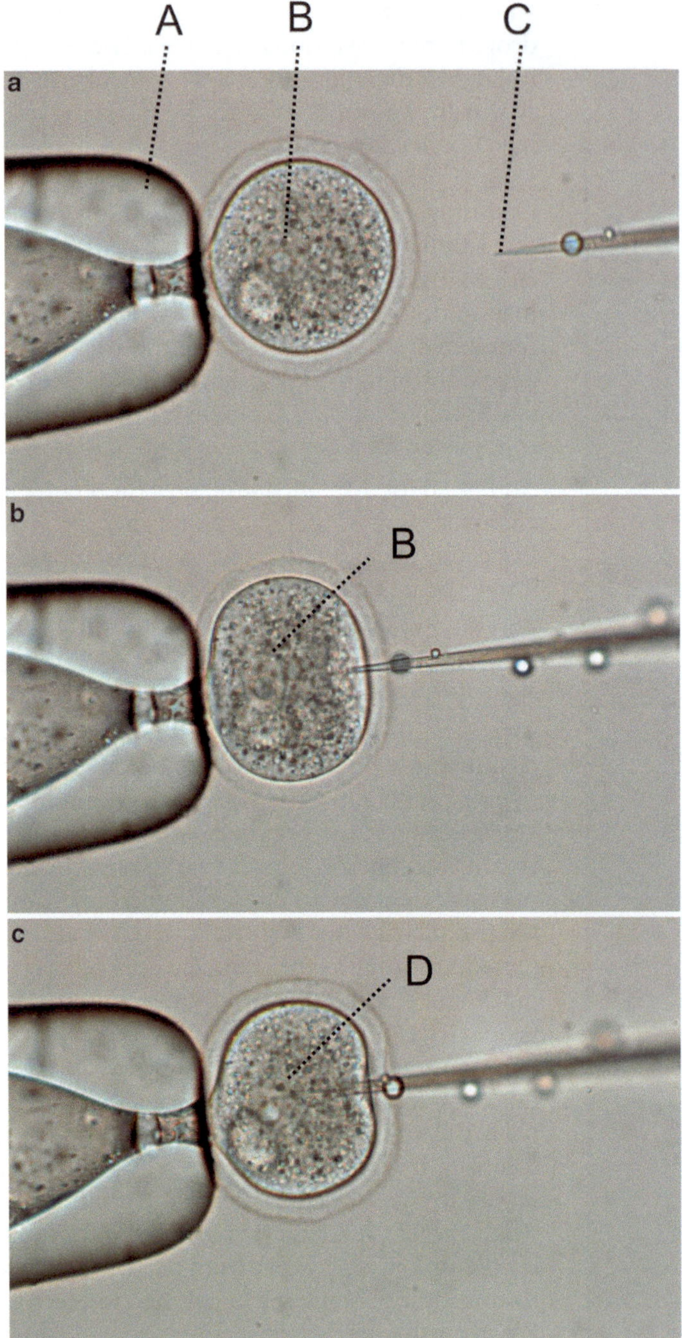

Fig. 5 Pronuclear microinjection. (**a**) A zygote is first grabbed by the holding pipette (A). The focus of the microscope is adjusted so that one of the two pronuclei (B) is in sharp focus. Next, the micromanipulator is adjusted to raise or lower the microinjection needle so that the tip of the needle (C) is in the same focal plan as the pronucleus. (**b**) The needle is inserted into the zygote with a quick motion. (**c**) When the needle tip is inside the pronucleus, the foot peddle of the Transjector is pushed to deliver DNA solution into the nucleus. A sudden expansion of the nuclear envelope (D) is an indication that the embryo has been successfully microinjected. Then, the needle is gently but quickly withdrawn

pipette to separate it from the uninjected eggs (*see* **Note 7**). When all eggs in the chamber are injected, they are transferred out of the chamber and placed in a drop of M16 (*see* **Note 8**) in the CO_2 incubator. Then, a new batch of eggs is transferred to the injection chamber for the next round of microinjection. After all the eggs are injected, they are washed through two or three M16 drops and then cultured in M16 medium overnight to allow them to develop into 2-cell stage embryos.

3.3 Oviduct Transfer (See Note 9 for Timing of Embryo Transfer)

1. On the day of microinjection, estrus Swiss Webster females are selected and set up with vasectomized male mice for mating (*see* Subheading 3.1, **step 2**, for the selection method).
2. The next morning, check the mice for vaginal plugs. The plugged mice can be used as recipient mothers for embryo transfer.
3. Examine the microinjected embryos using a stereo microscope. Count the number of embryos that have reached the 2-cell stage of development. Transfer these 2-cell embryos into a M2 drop.
4. Anesthetize the recipient mothers by injecting (i.p.) 2.5 % Avertin solution at a dose of 0.017 ml/g of body weight. While waiting for the injected mice to succumb (usually a few minutes), load the embryo transfer pipette with 13 embryos. Because the embryos are barely visible in the transfer pipette, as shown in Fig. 3b, two air bubbles are sucked into the transfer pipette to mark the boundaries of the suspended embryos.
5. Five minutes after the Avertin injection, check the depth of anesthesia by gently pinching one of the hind paws. If the animal can still respond to this gentle pinch, a supplemental dose of Avertin should be injected. Normally the supplementary dose is approximately a third of the original dose. After the animal is fully anesthetized, hair is removed from a generous area on the dorsal lumbar back using a small hair clipper. The clipped area is disinfected utilizing alternating applications of Betadine and 70 % alcohol.
6. Using dissecting scissors, cut a dorsal midline incision (~1 cm long) in the cleaned skin area (Fig. 6). Because mouse skin is only loosely connected to the body wall, the midline incision can be moved with a pair of iris forceps to either side of the paralumbar to find the ovary and associated fat pad, which are vaguely visible through the muscular body wall because they are paler than the surrounding internal organs. A 5–10-mm incision is made in the body wall by first making a small cut and then extending it using the back of the scissors' blades.
7. Grasp the fat pad that is associated with the ovary with a pair of iris forceps and gently pull out the fat pad, the ovary, the oviduct, and a segment of the uterine horn through the incision.

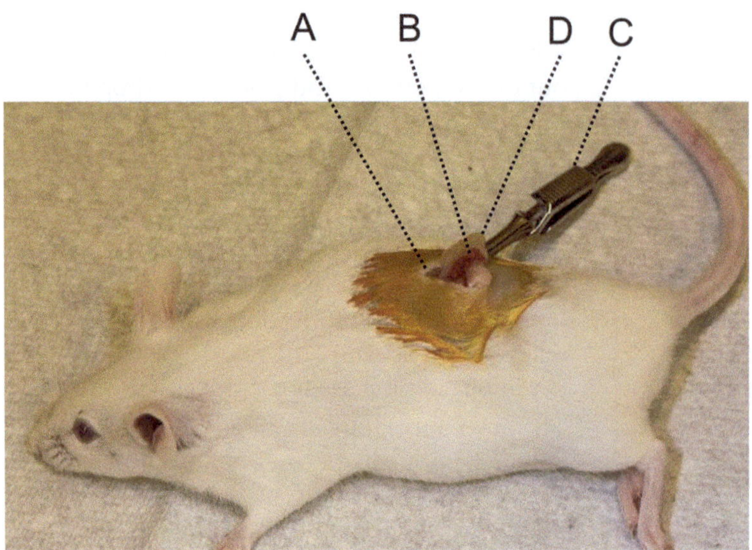

Fig. 6 Oviduct transfer surgical site. A midline incision (A) is made in the dorsal lumbar area. Then, the reproductive organs (including ovary, oviduct, and a segment of the uterine horn) are exteriorized through a small cut in the body wall (B). These organs are fixed in place by using a Dieffenbach clip (C) clamped to the fat pad (D) associated with the reproductive organs

Clamp the fat pad with a Dieffenbach clip to hold the organs in place. Under the dissecting microscope, find the oviduct and ampulla. Use a 29-gauge needle to punch a small hole in the oviduct wall between the ampulla and the ovary. Carefully insert the tip of the transfer pipette into the hole with the pipette opening pointing toward the ampulla. Gently blow into the mouthpiece to expel the two air bubbles and the embryos sandwiched between them. The presence of two air bubbles in the ampulla is a good indication that all embryos have been successfully implanted.

8. Remove the Dieffenbach clip and carefully push the reproductive organs back into the abdominal cavity. Close the abdominal wall with a cruciate suture utilizing absorbable 5-0 Vicryl suture. Drop one or two drops of 0.25 % bupivacaine solution on the muscle at the surgical site using a 26-gauge needle. Bupivacaine solution is a long-lasting local anesthetic for reducing pain at the surgical site.
9. For two-sided transfer, move the skin incision to the other side of the mouse, and transfer another 13 embryos into the other oviducts following the same procedure described in **steps 6–8**.
10. Close the skin incision with stainless steel surgical wound clips. Place the mice in a pre-warmed mouse cage placed on a circulating water blanket or under a carefully adjusted infrared heating lamp. Normally, the mice will wake 30–60 min post-Avertin injection. They can then be returned to the animal room.

11. Wound clips are removed 10–14 days after surgery. Pups developed from injected embryos will be born about 19 days after the embryo transfer procedure. The offspring will be weaned when they are 18–21 days old. At weaning, ear-tags are applied and the tail tips are clipped for identification and genotyping.

4 Notes

1. Mouse strains: Transgenic mice have been successfully generated using a variety of mouse strains. Generally, outbred strains, such as CD-1 and Swiss Webster, and hybrid strains, such as B6CBAF1, are very efficient, partly because they show good hormone-induced superovulation to yield a large number of injectable zygotes, and partly because hybrid embryos often can tolerate more trauma caused by the microinjection procedures. However, the undefined genetic backgrounds of these strains have greatly limited their usefulness for many transgenic studies. On the other hand, most inbred strains are inefficient for producing transgenic mice. FVB is an excellent inbred strain for generating transgenic mice [10], but it is not widely used in biomedical research. The C57BL/6 strain is often desired for many areas of research, including lipoprotein metabolism and atherosclerosis. It is significantly more susceptible to atherosclerosis than other mouse strains, and therefore many existing knockout and overexpression models, such as the standard atherogenic apoE-KO and LDLr-KO mice from The Jackson Laboratory, are already in a C57BL/6 background. However, C57BL/6 mice are not ideal for pronuclear microinjection. Many researchers superovulate C57BL/6 mice, but in our hands, their superovulatory response to hormone treatment is very inconsistent. When they do respond, we find that the zygotes collected from superovulated females do not survive the microinjection procedure as well as zygotes harvested from naturally mated females. Therefore, we only use natural matings for collecting C57BL/6 eggs. Normally, we can collect four to eight injectable zygotes from each female. Despite this relatively low efficiency, we still routinely generate transgenic mice using the C57BL/6J inbred strain. Compared with the time-consuming and expensive backcrossing, the extra effort upfront is well worth it.

   For outbred and hybrid mouse strains, the optimal age for superovulation is 4–7 weeks. For C57BL/6J females, 3–4 weeks of age is best for superovulation, but we prefer to use older (5–10 weeks) females for natural mating. For stud and vasectomized males, we use 2–12-month-old males.
2. Animal room environmental effects: The environmental conditions of the animal room can dramatically influence the out-

come of transgenic animal production. Most animal facilities have strict rules to regulate the temperature, light/dark cycle, noise level, and humidity to levels that far exceed the conditions for human living (hotels, homes, and working places). However, these strict rules do not automatically eliminate all the adverse conditions that interfere with mouse reproduction. Some mechanical and human errors are inevitable. For examples, someone may forget to turn off the blower or lights of the changing hood during the night, or the automatic light/dark control may be improperly set. Currently, racks with ventilation for each individual cage are increasingly being used. While they are useful for reducing cross-contamination among different animals, they also introduce more variables among different cages. Setting the air exchange rate too high not only increases noise and vibration but also reduces humidity in the cages because of the increased airflow. An inappropriate light/dark cycle, suboptimal humidity, and noise can significantly reduce both the number of zygotes harvested from each plugged female and the percentage of injected embryos that result in live births. Under certain unfavorable environments, even late-gestation embryos may be absorbed in utero. When the efficiency of transgenic mouse production is less than optimal, the housing environment is certainly worth careful checking.

3. We use a custom-made injection chamber for pronuclear microinjection because, for most microscopes, plastic tissue culture dishes are not optically acceptable for clearly visualizing pronuclei using DIC imaging. Dishes with a glass cover slip bottom (designed for confocal imaging) can be used, but the rim of the dish sometimes restricts the holding pipette and injection needle from accessing the dish at a small angle relative to the microscope stage. It should be mentioned that DIC imaging is possible for plastic dishes on some of the latest generation of inverted microscopes.

4. Avertin is an old-fashioned but very effective anesthetic. The solution is easy to make and stable when stored in the dark at −20 °C. However, some batches of this chemical may have adverse effect on animals. Therefore, it is important to test each new batch on several control animals to make sure it does not cause any health problems. Since the drug's adverse effects may take several days to show up, the test animals should be kept in the standard environment for 2–3 weeks. Ketamine (Ketaset)-xylazine (Rompun) mixture is a reliable and easy-to-use anesthetic for embryo transfer procedures, but ketamine is a controlled substance, which requires a drug lock box as well as a detailed log sheet for recording the amount used on each day in the United States.

5. We always use freshly pulled needles to perform microinjection. Old needles tend to collect dust and become sticky.
6. When using the Eppendorf CellTram Vario for controlling the pressure inside the holding pipette, the entire piston and attached tubing need to be completely filled with mineral oil. Any air bubbles will significantly reduce its controllability. AirTram from Eppendorf is more convenient because it does not need to be filled with oil, but its control is less precise.
7. When moving the embryos around the microinjection chamber, it is much easier to move the microscope stage, instead of moving the joystick of the holding pipette. When moving the stage, the holding pipette and microinjection needle will remain aligned to each other, and stay at the center of the field of view.
8. M2 medium can be used outside of CO_2 incubators because it contains HEPES buffer. However, the correct pH of M16 medium is dependent on the 5–6 % CO_2 in the incubator. Therefore, M16 medium needs to be equilibrated in the CO_2 incubator for several hours before using. When setting up drop culture dishes, it is important to cover the M16 drops with mineral oil very quickly because the small drops of medium can lose CO_2 and evaporate very quickly, which could lead to improper pH and ionic strength. Also, the mineral oil can hinder gas exchange in the CO_2 incubator, and therefore we often equilibrate the M16 drop culture dishes overnight before use.
9. We prefer to perform oviduct transfer the day after microinjection mainly because of convenience. Culturing embryos overnight is a convenient stop point for a very busy microinjection day. Of course, the microinjected zygotes can be implanted immediately after microinjection. In this case, the pseudopregnant foster mothers need to be prepared 1 day earlier than is described in the protocol. Occasionally, a shortage of pseudopregnant foster mothers occurs on the planned embryo transfer date. Under this circumstance, embryos can be further cultured for an additional 1 or 2 days in KSOM medium (KSOM medium is preferable to M16 for culturing embryos beyond the 2-cell stage of development). Embryos ranging from the 1-cell stage to the blastocyst stage can be transferred into the oviducts, but do not attempt to put embryos of any stage into the uteri of 0.5- or 1.5-day pregnant foster mothers, because the uteri are not ready to accept embryos yet.

Acknowledgments

This work was supported by the Intramural Research Program of the National Heart, Lung, and Blood Institute.

References

1. Jaenisch R (1976) Germ line integration and Mendelian transmission of the exogenous Moloney leukemia virus. Proc Natl Acad Sci USA 73:1260–1264
2. Gordon JW, Scangos GA, Plotkin DJ, Barbosa JA, Ruddle FH (1980) Genetic transformation of mouse embryos by microinjection of purified DNA. Proc Natl Acad Sci USA 77:7380–7384
3. Brinster RL, Chen HY, Trumbauer M, Senear AW, Warren R, Palmiter RD (1981) Somatic expression of herpes thymidine kinase in mice following injection of a fusion gene into eggs. Cell 27:223–231
4. Wagner TE, Hoppe PC, Jollick JD, Scholl DR, Hodinka RL, Gault JB (1981) Microinjection of a rabbit beta-globin gene into zygotes and its subsequent expression in adult mice and their offspring. Proc Natl Acad Sci USA 78:6376–6380
5. Costantini F, Lacy E (1981) Introduction of a rabbit beta-globin gene into the mouse germ line. Nature 294:92–94
6. Harbers K, Jahner D, Jaenisch R (1981) Microinjection of cloned retroviral genomes into mouse zygotes: integration and expression in the animal. Nature 293:540–542
7. Lin TP (1966) Microinjection of mouse eggs. Science 151:333–337
8. Lois C, Hong EJ, Pease S, Brown EJ, Baltimore D (2002) Germline transmission and tissue-specific expression of transgenes delivered by lentiviral vectors. Science 295:868–872
9. Lois C (2006) Generation of transgenic animals using lentiviral vectors. In: Pease S, Lois C (eds) Mammalian and avian transgenesis—new approaches. Springer, Heidelberg, Germany, pp 1–22
10. Taketo M, Schroeder AC, Mobraaten LE, Gunning KB, Hanten G, Fox RR, Roderick TH, Stewart CL, Lilly F, Hansen CT et al (1991) FVB/N: an inbred mouse strain preferable for transgenic analyses. Proc Natl Acad Sci USA 88:2065–2069

Chapter 11

Genotyping of Transgenic Animals by Real-Time Quantitative PCR with TaqMan Probes

Boris L. Vaisman

Abstract

Real-time quantitative PCR (qPCR) is a fast, sensitive, specific, and quantitative method for genotyping transgenic animals. Accurate quantitation of the number of transgenes helps to identify founders and to create and maintain pure lines of transgenic mice, thus reducing experimental variability. Here we describe an accurate method of genotyping using real-time quantitative PCR with primers and MGB TaqMan probes from Life Technologies. The first step in quantitating copy number is isolation of genomic DNA. To accurately compare the copies per genome (c/g) of a transgene in different mice, genomic DNA must be prepared by the same method for all the mice, with sample DNA and calibration standards dissolved in the same buffer. This chapter describes several "tried and true" methods, including an automatic system that isolates 16 samples at once in just 35–45 min, yielding DNA of excellent quality. Next, genomic DNA must be quantitated accurately so that similar amounts of DNA are added to each well. A fluorescent assay that is selective for dsDNA over RNA circumvents interference from RNA contamination and ensures more accurate DNA quantitation than A_{260} measurements. It is also very important to use appropriate calibration standards for accurate quantitation of transgene copy number. The best calibrator is the DNA fragment used for microinjection, mixed with normal mouse DNA in such a way that the transgene is present in a range of concentrations spanning the expected copy number in the transgenic mice. This chapter provides guidelines and sample calculations for preparing calibration standards that will accurately reflect the number of transgenes in the mice being tested. Finally, guidelines for preparing primers and TaqMan probes and techniques to prepare and run a 384-well plate smoothly and without errors are presented.

Key words Genotype or genotyping, Real-time PCR, DNA isolation, Transgenes, Calibration standards

1 Introduction

Genotyping is a critically important step in generating and maintaining transgenic animal lines. Potential founders obtained after DNA microinjections into fertilized eggs and subsequent implantation can have from zero to several hundred copies of microinjected DNA in their chromosomes. The DNA transgene is most frequently located in one chromosomal site but may have inserted independently into several (usually 2–3) different sites. Monitoring

Lita A. Freeman (ed.), *Lipoproteins and Cardiovascular Disease: Methods and Protocols*, Methods in Molecular Biology, vol. 1027, DOI 10.1007/978-1-60327-369-5_11, © Springer Science+Business Media, LLC 2013

the presence of the transgene and its number of copies per diploid genome (c/g) is used initially to reveal transgenic founders and then, by subsequent breeding, to create pure lines of transgenic animals with only one site of integration (if desired) with known numbers of the transgenes per genome. Genotyping will also help to reveal relatively rare cases of mosaic founders, which have a significantly lower transgene copy number in their tail DNA than their offspring [1]. The ability to quickly and accurately quantitate copy number permits the researcher to keep transgenic animal lines in a heterozygous and/or homozygous state and to study the effects of gene dose on phenotypic manifestations. Importantly, the ability to measure transgene copy number enables the researcher to create transgenic mice using mouse rather than human transgenes, as it is not difficult to distinguish the two endogenous copies per genome present in a wild-type diploid mouse from, say, 10–20 copies of a mouse transgene per genome present in a transgenic mouse, using the protocol provided below. Working with pure lines of transgenic mice which have the same copy number of the transgene, located in the same chromosomal locus, decreases the variability of the results and increases their statistical significance.

Several options exist for quantitative genotyping of transgenic mice, including Southern and dot-blot hybridization analysis [2] as well as real-time quantitative PCR (qPCR) [3]. Dot-blot and Southern analysis are time and labor consuming and may use radioactivity. Southern analysis is especially impractical when large number of samples must be analyzed. The TaqMan variation of real-time PCR presented in this chapter is a highly specific, sensitive, fast, nonradioactive approach, which can be used on different scales. The researcher has the option of using a manual approach or exploiting new robotic devices convenient for manipulating small volumes of liquid. In this chapter, we describe the main steps of real-time qPCR based on using MGB (minor groove binder) TaqMan probes [3].

2 Materials

2.1 DNA Isolation

For isolation of tail or blood DNA, any method which produces relatively pure high-molecular-weight DNA (with ratio 260 nm/280 nm near 1.8 and size 20–50 kb or more) can be used. Methods used successfully in our laboratory include:

1. Promega Maxwell 16 Instrument and Maxwell 16 DNA Purification Kit.
2. Blood/tail DNA isolation kits from, e.g., Qiagen (DNeasy), Promega (Wizard), or other companies.

3. Classical manual methods with proteinase K digestion, followed by phenol–chloroform purification [1]:
   (a) Digestion buffer:
      - 50 mM Tris–HCl pH 8.0.
      - 100 mM EDTA.
      - 0.5 % SDS.
   (b) Protease K, 10 mg/ml in water. Store at −20 °C.
   (c) Phenol saturated with 0.1 M Tris–HCl pH 8.0.
   (d) Phenol:chloroform (1:1).
   (e) 3 M sodium acetate, pH 6.0.
   (f) 100 % ethanol at room temperature.
   (g) 70 % ethanol.
   (h) TE buffer (pH 8.0).
4. University of Connecticut http://gttf.uchc.edu/protocols/tail_dna_extraction.html method:
   (a) 5 M NaCl.
   (b) 0.5 M EDTA pH 8.0.
   (c) 10 % SDS.
   (d) Protease K 10 mg/ml.
   (e) SE buffer.
      - 75 mM NaCl.
      - 25 mM EDTA.
      - 1 % SDS.
      - 200 μg/ml protease K.
   (f) 6 M NaCl (saturated), prewarmed to 55 °C before use.
   (g) Chloroform.
   (h) Isopropanol.
   (i) 70 % ethanol.
   (j) 10 mM Tris–HCl pH 8.5.

2.2 Determination of DNA Concentration

1. Life Technologies Quant-iT dsDNA Assay Kit (cat. no. Q33130).
2. Corning Inc. Costar 96-well plates (cat. no. 3915).
3. A plate reader with the ability to measure fluorescence at emission/excitation 485/535 nm.

2.3 Primers and MGB TaqMan Probes

Either ready-to-use Life Technologies TaqMan Gene Expression Assays or customized primers and TaqMan MGB probes designed with the help of Primer Express 3.0 software from Life Technologies. Primers and MGB TaqMan probes will be specific for each gene analyzed (*see* Subheading 3.3, below, for primer and probe design guidelines). Keep TaqMan assays, primers, and probes frozen.

2.4 Calibration Standards

Preparation of calibration standards will be specific for each experiment. *See* Subheading 3.4, below, for guidelines.

2.5 Preparing the Plan

No reagents needed.

2.6 Genomic DNA Solutions

1. Low-EDTA TE:
   - 10 mM Tris–HCl pH 7.4.
   - 0.1 mM EDTA.
2. 0.5 ml of each genomic DNA sample diluted to 0.5 μg/ml in low-EDTA TE. Place 0.5 ml of low-EDTA TE buffer into microcentrifuge tubes and add the appropriate volume of concentrated genomic DNA stock solution This can be calculated in advance as in spreadsheet #1 (*see* Subheading 3.5, below). Carefully mix and briefly spin tubes. The prepared DNA solutions can be kept frozen if necessary.

2.7 Real-Time qPCR

1. 7900HT Fast Real-Time PCR System (Life Technologies, part no. 4329001) with SDS 2.4 software, or any real-time PCR system which can work with TaqMan probes.
2. 384-well plates (Life Technologies, MicroAmp Optical 384-well Reaction Plate, part no. 4309849). PCR plates can be any size, which the qPCR system allows.
3. TaqMan Universal PCR Master Mix, 2×, No AmpErase (Life Technologies, part no. 4324018), which contains AmpliTaq Gold DNA Polymerase, dNTPs with dUTP, passive reference, and buffer. Keep frozen in 1-ml portions.
4. Primers and MGB TaqMan probes (*see* Subheading 2.3). Keep TaqMan assays, primers, and probes frozen until use.
5. MicroAmp Clear Adhesive Film (Life Technologies, part no. 4306311).
6. MicroAmp Adhesive Film Applicator (Life Technologies, part no. 4333183).
7. 8-Channel 384 Equalizer Pipette, 0.5–12.5 μl (Thermo Scientific, cat. no. 2139). The expandable pipette allows the user to take eight samples at once, from normal 1.5-ml test tubes in an 80-well rack, and to transfer them into eight wells of a 384-well plate.
8. Eppendorf Repeater Plus Dispenser (Eppendorf, cat. no. 022260201) with Eppendorf Combitips plus, 0.1-ml (Eppendorf, cat. no. 0030089510).

3 Methods

Important: The usual precautions for PCR against DNA cross-contamination should be followed throughout the entire procedure.

3.1 DNA Isolation

In mice, whole blood and tails are most routinely used as a source of DNA for genotyping. Blood can be safely taken from the retro-orbital sinus of a mouse when it has reached 2 months of age. In order to prevent coagulation, 4 μl of 0.2 M EDTA solution should be placed into the tube before adding 200 μl of blood.

Pieces of mouse tails can be clipped at the same time that mice have received identification markers. Usually it is convenient to do this at the day of weaning.

For isolation of tail or blood DNA, any method which produces relatively pure high-molecular-weight DNA (with ratio 260 nm/280 nm near 1.8 and size 20–50 kb or more) can be used. We provide several options that have been used successfully in our laboratory. It is important for all samples and standards to be purified by the same method if they are to be quantitatively compared side by side in the same assay (*see* **Note 1**).

1. The most convenient approach is to isolate genomic DNA using portable automatic systems, such as the Promega Maxwell 16 Instrument, which isolates 16 samples at once in just 35–45 min and produces DNA suitable for practically any application. Automated DNA purification systems are available from other suppliers as well. The yield from 5 to 7 mm of mouse tails using the Promega Maxwell 16 Instrument and the Maxwell 16 DNA Purification Kit is approximately 5–7 μg of DNA; from 200 μl of blood—around 4 μg. Automated systems should be used with the recommended kits according to manufacturer's instructions.
2. For blood/tail genomic DNA isolation, kits are available from Qiagen (DNeasy), Promega (Wizard), or other companies. For blood DNA in particular, these kits provide DNA of excellent quality for not just RT-PCR but subsequent Southern analysis, if desired.
3. The classical manual method with proteinase K digestion, following phenol–chloroform purification [1], gives excellent results, especially for tail DNA. This method, while relatively labor intensive, will produce abundant quantities of tail DNA suitable for Southern analysis as well as qPCR. The standard manual method, from reference [1], is:
   (a) For each mouse tail to be digested, prepare 500 μl digestion buffer:
      - 50 mM Tris–HCl pH 8.0.
      - 100 mM EDTA.
      - 0.5 % SDS.

(b) To a piece of mouse tail in a 1.5 ml microcentrifuge tube, add:
   - 500 μl digestion buffer.
   - 25 μl Protease K, 10 mg/ml.

(c) Incubate 55 °C overnight, keeping tail submerged.

(d) Add 500 μl Tris-buffer-saturated phenol and shake for 3 min.

(e) Centrifuge at top speed in a microcentrifuge 3 min.

(f) Transfer the supernatant (aqueous phase) to a fresh microcentrifuge tube, avoiding the interface.

(g) Extract with phenol:chloroform (1:1), centrifuge, and transfer the supernatant to a fresh tube as in **steps** (**d**)–(**f**), above.

(h) (Optional: repeat extraction 1× with pure chloroform).

(i) Add 50 μl (1/10 of obtained volume) 3 M sodium acetate, pH 5.0 to the supernatant (aqueous phase), and mix.

(j) Add 500 μl 100 % ethanol at room temperature and mix carefully.

(k) Centrifuge at top speed for 30 s in a microcentrifuge to precipitate DNA.

(l) Remove supernatant, taking care not to disturb the pellet.

(m) Wash pellet with 1 ml 70 % ethanol.

(n) Remove all traces of supernatant and briefly air-dry pellet.

(o) Add 100 μl of 10 mM Tris–HCl pH 7.4. Leave at room temperature overnight or heat at 55 °C while gently shaking to completely dissolve.

4. An alternative method that is rapid, although not automated, is a modification [4] of the method in **step 3**. This rapid manual method is presented on the Web site of the University of Connecticut http://gttf.uchc.edu/protocols/tail_dna_extraction.html.

(a) Digest 0.5–1 cm tail in 500 μl SE buffer/proteinase K, 55 °C, overnight.

(b) Prewarm 6 M NaCl (saturated NaCl) to 55 °C and add 170 μl to each tube of digested tails.

(c) Add 670 μl of chloroform and mix for 30–60 min by gentle rotation.

(d) Centrifuge at approximately 15,000 × g (12,000 rpm) for 5 min.

(e) Transfer the supernatant to a fresh tube.

(f) Add 1 volume of isopropanol at room temperature.

(g) Gently mix for 5 min.

(h) Pick DNA or centrifuge to collect DNA.
(i) Wash with 70 % ethanol for 1 h or overnight.
(j) Spin and briefly dry.
(k) Dissolve in 10 mM Tris–HCl pH 8.5.

The DNA purified by **option 4** is not as clean as DNA purified by **option 1**, **2,** or **3** above but is sufficiently pure for qPCR. PCR efficiency may be somewhat reduced due to presence of contaminants. If you are planning to use DNA for subsequent Southern analysis, be aware that some restriction enzymes that are sensitive to contaminants may not digest DNA purified by **option 4** well.

3.2 Determination of DNA Concentration

DNA solutions used in qPCR should have the same concentration in each well of the PCR plate. Thus, the original concentration of isolated DNA needs to be determined. It is convenient to use the Life Technologies Quant-iT dsDNA Assay Kit (cat. no. Q33130), Corning Inc. Costar 96-well plates (cat. no. 3915), and a plate reader with the ability to measure fluorescence at emission/excitation 485/535 nm for this purpose. The fluorescence signal is linear with DNA in the range of 2–1,000 ng. The method is highly sensitive, very fast, and selective for dsDNA over RNA.

To measure DNA concentration using the Quant-iT dsDNA Assay Kit:

1. Make the working solution provided by the kit by diluting the Quant-iT dsDNA BR reagent 1/200 in Quant-iT dsDNA BR buffer.
2. Aliquot 200 μl working solution to each microplate well.
3. For the standards, add 5 μl of each of the standards provided in the kit to a microplate well containing 200 μl working solution and mix well (*see* **Note 2**).
4. For the samples, add 15–300 ng of sample DNA (genomic DNA isolated from blood or tail) in 5 μl volume to the well containing working solution and mix well (*see* **Note 2**).
5. Read in a plate reader with the ability to measure fluorescence at an appropriate emission/excitation wavelength (510/527 nm recommended by manufacturer but 485/535 nm is acceptable).
6. Calculate the DNA concentration of the samples from a standard curve.

The results of the plate reader measurements can be quickly transformed into concentration data by using statistical software, such as GraphPad Prism from GraphPad Software.

3.3 Designing Primers and TaqMan MGB Probes for qPCR

A broad selection of ready-to-use assays (primers mixed with TaqMan MGB probes) is currently available on the market, including assays that can detect cDNA transgenes or genomic DNA transgenes (*see* **Note 3**). If a human cDNA was used as the transgene to make transgenic mice, it is possible to use for genotyping ready-to-use assays designed to analyze expression of that particular human gene. If a natural human gene was used as a transgene, assays designed for genomic DNA should be selected. The same is true if a BAC clone containing a fragment of genomic DNA was used.

By using software such as Primer Express from Life Technologies, it is possible to quickly design the desired set of primers and the corresponding MGB TaqMan probe. All designed assays will work efficiently at the same standard conditions of qPCR as pre-manufactured assays. Customized primers and probes are especially helpful for situations where no premanufactured assay is available, for example if the transgene was placed under a heterologous tissue-specific promoter that differs from the natural endogenous promoter for the transgene. In this case, for highly specific qPCR it is possible to place one primer in the promoter region, and the reverse primer within the transgene, with the probe placed between them as usual.

When selecting the DNA region for qPCR analysis and for primer and probe design, it is very important to follow the recommendations [3] described below:

1. The selection of the DNA region for genotyping
   (a) The sequence region to be used as the detection template should not have obvious secondary structures or long runs of the same nucleotides, as often happens in introns.
   (b) For detecting cDNA targets it is better to design the primers and probes so that they span exon-exon boundaries, rather than having the whole amplicon located in one of exons. This will decrease the risk of unspecific amplification and will also allow use of the same assay for future expression analysis of the transgene.
   (c) The selected sequence should be BLASTed against appropriate databases to be sure that there is no high degree of homology with other portions of mouse genome. BLAST can be performed by visiting http://www.ncbi.nlm.nih.gov/BLAST.
2. Amplicon design rules
   (a) 50–150 bp in length.
   (b) Whenever possible, the amplicon should not have a G/C content in excess of 60 %.
3. Primer design rules
   (a) T_m is 58–60 °C.
   (b) 20–80 % GC content.
   (c) Less than 2 °C difference in T_m between the two primers.

   (d) The total number of Gs and Cs in the last five nucleotides at the 3′ end of the primer should not exceed two.
   (e) Length 9–40 bases.
   (f) As close as possible to the probe without overlapping.
4. TaqMan MGB probe design rules
   (a) No Gs next to reporter (5′-end).
   (b) More Cs than Gs.
   (c) Probe T_m 10 °C higher than the primers.
   (d) Less than four contiguous Gs.
   (e) 13–20 bases long MGB.

Note again that software programs are available that can design correct primers and probes automatically, and that the researcher only needs to select the region of the construct which will be used for genotyping. Selected primers and probes should be checked by BLAST against appropriate databases in order to ensure that they will not cross-react with mouse genes and so are specific for the transgene.

3.4 Preparing Calibration Standards for qPCR

1. General considerations
   The precision of qPCR genotyping depends significantly on the calibration standards used in the assay. The range of the calibration curve should more than cover the expected copy number of the transgene, which can be from less than 1 (in the case of a mosaic founder) to several hundred c/g. The best available option for genotyping the founders is to utilize as a calibrator the DNA construct that was used at the microinjection step, mixed with normal mouse DNA in such a way that the resulting solutions will have the transgene in concentrations ranging from, say, 5–250 c/g. In addition, normal mouse genomic DNA must be included in the calibration standard PCR reaction to make their reaction conditions parallel to the reaction conditions for the transgenic mouse DNA. For a 384-well plate, each well, for standards as well as samples, routinely contains 4 ng mouse genomic DNA.

   Calibration standards should be prepared in low-EDTA TE (*see* **Note 4**) and must be in linear form (*see* **Note 5**). Multiple freeze–thaw cycles of calibration standards should be avoided (*see* **Note 6**).
2. Preparing the standards from the linear DNA fragment used for microinjection: equations and sample calculations

   We start with the linear DNA fragment used for microinjection at 10 ng/μl. How do we make standards that range from 5 to 250 copies per genome?

   First, note that “copies per genome” is equivalent to “number of molecules of the fragment per genome.”

Also note that each well of a 384-bp plate must contain 4.0 ng of mouse genomic DNA, as part of the standard PCR protocol.

Total diploid mouse genome is 6 × 10^9 bp. Suppouse our transgene has B bp length and C copies per diploid genome. In this case in any taken amount of total transgenic mouse DNA the portion of transgene will be (B × C)/(6 × 10^9). In 4 ng of total DNA we will have 4 × [(B × C)/(6 × 10^9)] ng transgene or 0.666 × B × C × 10^{-6} pg of transgene. Let us say the linear DNA fragment used for microinjection is 10,000 bp (i.e., 10 kb) and we have mouse with 5 c/g of the transgene. Then in 4 ng of taken total DNA we will have 0.666 × 10000 × 5 × 10^{-6} = 0.0333 pg transgenic DNA.

This is the total number of picograms of the linear fragment you would need to add to 4 ng of genomic mouse DNA to obtain a copy number of 5 copies calibration standard per genome.

More generally:

If M= number of ng mouse genomic DNA (usually 4 ng per well in a 384-well plate), C= number of copies of transgene fragment per genome desired in calibration standard, and B= size of linear fragment used for microinjection (in bp) to be placed in calibration standard.

Then the mass of linear fragment used for microinjection (in ng) of size B to add for a final c/g of N copies per genome is:

$$
\begin{aligned}
& M \times 152 \times C \times B \times 1.09674 \times 10^{-12} \\
& \text{ng} = M \times C \times B \times 1.667 \times 10^{-10} \qquad (1) \\
& \text{ng} = M \times C \times B \times 1.667 \times 10^{-7}\ \text{pg}
\end{aligned}
$$

This gives the number of picograms of the linear DNA fragment used for microinjection to add to the well.

Table 1, columns 1–4, exemplifies the use of Formula 1 to determine the picograms of fragment required to prepare calibration solutions with concentrations of 250, 125, 25, 5, and 0 c/g of a 10 kb linear DNA fragment, with 4 ng of total mouse genomic DNA (from a nontransgenic control) included in each well.The volume of the calibration standard (fragment plus 4 ng control mouse genomic DNA) is typically 8 μl for a 384-well plate, as the total volume of the qPCR reaction is typically 20 μl, but 12 μl are taken up by the primers, probe, and reaction mix. This leaves 8 μl left to add the linear fragment (the calibration standard) plus the 4 ng wt (nontransgenic) mouse genomic DNA ("carrier DNA") required in each calibration standard well.The final concentration of the carrier nontransgenic mouse genomic DNA in the calibration standard is 4 ng/8 μl, or 500 ng/ml. This concentration does not vary between calibration standards.

Table 1
Calculations for preparing calibration standards with 4 ng of total DNA in 8 μl solution

1	2	3	4	5	6	7	8
M (ng) (total mouse genomic DNA in well)	*C* (copies per genome; c/g)	*B* (size in bp)	Picograms of standard DNA in well (formula 11.1)	Volume of calibration standard per well	Final concentration of standard to be added to well (pg/μl)	Original concentration of standard DNA fragment (pg/μl)	Final dilution of standard DNA fragment
4	250	10,000	1.667 pg	8 μl	0.2083	10,000	48,000
4	125	10,000	0.833 pg	8 μl	0.1042	10,000	96,000
4	25	10,000	0.167 pg	8 μl	0.0208	10,000	480,000
4	5	10,000	0.033 pg	8 μl	0.0042	10,000	2,400,000
4	0	10,000	0 pg	8 μl	0	–	–

The concentration of the linear fragment in 8 μl of the calibration standard does vary. It can be calculated by dividing the picograms of standard DNA in the well (Column 4) by the volume of the calibration standard (8 μl; Column 5). The concentrations of linear fragment in 8 μl for the five different calibration standards in our example are shown in Column 6 of Table 1. Thus, each calibration standard must contain the same concentration of mouse genomic DNA (500 ng/μl) but different concentrations of the linear DNA fragment used for microinjection (e.g., Column 6). A convenient way to achieve this is simply to start with the linear DNA fragment used for microinjection (concentration 10 ng/μl, or 10,000 pg/μl) and just perform dilutions of the linear fragment into a 500 ng/μl mouse genomic DNA solution.

The fold-dilution required for the DNA fragments is:

$$\frac{\text{Original concentration of DNA standard}\left(\text{pg}/\mu\text{l}\right)}{\text{Final concentration of standard to be added to well}\left(\text{pg}/\mu\text{l}\right)}$$

In our example, dividing the original concentration of the linear DNA standard fragment used for microinjection (in pg/μl) (Table 1, Column 7) by the desired final concentration of this standard fragment in the 8 μl of calibration standard that will be added to the well (Table 1, Column 6) yields the final fold-dilution for the linear DNA fragment (Table 1, Column 8).

Note that the fold-dilutions are quite large and so serial dilutions (all in mouse genomic DNA, 500 ng/μl, in low-EDTA TE buffer) will be required. As an example, the first calibration standard (250 c/g) with final dilution 48,000-fold can be quickly prepared by three consecutive 36.35-fold dilutions of the linear transgenic DNA fragment ($36.35^3 = 48,000$). The less concentrated standards can be prepared by further serial dilutions. Prepared standards should be divided into small (50–100 μl) portions and kept frozen at −70 °C.

It is convenient to prepare the calculation table in Excel and keep the data in the columns 4, 5, 6, and 8 as a "formula," rather than as a "value." In this case one form can be used for any new situation.

3. Preparing calibration standards from previously genotyped mice DNA solutions for genotyping mice are to use offspring born from transgenic parents with previously established copy numbers. For example, assume you are keeping a line of transgenic mice which has 40 c/g in the homozygous state and 20 c/g in the heterozygous state. After mating heterozygous mice, you can expect zero, 20 or 40 c/g in the offspring. You can take DNA isolated from a previously genotyped homozygous animal with 40 c/g as a standard and create from it additional standards with 10 and 20 c/g by simple dilution of the original transgenic DNA with DNA isolated from normal mice, which will serve as a control for zero copies of the transgene.

3.5 Preparing a Plan for the qPCR Experiment

Prepare the following four Excel spreadsheets which will help in qPCR assay:

1. Spreadsheet for dilution of isolated mouse tail or blood DNA, which will contain the ID number of the samples, their DNA concentration, the gene selected for genotyping (target), and the volume of the DNA solution that needs to be taken for dilution.
2. Spreadsheet with the plan for location of the samples, standards, and no-target controls on the 384-well plate. This spreadsheet also should show the total number of wells for each particular detector (TaqMan Assay) and the location of these wells on the plate.

   In this protocol we will place reagents and tested DNA into the plate by columns (from 1 to 24) (*see* **Note 7**).
3. Spreadsheet with the tables for preparing qPCR working reagent mix for each of the qPCR TaqMan assays.

   Each PCR well on the plate should have ready for the qPCR reaction:

   (a) 10 μl of 2× TaqMan Universal PCR Master Mix.

   (b) Prefabricated TaqMan Gene Expression Assay (*see* **Note 3**) or mixture of customized primers and probe.

   (c) 8 μl of prepared DNA solution (4 ng).

   (d) Water to a total volume of 20 μl, if necessary.

Table 2
Preparing the qPCR working reagent mix from a prefabricated TaqMan assay

Component	Volume per tube	For *N* samples
Number of samples	1	100
Water (μl)	1	100
Master mix, 2× (μl)	10	1,000
TaqMan Gene Expression Assay, 20× (μl)	1	100
Total (μl)	12	1,200

The qPCR working reagent mix is a mix of all the necessary reagents except DNA that will be used with a given TaqMan Gene Expression Assay (or customized primer/MGB TaqMan probe set), ready to be pipetted into the 384-well PCR plate.

Table 2 shows an example of preparing the qPCR working reagent mix from a prefabricated TaqMan Gene Expression Assay.

ABI TaqMan MGB probes and primers are supplied premixed at a concentration of 18 μM for each primer and 5 μM for the probe. This is a 20× mix. Thus, the final concentration in the qPCR well will be 900 nM for the primers and 250 nM for the TaqMan MGB probe. If using customized primers and TaqMan MGB probes they should be prepared such that they have the same concentration of primers and probe as in the prefabricated mix (*see* Table 3 as an example).

4. Spreadsheet into which results of qPCR will be transferred from qPCR system after finishing the analysis.

   Prepare in advance a file for the qPCR in the software provided with the qPCR system, containing information about sample locations, calibration standards, no-target controls, and negative controls. Follow the manufacturer's protocol on this step. The conditions of the qPCR reaction will be the same for any ready-to-use TaqMan Assay or for any customized set of primers and probe, designed with the help of the Primer Express 3.0 software:

   First one cycle of:

   95 °C, 10 min

   Then 40 cycles of:

   95 °C, 15 s

   60 °C, 1 min

Table 3
Preparing the qPCR reagent mix from customized primers and TaqMan probe

Component	Volume per tube (μl)	For *N* samples (μl)
Number of samples	1	100
Water	0	0
Master mix, 2×	10	1,000
Forward primer, 36 μM	0.5	50
Reversed primer, 36 μM	0.5	50
TaqMan MGB probe (5 μM)	1	100
Total	12	1,200

3.6 Preparing DNA Solutions for qPCR

The total PCR reaction volume is 20 μl per well for 384-well plates. 8 μl of this is the DNA solution containing 4 ng of genomic DNA (i.e., 8.0 μl of 0.5 μg/ml genomic DNA). Thus, each stock genomic DNA sample, prepared in Subheading 3.1, should be diluted to 0.5 μg/ml in 0.5 ml low-EDTA TE buffer in advance as follows:

1. Place 0.5 ml of low-EDTA TE buffer into 1.5-ml microcentrifuge tubes.
2. Add the appropriate volume of original DNA stock solution, calculated in advance in spreadsheet no. 1.
3. Carefully mix and briefly spin tubes.

The prepared DNA solutions can be kept frozen if necessary.

3.7 Preparing the qPCR Working Reagent Mix

Follow the tables from spreadsheet no. 3 in preparing working reagent mix for each assay. Prepare the ready mixes on the day of use and keep on ice until use.

3.8 Preparing the qPCR Plate and Running qPCR

Placing several hundred samples on 384-well plate in a reasonable time and without mistakes presents some challenges. However, by using an Eppendorf Repeater Plus Dispenser and an 8-Channel 384 Equalizer Pipette, this task can be done manually in about 1–2 h.

1. Place on ice a 384-well qPCR plate.
2. Using an Eppendorf Repeater Plus Dispenser with 0.1-ml tips, dispense 12 μl of the prepared working reagent mixes into each of the wells, following the previously prepared spreadsheet no. 2.
3. As soon as one column is finished, place vertical marks at the top and the bottom of the column outside the rim of the well (*see* **Note 8**). Then proceed to the next column.
4. After reagent mix has been added to the wells that will be used, proceed with addition of DNA (calibration, controls, and the samples that will be tested) to the wells.

5. Using an 8-channel 0.5–12.5 μl 384 Equalizer Pipette and the prepared chart with locations of the samples on the well (spreadsheet no. 2), place into the wells 8 μl of the diluted DNA from step 3.6.3 (again, be sure the plate includes calibration standards and controls as well as sample DNA, all appropriately diluted).
6. Repeat by columns from 1 to 24. As soon as one column is finished, cross the previously placed vertical marks at the top and the bottom of the column.
7. When all DNA samples have been added, remove the plate from ice and cover it with MicroAmp Clear Adhesive Film. Firmly attach the film to the plate by using a MicroAmp Adhesive Film Applicator. Be careful and do not scratch the film.
8. Briefly shake the plate on a plate shaker (an inexpensive attachment to Vortex-Genie 2, Daigger, cat. no. EF22221D can be used).
9. Then, centrifuge the plate 10 s at 800–1,000 × *g* in an appropriate centrifuge, making sure that there are no air bubbles at the bottom of the wells (*see* **Note 9**).
10. Check that the bottom of the plate is clean from any debris so that the plate will not contaminate the PCR system.
11. Transfer the plate into the qPCR system and run the assay following the manufacturer's protocol.
12. After finishing qPCR, remove the plate from the machine and check that in all wells with samples, the adhesive cover effectively prevented samples from evaporating. Partial evaporation of the PCR media will change the PCR conditions and affect the results.

3.9 Analysis of the Results

The software provided with the real-time qPCR system will analyze the obtained data and calculate the results, based on the user-provided information about the standards. Follow the manufacturer's manual for this step. Wells with zero copies of the target or empty tubes will not have the typical sigmoid form of the amplification curve. Sometimes the software will calculate a Ct value and some quantitative data which deviates significantly from what is possible to expect. Amplification plots should be checked and results for wells without amplification curve should be discarded. Also check the quality of the calibration curves. The square of the coefficient of regression (R^2) should be more than 0.98 [3]. The slope of the standard curve should be between −3.1 and −3.6. In this case, reaction efficiency will be 90–110 %. The formula for calculating efficiency is [3]:

$$\text{Efficiency} = \left[10^{(-1/\text{slope})}\right] - 1$$

To obtain a percentage, this number should be multiplied by 100.

The Life Technologies 7900 software SDS 2.4 calculates the results for each detector by rows (from top to bottom) and is able to export the results to an Excel file in this form of the respective tables. Because of this obstacle it is often necessary to transform results presented by rows to ones in which data are placed according to the sequences of the wells in the plate columns. If the software for the qPCR system does not provide this option, the transformation can be easily obtained by using the "Sort" option of the Excel.

Here is an example of two columns (32 samples), where the same detector was used. In this case the software SDS 2.4 presented the results in the form of a table with 32 rows in which samples will be placed in the following order from 1 to 32:

	Column 1	Column 2
A	1	2
B	3	4
C	5	6
D	7	8
E	9	10
F	11	12
G	13	14
H	15	16
I	17	18
J	19	20
K	21	22
L	23	24
M	25	26
N	27	28
O	29	30
P	31	32

However, we need the results presented as shown in the following table:

	Column 1	Column 2
A	1	17
B	2	18
C	3	19
D	4	20
E	5	21

(continued)

(continued)

	Column 1	Column 2
F	6	22
G	7	23
H	8	24
I	9	25
J	10	26
K	11	27
L	12	28
M	13	29
N	14	30
O	15	31
P	16	32

In order to rearrange the data from the upper table to the order shown in the lower table, it is necessary to insert a column to the left of each column in the table, with numbers corresponding to the sequence in which the software is placed our samples, namely,

1, 17, 2, 18, 3, 19, 4, 20, and so on up to the last two samples, 16 and 32. Then use the Excel option "Sort smallest to largest" and samples will be placed in order from 1 to 32. The results can then be transferred into the final table with results (spreadsheet no. 4).

The same approach should be used for any number of the columns. For three columns, the first six numbers will be 1, 17, 33, 2, 18, and 34 and so on up to 16, 32, and 48. For four columns, the first eight numbers will be 1, 17, 33, 49, 2, 18, 34, and 50 and so on up to 16, 32, 48, and 64. From these examples the principle of calculations is evident for all other cases.

4 Notes

1. For qPCR genotyping, it is important for all samples of DNA used in an assay as well as the DNA used as calibration standards to have the same quality. Thus, all samples of DNA must be isolated by the same method. The DNA to be analyzed and the calibration standards should be dissolved in the same buffer. This ensures that differences in purity and size will be minimal and will not affect the efficiency of amplification and the calculated number of transgene c/g.
2. Using multichannel expandable electronic pipettors like Matrix EXP or Equalizer (Thermo Scientific, *see* below) allows rapid

transfer of DNA solutions as well as calibration standards included in the kit from microcentrifuge test tubes placed in an 80-well rack (Daigger, cat. no. EF29025A) to 96-well plates with wells containing working solution.

3. Life Technologies has more than 1,000,000 ready-to-use TaqMan Gene Expression Assays available. The following table from Life technologies provides an explanation about all types of assays which they manufacture:

Assay suffix	Assay placement
"_m"	An assay whose probe spans an exon-exon junction and will not detect genomic DNA
"_s"	An assay whose primers and probes are designed within a single exon; such assays will, by definition, detect genomic DNA
su "_g"	Although the assay probe spans an exon-exon junction, the assay may detect genomic DNA if present in the sample
"_mH", "_sH", or "_gH"	The assay was designed to a transcript belonging to a gene family with high sequence homology. The assays have been designed to give between 10 Ct and 15 Ct difference between the target gene and the gene with the closest sequence homology. This means that an assay will detect the target transcript with 1,000–30,000-fold greater discrimination (sensitivity) than the closest homologous transcript, if they are present at the same copy number in a sample

Keep in mind that Life Technologies does not provide precise information about their assays. In case of doubt, it is often more simple to design and order your own assay, where every nucleotide is known.

4. The nontransgenic mouse DNA used as a control ("carrier DNA") must be dissolved in TE with low EDTA, as higher levels of EDTA can interfere with the PCR reaction.

5. The standard DNA used in calibration solutions must be in linear form, as circular and linear DNA have different amplification efficiencies.

6. Note that each freeze–thaw cycle will gradually partially degrade prepared DNA standards and will decrease its efficiency of amplification in qPCR. As a result, the calculated number of copies of the transgene will be larger than the actual number. Thus, we recommend aliquoting standards into relatively small portions, in order to limit the number of freeze–thaw cycles to about five times, maximum. If necessary, standards should be prepared again.

7. Keep in mind that the ABI SDS 2.4 software calculates results by rows (from A to P) and is able to export them in Excel file in the form of the table with the same sequence of rows. However, when 384-well plates are loaded manually with the aid of a 0.5–12.5 μl 8-channel 384 Equalizer Pipette, the risk of mistakes is lessened, and it is more convenient to place samples by 16-well columns (just 8 × 2), than by 24-well rows. If this method is used, then the results presented in a table by rows after the run must be converted to the same sequence as samples were placed (by column). *See* Subheading 3.9 for explanation. In this protocol we will place reagents and DNA samples into the plate by columns (from 1 to 24).
8. Do not use a plastic ruler to cover wells to mark them as filled, as the contents will transfer to the ruler due to electrostatic interactions.
9. In an emergency, you can aliquot everything (working reagent mix plus DNA) into the plates and freeze and run the next day. The Taq Gold enzyme present in the mix is sequestered in inactive form by binding to anti-Taq antibody, which is removed only by heating.

References

1. Nagy A, Gertenstein M, Vintersten K, Behringer R (2003) Chapter 12. Manipulating the mouse embryo: a laboratory manual. Cold Spring Harbor Laboratory Press, Cold Spring Harbor, NY
2. Sambrook J, Russell DW (2001) Molecular cloning: a laboratory manual, vol 1. Cold Spring Harbor Laboratory Press, Cold Spring Harbor, NY, pp 6.33–6.64
3. Life Technologies. Real-time PCR handbook
4. Mullenbach R, Lagoda PJL, Welter C (1989) An efficient salt-chloroform extraction of DNA from blood and tissues. Trends Genet 5:391

Chapter 12

Generation of General and Tissue-Specific Gene Knockout Mouse Models

Xian-Cheng Jiang

Abstract

Knockout technology has established the functions of many genes affecting plasma lipid and lipoprotein levels and the development of atherosclerosis. However, many genes remain to be characterized. The ability to produce mice lacking whole-body expression of a given gene is still one of the most powerful techniques available for determining gene function. A complementary approach, underutilized yet vitally important to understanding lipoprotein metabolism, is the ability to create mice with gene deficiency only in a specific tissue. Liver, intestine, and macrophages are the major tissues and cells involved in lipoprotein metabolism and atherosclerosis, and additional tissues such as adipose tissue and brain are also of interest. Thus, feasible approaches to prepare general and tissue-specific gene knockout mouse models are necessary. Here, we describe our general procedure for generating whole-body knockout mice, using as an example the preparation of general (whole-body) phospholipid transfer protein (PLTP) gene knockout mice. We also describe several approaches to generating liver, intestine, and myeloid cell-specific tissue-specific knockout mice, using as an example the preparation of tissue-specific knockout mice for the subunit 2 of serine palmitoyltransferase (SPT), a key enzyme for sphingomyelin de novo synthesis. Bone marrow transplantation is an alternative means of creating myeloid cell-specific knockout mice. The general principles and techniques described here apply to the establishment of other gene knockout mouse models as well. The ability to manipulate gene expression in specific tissues as well as throughout the entire body of the mouse is anticipated to yield novel insights into lipid and lipoprotein metabolism and the development of atherosclerosis.

Key words General gene knockout, Liver-specific knockout, Intestine-specific knockout, Myeloid cell-specific knockout, Phospholipid transfer protein, Serine palmitoyl-CoA transferase, Lipid metabolism

1 Introduction

Plasma lipoprotein metabolism plays critical roles in the development of atherosclerosis. The liver is the major site of lipoprotein synthesis, secretion, and degradation. Very-low-density lipoprotein (VLDL) and high-density lipoprotein (HDL) are the two main classes of lipoprotein particles that are synthesized and secreted by the liver. The specific protein moieties [apolipoproteins (apo)] of these lipoprotein particles are also synthesized in the liver.

Lita A. Freeman (ed.), *Lipoproteins and Cardiovascular Disease: Methods and Protocols*, Methods in Molecular Biology, vol. 1027, DOI 10.1007/978-1-60327-369-5_12, © Springer Science+Business Media, LLC 2013

ApoB-100 [1] and apoA-I [2] are the major apolipoproteins of VLDL and HDL, respectively.

The small intestine is the site for absorption of dietary lipids and their secretion into the blood. Lowering plasma cholesterol is important because its levels closely relate to the cardiovascular and metabolic disorders. About 30 % of plasma cholesterol is derived by intestinal absorption [3]. It has been estimated that a 36–37 % reduction in plasma cholesterol levels could be achieved by total inhibition of cholesterol absorption [3]. Absorption is a multistep process in which cholesterol is micellized by bile acids and phospholipids in the intestinal lumen, taken up by the enterocytes, assembled into lipoproteins, and transported to the lymph and the circulation.

Foam cell formation due to excessive accumulation of cholesterol by macrophages is a pathological hallmark of atherosclerosis [4]. Macrophage scavenger receptor class A is implicated in the deposition of cholesterol in arterial walls during atherogenesis, through receptor-mediated endocytosis of chemically modified low-density lipoproteins [5, 6]. One member of the scavenger receptor class B family, CD36, is also involved in macrophage foam cell formation [7]. Macrophages cannot limit their uptake of cholesterol and therefore depend on cholesterol efflux pathways for preventing their transformation into foam cells. Several ABC-transporters, including ABCA1 [8] and ABCG1 [9], facilitate the efflux of cholesterol from macrophages.

It is conceivable that liver-, intestine-, and macrophage-controlled lipid metabolism-related gene expressions specifically would influence the levels of plasma lipids. In this chapter, we will describe the general and tissue-specific gene knockout approaches for lipid metabolism-related proteins. These general principles are applicable to other knockout mouse preparations. We will use phospholipids transfer protein (PLTP) and serine palmitoyltransferase (STP) subunit 2 as two examples to describe the procedure for general and tissue-specific knockout mouse preparations.

2 Materials

2.1 Isolation of MEF

1. Mouse. Any mouse line with a neomycin-resistant gene insertion can be used to make MEFs.
2. 70 % ethanol.
3. Forceps and scissors. Sterilize by autoclaving or dipping in 70 % ethanol prior to use.
4. Ca/Mg-free PBS.
5. 50-ml screw-cap tube, sterile.
6. Dulbecco's modified Eagle's medium (DMEM; Gibco, Cat #11965-118).

7. BSA, 10 mg/ml, tissue culture grade. Store at 4 °C.
8. DMEM with 1 mg/ml BSA.
9. 10-ml syringes.
10. 16G needle.
11. 18G needle.
12. 0.25 % Trypsin with 0.02 % (1 mM) EDTA in PBS or DMEM.
13. DNase I (25 units/μl) or 10 mg/ml frozen stock in PBS, specific activity 2,500 units/mg, Sigma D-4527.
14. Rotator at 37 °C.
15. DMEM (high glucose).
16. Fetal calf serum (Hyclone).
17. Pen/Strep (100×).
18. 2-Mercaptoethanol.
19. MEM nonessential amino acids (100×).
20. Glutamine-200 mM (100×).
21. Leukemia inhibitory factor (1,000 units/ml final concentration).
22. ES media: Per 500 ml, combine:
    (a) 410 ml DMEM (high glucose).
    (b) 75 ml Fetal calf serum (Hyclone).
    (c) 5 ml Pen/Strep (100×).
    (d) 5 μl 2-mercaptoethanol.
    (e) 5 ml MEM nonessential amino acids (100×).
    (f) 5 ml Glutamine-200 mM (100×).
    (g) 50 μl leukemia inhibitory factor (1,000 units/ml final concentration).
23. Treat 8×10-cm tissue culture plates (Corning) with sterile 0.1 % gelatin. Swirl 2–3 ml gelatin to fully cover the plate and let it stand for 5–10 min. Aspirate the gelatin solution and discard. There is no need to dry the plates following treatment.
24. CO_2 incubator (5 % CO_2, 37 °C).
25. 2 mg Mitomycin C (Sigma M-0503).
26. 1× D-PBS.
27. 1× mitomycin medium: Dissolve 2 mg Mitomycin C (Sigma M-0503) in 2 ml 1× D-PBS, and add this 2-ml mixture to 200 ml ES cell medium.

    Number of MEF feeder cells needed for each plate or well ($1 \times 10^5/cm^2$)

    (a) 150-mm dish (88 cm^2): 9×10^6 cells.
    (b) 100-mm dish (40 cm^2): 4×10^6 cells.

(c) 60-mm dish (14 cm^2): 1.4×10^6 cells.
(d) 35-mm well (4.5 cm^2): 5×10^5 cells/well (e.g., 6-well plate).
(e) 25-mm well (2.4 cm^2): 2.5×10^5 cells/well (e.g., 12-well plate).
(f) 1.5-cm well (0.9 cm^2): 9×10^4 cells/well (24-well plate or Nunc 4-well plate).
(g) 0.7-cm well of 96-well plates (0.2 cm^2): 2×10^4 cells/well.

2.2 ES Cell Growing, Transfection, and Selection

1. ES cells (ES-E14TG29, ATCC).
2. ES medium.
3. T-25 flask containing a monolayer of inactivated MEF cells.
4. Inactivated MEF cells in 10 cm plate.
5. 1.8 ml of transfection buffer (20 mM HEPES, pH 7.05, 137 mM NaCl, 5 mM KCl, 0.7 mM Na_2HPO_4, 6 mM dextrose).
6. 25 μg of linearized DNA (targeting vector, see below).
7. Electroporator (Bio-Rad Gene Pulser Electroporator, Bio-Rad).
8. Electroporation cell (Bio-Rad, Cat. No. 165-2088).
9. 15 ml conical tube containing 10 ml fresh and warm ES medium.
10. Selection medium containing G418 (300–400 μg/ml).
11. Trypsin.
12. 96-well round-bottom plate.
13. 24-well plate which is coated with a monolayer of inactivated MEF cells and contains 0.5 ml ES cell medium without G418.
14. Fresh ES cell medium without G418.
15. 6-well plate coated with a MEF cell monolayer.
16. Cell freezing media.
17. Cryogenic vials for freezing cells.
18. Liquid nitrogen.

2.3 DNA Extraction from ES Cells

1. 2× Cell lysis buffer (20 mM Tris–HCl, pH 7.5, 20 mM EDTA, 1 % SDS, 400 μg/ml proteinase K).
2. Phenol.
3. Chloroform.
4. NH_4OAc (7.5 M).
5. Isopropanol.
6. 70 % Ethanol.

7. Speed vacuum centrifuge.
8. 1× TE buffer.

2.4 Blastocyst Isolation, Targeted ES Injection, and Transplantation

2.4.1 Animal Setting

1. Ten C57BL/6J female mice (4–6 weeks, maximum pregnant mare's serum (PMS) responding age) for harvesting blastocysts.
2. Ten C57BL/6J male mice (8–20 weeks) for mating.
3. Ten to fifteen Swiss Webster female mice (7–15 weeks) for preparing pseudopregnant females.
4. Ten to fifteen vasectomized male mice (sterile stud male mice, 8–24 weeks) for mating with Swiss Webster females.

2.4.2 Materials

1. Stereomicroscope, *see* Subheading 3.
2. PMS, 5 units/C57BL/6J female mouse.
3. Human chorionic gonadotropin 5 units/C57BL/6J female mouse.
4. ES cell medium.
5. Humidified incubator with 5 % CO_2 at 37 °C.
6. Injection pipette.
7. Holding pipette.
8. DMEM plus 10 % FBS.
9. Light paraffin oil.
10. 2.5 % Avertin (0.016 ml/g).
11. Blunt fine forceps.
12. 26G needle.
13. Serrefine clamp.
14. Needles and thread for sewing up mouse.

2.5 PLTP KO (Conventional) Mouse Preparation

2.5.1 Construction of PLTP Gene Replacement Vector

1. Targeting vector (NeoR gene in reverse direction).
2. Primer pairs for short arm.
3. Primer pairs for long arm.

Important: The long and short arms must not contain a site for the enzyme that will be used to linearize the vector/make the fragment.

4. PacI or other restriction enzyme for linearizing targeting vector.
5. Restriction enzyme to screen for homologous integrants.
6. Probe to screen for homologous integrants.
7. Southern blot apparatus.
8. Two primers, one located outside of targeting vector and the other at the 3′ end of the neomycin-resistant gene to perform PCR (*see* Fig. 1a).

2.5.2 PLTP KO Mouse Preparation

1. Upstream probe for Southern analysis.
2. cDNA probe for Northern analysis.

2.6 Serine Palmitoyltransferase subunit 2 Tissue-Specific Mouse Preparation

"Sptlc2-Flox" is presented in this chapter as a model for the user to subunit 2 follow in creating gene replacement vectors and tissue-specific mice for their own gene of interest. Specific materials used will vary depending on the gene. *See* Subheading 3.6.1 for further guidelines.

2.6.1 Construction of "Sptlc2-Flox" Gene Replacement Vector

2.6.2 Preparation of Sptlc2$^{Flox/Frt}$ Mice

1. Mouse blastocysts with 129 (an inbred mouse strain) background.
2. C57BL/6 females.

2.6.3 Preparation of Sptlc2$^{\Delta Neo}$ Mice

1. Heterozygous Sptlc2$^{Flox-FRT}$ mice.
2. Flp transgenic mice (the Jackson Laboratory (Maine)).

2.6.4 Preparation of Sptlc2-Flox Mice

3. An internal 250 bp probe for Southern blot analysis (Fig. 3a).

2.7 Preparation of a Liver-Specific Sptlc2 KO Mouse

2.7.1 Adenovirus-Cre Method

1. Adenovirus (AdV)-Cre (ViraQuest, Inc).
2. Sptlc2-Flox mice.
3. WT mice.

2.7.2 Pharmacological Induction of a Cre Transgene

1. Mx1-Cre transgenic (Jackson Laboratories (Maine)).
2. Polyinosinic-polycytidylic ribonucleic acid (pIpC) (Sigma).
3. Sptlc2-Flox mice.
4. WT mice.

2.7.3 Albumin Promoter/Enhancer-Cre Transgenic Mice

1. Albumin promoter/enhancer-Cre transgenic mice (Jackson Laboratory).
2. Sptlc2-Flox mice.
3. WT mice.

2.8 Intestine-Specific Gene KO Mice

1. Vil-Cre transgenic mice (Jackson Laboratory).
2. Sptlc2-Flox mice.
3. WT mice.

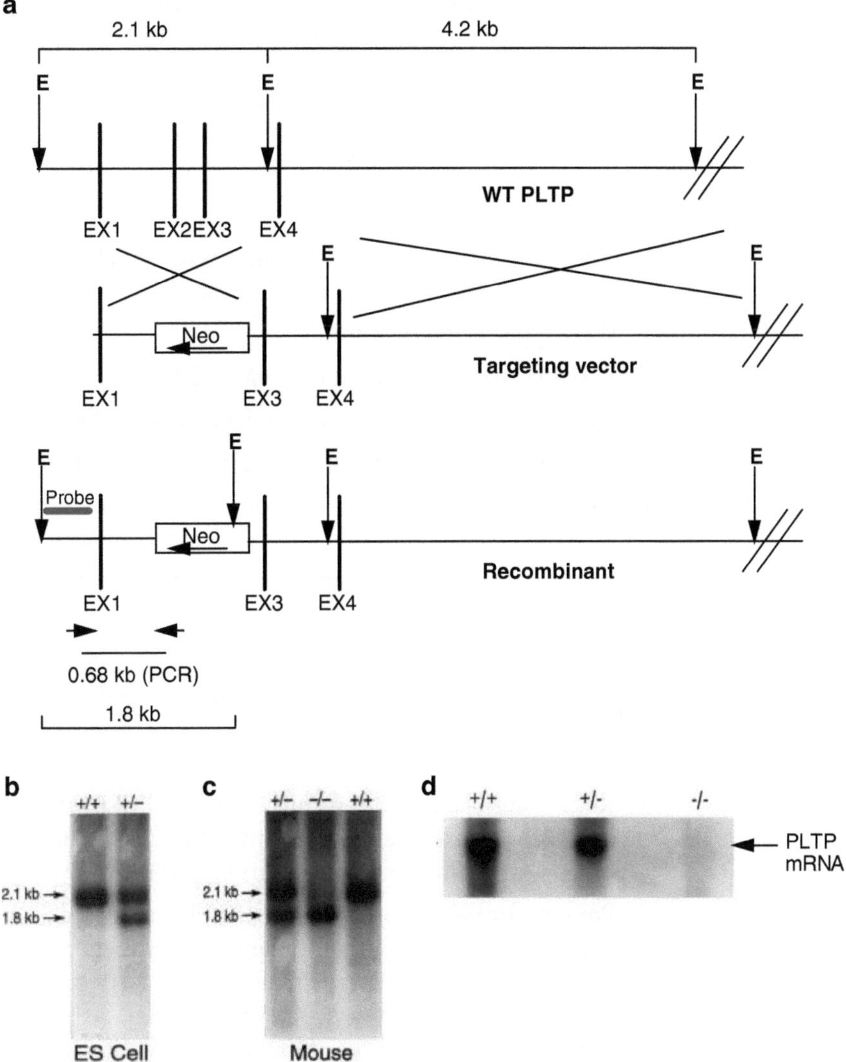

Fig. 1 PLTP KO mouse preparation. (**a**) Strategy used to disrupt the mouse PLTP gene. The *top line* represents the map of the endogenous murine PLTP gene and its flanking sequence. The *middle line* represents the vector used to target the PLTP locus. The *bottom line* shows the predicted organization of the locus after homologous recombination. A probe and a pair of PCR primers indicated in this line were used to confirm the integrity of site-specific integration. (**b**) Southern blot analysis of ES cells, and (**c**) mouse tail-tip genomic DNA, respectively, digested with EcoRI and hybridized with the probe. Normal ES cell DNA and control mice DNA with 2.1-kb signal only (+/+); targeted ES cell DNA and heterozygous deficient mice with 2.1-kb and 1.8-kb signal (+/–); homozygous deficient mice with 1.8-kb signal only (–/–). (**d**) Northern blot of poly(A)$^+$ RNA (2 μg) from lung probed with a mouse PLTP cDNA (666 bp, nucleotides 3–669). *ES* embryonic stem, *neo* neomycin-resistant gene, *PLTP* phospholipid transfer protein

2.9 Preparation of Macrophage-Specific Gene KO Mice

2.9.1 LysM-Cre Transgenic Mouse Approach (myeloid cell-specific)

1. LysM-Cre transgenic mice which express Cre recombinase from the endogenous lysozyme locus (Jackson Laboratory).
2. Sptlc2-Flox mice.
3. WT mice (myeloid cell-specific).

2.9.2 Bone Marrow Transplantation Approach (hematopoietic cell-specific)

1. Male donor mice with whole-body knockout of gene of (hematopoietic cell-specific) interest.
2. Wild-type recipient mice.
3. DMEM with 5 % FBS and 1 mM HEPES.
4. Irradiator capable of delivering irradiated 1,000 rad (10 Gy).
5. PCR primers for monitoring the process of cell replacement.
6. Genomic DNA isolation kit from mouse white blood cells.

3 Methods

3.1 Isolation of MEF

ES cells (ES-E14TG29, ATCC) must grow on top of a monolayer of mouse embryonic fibroblasts (MEF).

Procedure

1. Kill the mouse and dip it into 70 % ethanol for 5 min. Dissect out 13.5-day mouse embryos into PBS. Scoop out the internal organs (anything that appears to be red in color).
2. Put ten carcasses into a 50-ml sterile tube, then add 5 ml (DMEM with 1 mg/ml BSA), then disaggregate the embryos into small pieces by sucking embryos in and out of a 10-ml syringe with a 16G needle attached to it, repeat three times, then use an 18G needle to repeat this procedure for another three times.
3. Add 2 ml 0.25 % trypsin/EDTA and 10 μl DNase I (25 units/μl or 10 mg/ml frozen stock in PBS, specific activity 2,500 units/mg, Sigma D-4527) into this tube and shake the tube on a rotator at 37 °C for 15 min.
4. Add 5 ml ES medium to the tube and try to break up the embryo block further by pipetting with a 10-ml pipette several times.
5. Add 40 ml ES medium and mix gently. Let the tube sit still for 5 min for large embryo pieces to settle to the bottom.
6. Dispense the cells to 10-cm gelatinized tissue culture plates, incubate in a CO_2 incubator overnight, then change medium next morning. The plates should be confluent in 24–36 h.
7. Split the plates 1:5 to make passage #2 MEF cells and either make the fibroblasts into feeders by mitomycin C treatment or freeze them in one vial/10 cm confluent plate for future use (*see* **Note 1**).

8. Mitomycin C inactivation of MEF.
   (a) Dissolve 2 mg mitomycin C (Sigma M-0503) in 2 ml 1× D-PBS and add this 2 ml mixture to 200 ml ES cell medium. This will make 1× mitomycin C medium.
   (b) Treat MEF with 1× mitomycin C medium at 37 °C for 2–3 h. Remove mitomycin C medium and wash MEF three times with 1× D-PBS to wash away any residual mitomycin C.
   (c) Trypsinize MEF and aliquot the appropriate amount of mitomycin C-treated feeder cells to plates. Plates should be ready after 2 h.

   Number of MEF feeder cells needed for each plate or well ($1 \times 10^5/\text{cm}^2$)
   - 150-mm dish (88 cm^2): 9×10^6 cells.
   - 100-mm dish (40 cm^2): 4×10^6 cells.
   - 60-mm dish (14 cm^2): 1.4×10^6 cells.
   - 35-mm well (4.5 cm^2): 5×10^5 cells/well (e.g., 6-well plate).
   - 25-mm well (2.4 cm^2): 2.5×10^5 cells/well (e.g., 12-well plate).
   - 1.5-cm well (0.9 cm^2): 9×10^4 cells/well (24-well plate or Nunc 4-well plate).
   - 0.7-cm well of 96-well plates (0.2 cm^2): 2×10^4 cells/well.

3.2 ES Cell Growing, Transfection, and Selection

The whole procedure takes about 20–25 days.

Day 1. Thaw ES cells. Place cells into a T-25 flask containing a monolayer of inactivated MEF cells.

Day 2. Feed ES cells.

Day 3. Expand ES cells. Place cells on inactivated MEF cells in 10 cm plate.

Day 4. Feed ES cells.

Day 5. Collect ES cells by trypsinization from a 10-cm plate. Resuspend the cells in 1.8 ml of transfection buffer (20 mM HEPES, pH 7.05, 137 mM NaCl, 5 mM KCl, 0.7 mM Na_2HPO_4, 6 mM dextrose). Gently mix 25 μg of linearized DNA (targeting vector, see below) with 900 μl of the cells and place in the electroporation cell and transfect. We routinely use 250 V, 500 microF for the Bio-Rad GenePulser. Immediately remove the transfected cells to a 15-ml conical tube containing 10 ml fresh and warm ES medium, and then add it on top of MEF cells (monolayer).

Days 6–12. Feed transfected ES cells with selection medium containing G418 (300–400 μg/ml).

Day 13. Select ES colonies that are undifferentiated. The best colonies to select are rounded or oval in shape with a phase bright edge, and often a dark necrotic center. Colonies that are differentiated may appear flat, and be surrounded by "cobblestones" of fibroblast-like cells. These differentiated colonies are not picked. Place 50 μl trypsin in each well of a 96-well round-bottom plate. Using a 20 μl Pipetman, pick individual colonies from the 10-cm plate and place each colony in a well of the 96-well plate. After the ES cells of each clone are dissociated, they are transferred to a well of a 24-well plate which is coated with a monolayer of inactivated MEF cells and contains 0.5 ml ES cell medium without G418. Repeat the selection process until about 200 colonies are picked.

Days 14–16. Feed all 24-well plates of selected ES clones with fresh medium without G418.

Day 17. Expanding the ES clones. Transfer one well ES cells from the 24-well plate to one well of a 6-well plate coated with a MEF cell monolayer. Repeat for each well of the 24-well plate. Incubate with fresh medium without G418.

Days 18 and 19. Feed all 6-well plates of selected ES clones with fresh medium without G418. The healthy ES cells should be 95–100 % confluent at day 19.

Day 20. Freezing cloned ES cells and collecting DNA for screening. Place 500 μl trypsin in each well of a 6-well plate and incubate in tissue culture hood for 5 min. Half of the dissociated ES cells (about 250 μl) are collected into an Eppendorf tube for DNA extraction (*see* below). The remaining half are quickly mixed with 500 μl of cell freezing media and stored in a capped tube (cryogenic vial) in liquid nitrogen.

3.3 DNA Extraction from ES Cells

1. Add 250 μl of 2× cell lysis buffer (20 mM Tris–HCl, pH 7.5, 20 mM EDTA, 1 % SDS, 400 μg/ml proteinase K) to above 250 μl of ES solution and incubate at 52°C from 1 h to overnight.
2. Extract the DNA once with phenol and once with chloroform.
3. Add 250 μl NH_4OAc (7.5 M) and 750 μl isopropanol and mix thoroughly.
4. Spin to pellet the DNA and then wash the pellet once with 70 % ethanol.
5. Dry the pellet in a speed vacuum centrifuge.
6. Resuspend the pellet in 100 μl 1× TE buffer and dissolve at 52 °C for 1–2 h before restriction enzyme digestion or PCR analysis.

3.4 Blastocyst Isolation, Targeted ES Injection, and Transplantation

3.4.1 Animal Setting

1. Ten C57BL/6J female mice (4–6 weeks, maximum pregnant mare's serum (PMS) responding age) for harvesting blastocysts.
2. Ten C57BL/6J male mice (8–20 weeks) for mating.
3. Ten to fifteen Swiss Webster female mice (7–15 weeks) for preparing pseudopregnant females.
4. Ten to fifteen vasectomized male mice (sterile stud male mice, 8–24 weeks) for mating with Swiss Webster females.

3.4.2 Procedure

Day 1: At 1:00 p.m., intraperitoneal injection of PMS (mimic FSH in inducing superovulation), 5 units/C57BL/6J female mouse.

Day 3: At 1:00 p.m., intraperitoneal injection of human chorionic gonadotropin (HCG, mimics LH in inducing superovulation), 5 units/C57BL/6J female mouse. Set up mating—C57BL/6J female with C57BL/6J male—for harvesting blastocysts.

Day 6: Mate Swiss Webster females with vasectomized males to prepare pseudopregnant females.

Day 7: At 9:00 a.m., check for plugged Swiss Webster females, which will be used as pseudopregnant blastocyst recipient females.

At 9:30 p.m., collect blastocysts from mated C57BL/6J females for blastocyst harvesting. The blastocysts are incubated in ES medium in a humidified incubator with 5 % CO_2 at 37 °C until they are sufficiently expanded.
Collect targeted ES cells (*see* Subheading 3.2, Day 13).

Draw approximately 10–15 ES cells into the injection pipette and position them near the tip in a minimal amount of medium. Then, immobilize a single blastocyst by applying suction to the holding pipette. Insert the transfer pipette containing ES cells into the blastocyst and blow on the pipette until the ES cells enter the blastocyst. Incubate the injected embryos in a humidified incubator with 5 % CO_2 at 37 °C in microdrops of DMEM plus 10 % FBS under light paraffin oil. Three hours later, they are ready for uterine transfer.

Uterine transfer: weigh the pseudopregnant blastocyst recipient mouse and then anesthetize by injecting it intraperitoneally with 0.016 ml/g of 2.5 % avertin. On the lateral side of the mouse, make a small incision of skin and underlying muscle, and then pull out the ovary, oviduct, and uterus. Hold the top of the uterus gently with blunt fine forceps and use a 26G needle to make a hole in the uterus. Make sure that the needle has entered the lumen. Pull out the needle and insert approximately 5 mm of the prepared transfer pipette containing the blastocysts. Blow gently on the transfer pipette until the air bubble closest to the blastocysts is at the tip of the pipette and all of the blastocysts have been expelled. Unclip the Serrefine clamp and remove the mouse from the stage of the stereomicroscope. Use blunt fine forceps to pick up the fat pad and place the uterus, oviduct, and ovary back inside the body wall. Sew up the body wall.

3.5 PLTP KO (Conventional) Mouse Preparation

3.5.1 Construction of PLTP Gene Replacement Vector

The overall strategy for PLTP gene targeting was to replace intron 1, exon 2, and part of intron 2 with the neomycin-resistant gene (Fig. 1a). Because exon 2 contains the translation initiation codon, ATG, as well as the signal peptide and the first 16 amino acids of mature PLTP, deletion of exon 2 would be expected to create a null PLTP mouse model. The short arm (530 bp) and long arm (5.2 kb) of the targeting vector were prepared from genomic DNA by PCR using two pairs of primers.

First pair:

5′-CCACGTGACCACACTACTAAG-3′ (located in first exon)

5′-TGGTCATGCAACTAGAACGG-3′ (located in the end of intron 1)

Second pair:

5′-GGGACCTGAGTGTCTCCATG-3′ (located in intron 2)

5′-CGGAATTCCATCTCGAGGTTGCCGT-3′ (located in intron 6).

The direction of the neomycin resistance gene (Neo) expression is opposite of that of PLTP in the targeting vector (*see* **Note 2**). Embryonic stem (ES) cells were electroporated by PacI-linearized targeting vector (PLTP targeting vector; note that the Pac1 site is on the multicloning site of the vector and that there is no Pac1 site in the region of interest) and screened by sequential selection with G418. To screen for homologous integrants, genomic DNA from ES cells was digested with EcoRI, and a 5′ PLTP flanking probe (*see* **Note 3**) was used to analyze Southern blots (Fig. 1a–c). The addition of a 1.8-kb signal to the endogenous 2.1-kb signal indicated site-specific integration at the PLTP locus. Two primers, one located outside of targeting vector and the other at the 3′ end of the neomycin-resistant gene (Fig. 1a) (*see* **Note 4**), were also used to perform PCR.

3.5.2 PLTP KO Mouse Preparation

The correctly targeted cells were then injected into C57BL/6J host blastocysts. Fifteen chimeras (*see* **Note 5**) were generated (12 male, 3 female), and eight of the resulting males transmitted the disrupted PLTP allele through the germline. The resulting heterozygous mice were crossed, and the targeted allele segregated in a Mendelian fashion. Of 96 progeny analyzed from heterozygous crosses by Southern blotting of tail-tip DNA, 25 (26 %) of the progeny were wild type, 51 (53 %) heterozygous, and 20 (21 %) homozygous for the disrupted allele (Fig. 1c). Homozygous crosses yielded viable progeny. A Northern blot of RNA prepared from mouse lung was probed with a mouse PLTP cDNA probe (666 bp of 5′ end) and showed that there was no 1.7-kb PLTP message in homozygous deficient animals (PLTP$^{-/-}$) and that there was a reduction of the message in heterozygous deficient animals (PLTP$^{+/-}$) (Fig. 1d).

3.6 Serine Palmitoyltransferase subunit 2 Tissue-Specific Mouse Preparation

Serine palmitoyltransferase (SPT) is the rate-limiting enzyme in the subunit 2 biosynthesis of sphingolipids [10]. Mammalian SPT contains two subunits, Sptlc1 and Sptlc2, encoding 53- and 63-kDa proteins, respectively [11, 12]. Both subunits are necessary for SPT activity. SPT is known to play an important role in the metabolism of SM and atherosclerosis [13, 14]. Since conventional homozygous Sptlcl and Sptlc2 KO mice are embryonic lethal

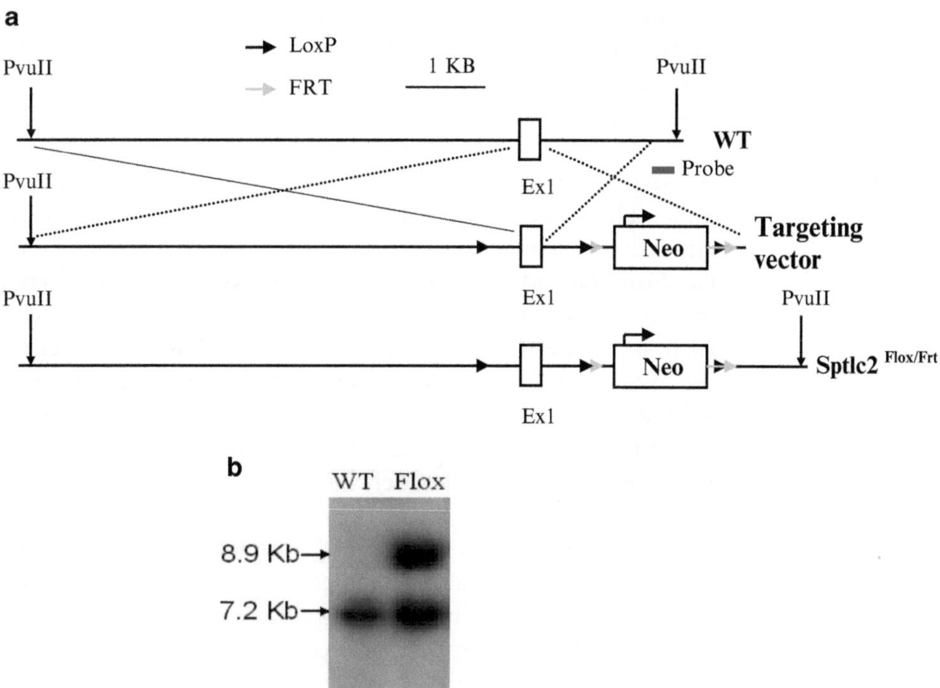

Fig. 2 Sptlc2$^{Flox-FRT}$ mouse preparation. (**a**) Strategy. A 300 bp fragment was used as a probe for Southern blot analysis to check homologous recombination. (**b**) Southern blot of WT and Sptlc2$^{Flox-FRT}$ mice. The WT mouse has a 7.2 kb PvuII digested signal; the heterozygous Flox mouse has 7.2 and 8.9 kb signals

[15], we planned to use a tissue-specific KO approach to study the relationship between SPT and SM metabolism. We chose Sptlc2 as a tissue-specific target.

The concept of making a tissue-specific knockout is straightforward. First, a typical targeting construct containing a neomycin resistance gene is made as usual, with the key differences being that the region to be excised is flanked by Flox sites; the adjacent neo gene is flanked by FRT and Flox sites; and no endogenous sequences are deleted (*see* Fig. 2). After targeted recombination as usual, the Flox/NeoFRT mice are then mated with mice that express FLP under control of the human EF-1α gene promoter, which excises the genomic region between the FRT sites (i.e., the neoR gene) in all tissues of the body. This leaves the gene in an almost wild-type state, with only FLOX sites flanking the region to be excised. (Ideally these FLOX sites are placed in intronic regions, where they will have minimal impact on gene function.) To excise the portion of the gene flanked by FLOX sites only in a specific tissue, Cre recombinase is expressed in a tissue-specific manner—e.g., by introducing adenovirus expressing Cre recombinase into the liver, or crossing with mice that express Cre recombinase under a tissue-specific promoter and/or enhancer, such as the albumin promoter/enhancer for liver, villin for intestine, and LysM for macrophages.

Below we describe the creation of mice with selective inactivation of the Sptlc gene in liver, intestine, and macrophages. The general strategy can be used to create tissue-specific inactivation of any gene.

3.6.1 Construction of the Sptlc2-Flox Gene Replacement Vector

A Loxp site was inserted in the 5′ flanking region of the Sptlc2 gene. The targeting vector was Sptlc2$^{Flox/FRT}$ (Fig. 2a). A neomycin-resistant gene (Neo) cassette, flanked with a pair of Loxp sites (*see* **Note 6**) and a pair of FRT sites (*see* **Note 7**), was inserted into intron 1 (Fig. 2a), creating the targeting vector. A probe complementary to a 330 bp fragment of the gene 3′ to the targeting vector was used first to detect successfully targeted ES cells, and subsequently in screening for positive mice. The recombinant sequence contains sites allowing for subsequent creation of mice with tissue-specific Sptlc2 deficiency.

3.6.2 Preparation of Sptlc2$^{Flox/Frt}$ Mice

Four positive ES cells with C57BL/6 background were identified out of 210 screened and were injected into mouse blastocysts with 129 background. Chimeric males were mated with C57BL/6 females, and F1 generation mice containing the Sptlc2$^{Flox/Frt}$ genotype were identified (Fig. 2b).

3.6.3 Preparation of Sptlc2$^{\Delta Neo}$ Mice

To delete the Neo cassette in the Sptlc2$^{Flox/Frt}$ mouse, we crossed heterozygous Sptlc2$^{Flox\text{-}FRT}$ mice with Flp transgenic mice (Fig. 3) [16, 17], which we obtained from the Jackson Laboratory (Maine). To facilitate screening for the successful deletion, we designed an internal 250 bp probe for Southern blot analysis (Fig. 3a). We have obtained three mosaic mice with more than 60 % of Sptlc2$^{\Delta Neo}$ (Fig. 3b) (*see* **Note 8**). We then crossed them to WT mice to establish 100 % of Sptlc2$^{\Delta Neo}$ mice (heterozygous) (Fig. 3c).

3.6.4 Preparation of Sptlc2-Flox Mice

Heterozygous Sptlc2$^{\Delta Neo}$ mice were crossed with each other. We obtained 20 Sptlc2-Flox (homozygous for Sptlc2$^{\Delta Neo}$) mice (Fig. 3d).

3.7 Preparation of a Liver-Specific Sptlc2 KO Mouse

3.7.1 Liver-Specific Sptlc2 Inactivation by Injecting Adenovirus-Cre into Sptlc2-Flox Mice

To investigate whether a Sptlc2-KO gene could be created specifically in the liver, we injected AdV-Cre (ViraQuest, Inc) (2×10^{11} particles/animal) i.v. into five Sptlc2-Flox and five WT mice. Twenty-one days later, we sacrificed the animals and isolated liver genomic DNA and total RNA. We demonstrated that recombination of the floxed Sptlc2 alleles (and thereby inactivation of the Sptlc2 gene) could be achieved in essentially 100 % of the hepatocytes after intravenous injection of a recombinant adenovirus carrying the Cre recombinase into the animals (Fig. 3e). Because expression of virally transferred exogenous genes, in this case the Cre recombinase, is almost exclusively limited to the liver [18],

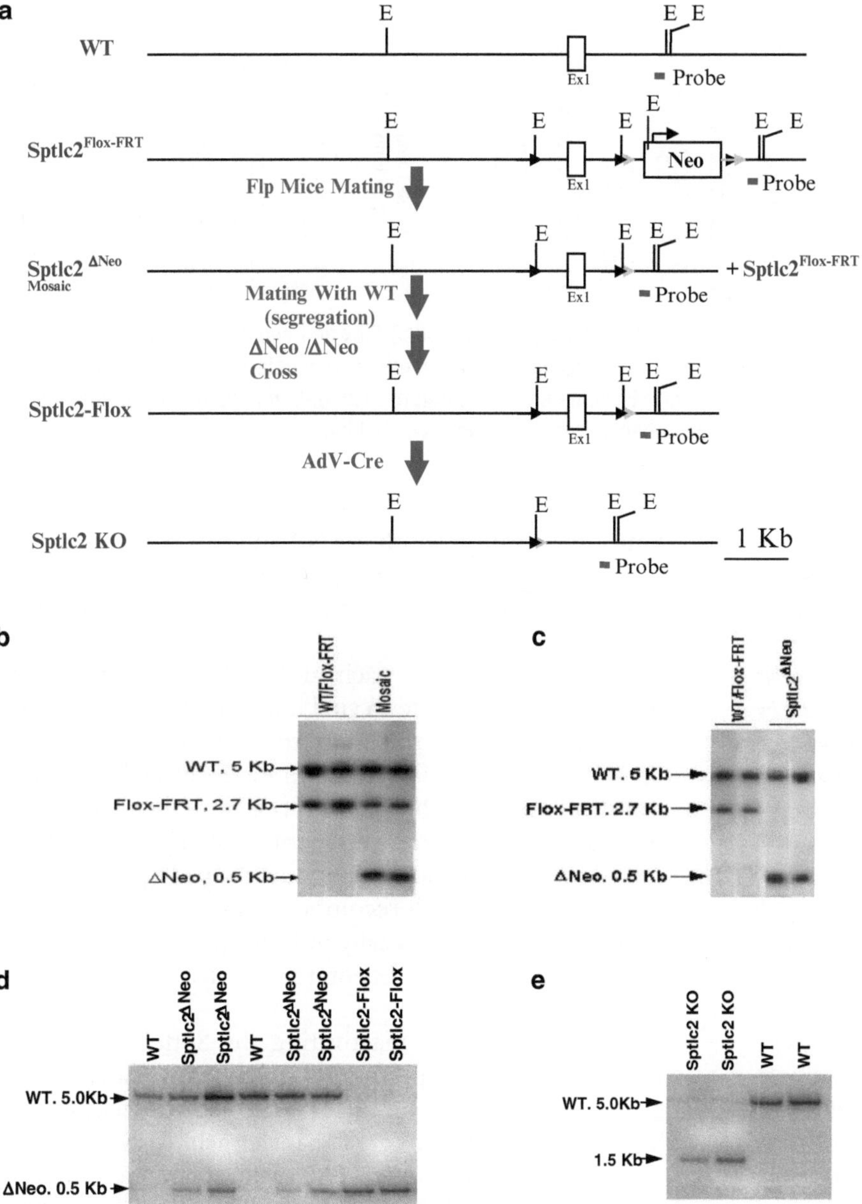

Fig. 3 Sptlc2-Flox (homozygous Sptlc2$^{\Delta Neo}$) mouse preparation. (**a**) Strategy; (**b**) Southern blot analysis for mosaic mice. Heterozygous Sptlc2$^{Flox\text{-}FRT}$ mice are crossed with heterozygous Flp recombinase transgenic mice. A 250 bp internal fragment is used as a probe for Southern blot analysis. The heterozygous mouse (WT/Flox-FRT) has 5.0 and 2.7 kb EcoRI digested signals and the mosaic (WT + Flox-FRT + ΔNeo) mouse has 5.0, 2.7, and 0.5 kb signals. (**c**) Southern blot analysis for heterozygous Sptlc2$^{\Delta Neo}$ mice. The mosaic mice were crossed with WT mice. The heterozygous mouse (WT/Flox-FRT) has 5.0 and 2.7 kb EcoRI digested signals and the Sptlc2$^{\Delta Neo}$ (WT + ΔNeo) mouse has 5.0 and 0.5 kb signals. (**d**) Southern blot analysis for Sptlc2-Flox (homozygous Sptlc2$^{\Delta Neo}$) mice. The WT mouse has a 5.0 kb EcoRI digested signal; the heterozygous Sptlc2$^{\Delta Neo}$ (WT + ΔNeo) mouse has 5.0 and 0.5 kb signals; the Sptlc2-Flox mouse has a 0.5 kb signal. (**e**) Southern blot analysis for liver Sptlc2 after AdV-Cre treatment. The complete Sptlc2 liver KO mice have a 1.5 kb EcoRI digested signal; the WT mice have a 5.0 kb digested signal. *E* EcoRI

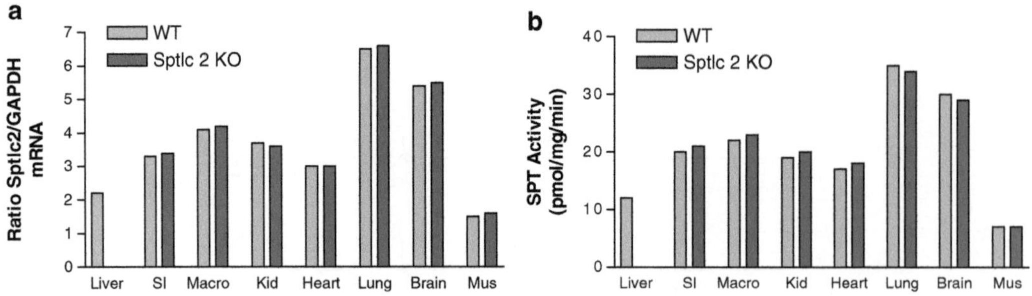

Fig. 4 Sptlc2 mRNA and SPT activity analyses. Two Sptlc2-Flox and two WT mice were injected with AdV-Cre. Twenty-one days later, tissues and cells were isolated. mRNA was determined by real-time PCR (**a**) and SPT activity was determined using [^{3}H]serine as a substrate (**b**). The data represent an average of two animals. *SI* small intestine, *Macro* macrophages from bone marrow, *Kid* kidney, *Mus* muscle

Sptlc2 inactivation (both mRNA and SPT activity) was highly liver specific. We also measured Sptlc2 mRNA and SPT activity levels in different tissues and found the depletion is liver specific (Fig. 4).

3.7.2 Liver-Specific Sptlc2 Inactivation by Pharmacological Induction of a Cre Transgene in Sptlc2-Flox Mice

As an independent approach, and to allay concerns about potential hepatotoxic effects due to viral infection, hepatic expression of Cre can be controlled by the interferon-inducible Mx1 promoter [19]. Sptlc2-Flox (or any floxed gene) mice can be bred with Mx1-Cre transgenic animals [20]. Cre expression is then induced in the liver with a series of three intraperitoneal injections of 250 μg of polyinosinic-polycytidylic ribonucleic acid (pIpC) (Sigma) [21]. Thus, Sptlc2-Flox will recombine only in liver, and inactivation of Sptlc2 will occur specifically in liver. Mx1-Cre transgenic mice are available from Jackson Laboratories (Maine) (*see* **Note 9**).

3.7.3 Liver-Specific Sptlc2 Inactivation by Crossing Sptlc2-Flox Mice with Albumin Promoter/Enhancer-Cre Transgenic Mice

A third approach to inactivating the Sptlc2 gene specifically in liver is to cross Sptlc2-Flox mice with albumin promoter/enhancer-Cre transgenic mice (available from Jackson Laboratory). The albumin promoter/enhancer will drive Cre expression specifically in liver, Sptlc2-Flox will recombine only in liver, and inactivation of Sptlc2 will occur specifically in liver. This approach will work only for genes that are not required in liver during development (*see* **Note 10**).

3.8 Intestine-Specific Sptlc2 Inactivation by Crossing Sptlc2-Flox Mice with Villin 1-Cre Transgenic Mice

Sptlc2-Flox (or other floxed genes) can be inactivated specifically in the intestine by crossing Sptlc2-Flox mice with Vil-Cre transgenic mice (Jackson Laboratory). The transgenic construct for the Vil-Cre transgenic mice contains the coding domain of Cre recombinase, driven by a 12.4 kb fragment of the mouse villin 1 gene and containing a metallothionein polyadenylation site sequence. Recombination occurs in villi and crypt cells of the small and large intestines, closely patterning the endogenous villin 1 gene expression [22, 23]. The animals have C57BL/6 background.

3.9 Preparation of Macrophage-Specific Gene KO Mice

3.9.1 Myeloid-Specific Sptlc2 Inactivation by Crossing Sptlc2-Flox Mice with LysM-Cre Transgenic Mice

Sptlc2-Flox mice (or mice with other floxed genes) can be crossed with LysM-Cre transgenic mice which express Cre recombinase from the endogenous lysozyme locus. When crossed with Sptlc2-Flox mice, Cre-mediated recombination results in deletion of the targeted gene in the myeloid cell lineage, including monocytes, mature macrophages, and granulocytes [24, 25]. The LysM-Cre transgenic mice are available from Jackson Laboratory (Maine) (*see* **Note 11**).

3.9.2 Bone Marrow Transplantation from Whole-Body Knockout Mice to a Acceptor Mice to Prepare Hematopoietic Cell-specific Gene Knock out Mice

Bone marrow contains hematopoietic progenitor cells, so transplanting bone marrow from a whole-body knockout mouse into an irradiated wild-type mouse will generate a hematopoietic cell- specific gene knockout animal. Briefly, bone marrow cells for transplantation into the irradiated mice are prepared by flushing both femurs of male donor mice. The donor mice are either apoE KO/PLTP KO or apoE KO (control) (this is an example). Donor cells are washed and suspended in DMEM with 5 % FBS and 1 mM HEPES. The cell concentration is 5×10^6 cells/ml. Twenty apoE$^{-/-}$ mice are lethally irradiated with 1,000 rad (10 Gy). Ten of these animals are transplanted with PLTP KO/apoE KO mouse bone marrow cells (5×10^6 cells), and the other ten with apoE KO mouse ones, via the femoral vein within 3 h of irradiation. We monitor the process of cell replacement by PCR, using genomic DNA from mouse white blood cells as a template [26]. Full details can be found in ref. 26.

4 Notes

1. We prefer to use MEF at Passage #3 to #4 as feeders.
2. It is better to put the direction of the Neo expression opposite to that of PLTP in the targeting vector. This approach blocks the opportunity of getting a fusion protein which still has PLTP activity.
3. The probe must be outside of the targeting vector; otherwise, it is not possible to distinguish whether the incorporation of the target vector is through homologous recombination or random insertion.
4. One primer must be located outside of the targeting vector; otherwise, it will not be possible to distinguish whether the incorporation of the target vector is through homologous recombination or random insertion.
5. We utilized ES from 129 mice. Based on fur color, we can determine whether there are chimeras or not. The range should be from no chimeras (100 % black) to 100 % chimeras (100 % agouti).

6. Three Loxp sites are necessary to delete Exon 1 of Sptlc2. A neomycin-resistant gene (Neo) cassette is necessary for positive ES cell clone screening. It is better to put the direction of Neo expression opposite to that of Sptlc2 in the targeting vector, for the same reason as in **Note 2**.
7. A pair of FRT sites is necessary to delete the Neo cassette. Based on our experience, the homozygous Sptlc2$^{Flox\text{-}FRT}$ mice are embryonic lethal, since the mice cannot tolerate a Neo cassette inserted into intron 1 of the Sptlc2 gene. This may also be true for some other genes.
8. The efficiency of Flp recombinase will not be 100 %, so we will obtain mosaic mice which need to undergo further segregation. From our experience, the efficiency can be as high as 60–70 %.
9. The Mx-1-Cre approach can completely solve the problem of embryonic lethality for certain gene deficiencies, since the gene depletion can be achieved in adult mice.
10. If some genes are absolutely necessary for embryo development from day 1, this approach will not work. It will be necessary to use the Mx-1-Cre approach.
11. The only problem with this approach is that the mice are myeloid-specific KO animals, but not pure macrophage-specific animals.

References

1. Siuta-Mangano P, Janero DR, Lane MD (1982) Association and assembly of triglyceride and phospholipid with glycosylated and unglycosylated apoproteins of very low density lipoprotein in the intact liver cell. J Biol Chem 257:11463–11467
2. Banarjee D, Redman CM (1984) Biosynthesis of high density lipoprotein by chicken liver: conjugation of nascent lipids with apoprotein A1. J Cell Biol 99:1917–1926
3. Gylling H, Miettinen TA (1995) The effect of cholesterol absorption inhibition on low density lipoprotein cholesterol level. Atherosclerosis 117:305–308
4. Ross R (1993) The pathogenesis of atherosclerosis: a perspective for the 1990s. Nature 362:801–808
5. Freeman M, Ashkenas J, Rees KJG, Kingsley DM, Copeland N, Jenkins NA, Krieger M (1990) An ancient, highly conserved family of cysteine-rich protein domains revealed by cloning type I and type II murine macrophage scavenger receptors. Proc Natl Acad Sci USA 87:8810–8814
6. Kodama T, Freeman M, Rohrer LJ, Zabrewky J, Matsudaira P, Krieger M (1990) Type I macrophage scavenger receptor contains alpha-helical and collagen-like coiled coils. Nature 343:531–535
7. Rahaman SO, Lennon DJ, Febbraio M, Podrez EA, Hazen SL, Silverstein RL (2006) A CD36-dependent signaling cascade is necessary for macrophage foam cell formation. Cell Metab 4:211–221
8. Cavelier C, Lorenzi I, Rohrer L, von Eckardstein A (2006) Lipid efflux by the ATP-binding cassette transporters ABCA1 and ABCG1. Biochim Biophys Acta 1761:655–666
9. Jessup W, Gelissen IC, Gaus K, Kritharides L (2006) Roles of ATP binding cassette transporters A1 and G1, scavenger receptor BI and membrane lipid domains in cholesterol export from macrophages. Curr Opin Lipidol 17:247–257
10. Merrill AH, Jones DD (1990) An update of the enzymology and regulation sphingomyelin metabolism. Biochim Biophys Acta 1044:1–12

11. Weiss B, Stoffel W (1997) Human and murine serine-palmitoyl-CoA transferase—cloning, expression and characterization of the key enzyme in sphingolipid synthesis. Eur J Biochem 249:239–247
12. Hanada K, Hara T, Nishijima M, Kuge O, Dickson RC, Nagiec MM (1997) A mammalian homolog of the yeast LCB1 encodes a component of serine palmitoyltransferase, the enzyme catalyzing the first step in sphingolipid synthesis. J Biol Chem 272:32108–32114
13. Hojjati MR, Li Z, Zhou H, Tang S, Huan C, Ooi E, Lu S, Jiang XC (2005) Effect of myriocin on plasma sphingolipid metabolism and atherosclerosis in apoE-deficient mice. J Biol Chem 280:10284–10289
14. Park TS, Panek RL, Mueller SB, Hanselman JC, Rosebury WS, Robertson AW, Kindt EK, Homan R, Karathanasis SK, Rekhter MD (2004) Inhibition of sphingomyelin synthesis reduces atherogenesis in apolipoprotein E-knockout mice. Circulation 110:3465–3471
15. Hojjati MR, Li Z, Jiang XC (2005) Serine palmitoyl-CoA transferase (SPT) deficiency and sphingolipid levels in mice. Biochim Biophys Acta 1737:44–51
16. Farley FW, Soriano P, Steffen LS, Dymecki SM (2000) Widespread recombinase expression using FLPeR (flipper) mice. Genesis 28:106–110
17. Jones JR, Shelton KD, Magnuson MA (2005) Strategies for the use of site-specific recombinases in genome engineering. Methods Mol Med 103:245–257
18. Herz J, Gerard RD (1993) Adenovirus-mediated transfer of low density lipoprotein receptor gene acutely accelerates cholesterol clearance in normal mice. Proc Natl Acad Sci USA 90:2812–2816
19. Kühn R, Schwenk F, Aguet M, Rajewsky K (1995) Inducible gene targeting in mice. Science 269:1427–1429
20. Gu H, Zou YR, Rajewsky R (1993) Independent control of immunoglobulin switch recombination at individual switch regions evidenced through Cre-loxP-mediated gene targeting. Cell 73:1155–1164
21. Ramirez-Solis R, Davis AC, Bradley A (1993) Gene targeting in embryonic stem cells. Methods Enzymol 225:855–878
22. Kucherlapati MH, Nguyen AA, Bronson RT, Kucherlapati RS (2006) Inactivation of conditional Rb by Villin-Cre leads to aggressive tumors outside the gastrointestinal tract. Cancer Res 66:3576–3583
23. Adachi M, Kurotani R, Morimura K, Shah Y, Sanford M, Madison BB, Gumucio DL, Marin HE, Peters JM, Young HA, Gonzalez FJ (2006) PPAR{gamma} in colonic epithelial cells protects against experimental inflammatory bowel disease. Gut 55:1104–1113
24. Boesten LS, Zadelaar AS, van Nieuwkoop A, Hu L, Jonkers J, van de Water B, Gijbels MJ, van der Made I, de Winther MP, Havekes LM, van Vlijmen BJ (2006) Macrophage retinoblastoma deficiency leads to enhanced atherosclerosis development in ApoE-deficient mice. FASEB J 20:953–955
25. Grivennikov SI, Tumanov AV, Liepinsh DJ, Kruglov AA, Marakusha BI, Shakhov AN, Murakami T, Drutskaya LN, Forster I, Clausen BE, Tessarollo L, Ryffel B, Kuprash DV, Nedospasov SA (2005) Distinct and nonredundant in vivo functions of TNF produced by t cells and macrophages/neutrophils: protective and deleterious effects. Immunity 22:93–104
26. Liu R, Hojjati MR, Devlin CM, Hansen IH, Jiang XC (2007) Macrophage phospholipid transfer protein deficiency and ApoE secretion: impact on mouse plasma cholesterol levels and atherosclerosis. Arterioscler Thromb Vasc Biol 27:190–196

Chapter 13

Adeno-associated Viruses as Liver-Directed Gene Delivery Vehicles: Focus on Lipoprotein Metabolism

William R. Lagor, Julie C. Johnston, Martin Lock, Luk H. Vandenberghe, and Daniel J. Rader

Abstract

Adeno-associated viral vectors have proven to be excellent gene delivery vehicles for somatic overexpression. These viral vectors can efficiently and selectively target the liver, which plays a central role in lipoprotein metabolism. Both liver-expressed as well as non-hepatic secreted proteins can be easily examined in different mouse models using this approach. The dosability of adeno-associated viral (AAV) vectors, as well as their potential for long-term expression, makes them an excellent choice for assessing gene function in vivo. This section will cover the use of AAV to study lipoprotein metabolism—including vector design, virus production and purification, and viral delivery, as well as monitoring of transgene expression and resulting phenotypic changes. Practical information is provided to assist the investigator in designing, interpreting, and troubleshooting experiments.

Key words Adeno-associated virus, AAV, Adenovirus, Liver, Gene delivery, Somatic overexpression, Mice, Lipoproteins, Lipids, Atherosclerosis

1 Introduction

Transgenic mice have yielded valuable insights into lipid metabolism and atherosclerosis—particularly in cases where a missing human gene can be expressed in mice. Perhaps the most notable examples of this technology as it relates to lipoprotein metabolism are the introduction of cholesteryl ester transfer protein (CETP), human apolipoprotein A-I (ApoA-I), and human apolipoprotein B (ApoB). These mouse models have proved invaluable by providing a more humanized lipoprotein environment in which to study lipid metabolism. Although significant improvements have been made, the generation of transgenic mice is still expensive and time-consuming, often derailing investigators from pursuing other important questions in their area of research interest. In addition, technical issues such as the mixed genetic background of the mice, passenger genes, and transgene copy number can greatly confound data

Lita A. Freeman (ed.), *Lipoproteins and Cardiovascular Disease: Methods and Protocols*, Methods in Molecular Biology, vol. 1027, DOI 10.1007/978-1-60327-369-5_13,

interpretation. Somatic overexpression with viral vectors offers an attractive alternative to transgenic mice. Viral vectors allow the investigator to initiate expression in a temporal and tissue-specific manner. The dose of virus can easily be titrated over a wide range to achieve the desired level of expression. This precisely controlled expression circumvents the problems of variable transgene copy number and passenger genes. Furthermore, viral vectors allow the researcher the flexibility to readily move between different mouse models without time-consuming and expensive breedings. This chapter will focus on the powerful approach of liver-directed gene transfer as a method of studying lipoprotein metabolism in mice.

The liver plays a central role in lipid metabolism and whole-body cholesterol balance. This organ is a major site of de novo synthesis of cholesterol and triglycerides that are secreted for use by peripheral tissues. Plasma cholesterol levels are largely dependent upon the rate of production as well as uptake by the liver. Through the action of the LDL receptor, the liver is responsible for clearance of atherogenic ApoB-containing lipoproteins from the circulation. In addition, the liver can accept macrophage-derived cholesterol from the atheroprotective HDL particles, as well as act as a gatekeeper for cholesterol removal from the body. Defects in key liver genes are the basis for most of the currently known inherited lipid disorders. Replacement of these defective genes has been achieved in animal models of atherosclerosis and holds great promise for future gene therapy in humans. Likewise, liver-directed somatic overexpression has also yielded important insights into the basic biology of lipoprotein metabolism and atherosclerosis.

1.1 Liver-Directed Gene Transfer

Several viral vectors have been used successfully to somatically overexpress genes in liver including most notably adenovirus and adeno-associated virus (AAV). Adenoviral vectors have an extensive history and their in vivo functionality has been thoroughly characterized. Adenovirus is a reliable and well-established method for somatic overexpression, but does come with significant limitations. Adeno-associated viruses (AAVs) are rapidly gaining acceptance as a highly efficient gene delivery tool that can selectively target the liver. AAVs offer many advantages over adenoviral vectors, most notably—low immunogenicity and the potential for sustained high-level expression. AAVs also hold great promise as vectors for human gene therapy, and several are currently in clinical trials. We expect AAV-mediated overexpression to supplant adenoviral gene transfer as the preferred methodology for probing questions related to lipoprotein metabolism and atherosclerosis in mice. As such, we will briefly discuss the use of adenovirus to provide historical perspective, and then move on to the use of AAV for liver-directed somatic overexpression studies.

1.2 Adenovirus

Adenovirus is a large non-enveloped, double-stranded DNA virus that is capable of infecting both dividing and nondividing cells. The adenoviral genome is 36–38 kb in size with inverted terminal repeats (ITRs) on each end of the DNA. The ITRs play an important role in DNA replication, and are required for completion of the viral life cycle. This large DNA genome is packaged into the viral capsid with the help of a histone-like protein. Three major structural proteins known as hexon, penton base, and knobbed fiber combine to form the adenoviral capsid. Twelve copies of the hexon trimer join together to form each of the 20 triangular faces of the icosahedral capsid. The vertices of the capsid each consist of five copies of the penton base, and three of the knobbed fiber. The adenoviral capsid is very large with an estimated mass for Ad2 of 150×10^6 Da [1]. The size of the adenoviral genome makes this virus able to accept large transgenes. In general, DNA cassettes not exceeding 105 % of the size of the adenoviral genome will package with high efficiency. Modified adenoviruses are widely used as gene delivery vehicles for animal studies.

Adenovirus was the first viral vector used to efficiently deliver genes to the liver in mice. Stratford-Perricaudet et al. [2] used a recombinant adenovirus to effectively cure newborn mice of ornithine transcarbamylase deficiency. Sustained expression was observed out to a year following administration, clearly demonstrating the power of viral gene transfer. A few years later, adenovirus was used to overexpress the LDL receptor (LDLR) gene in mice, markedly improving clearance of ^{125}I-labeled LDL from the circulation [3]. This initial finding was followed up with the successful reversal of hypercholesterolemia in LDLR-deficient mice [4]. Since that time, adenovirus has been used to deliver ApoA-I [5], apolipoprotein B mRNA-editing protein (ApoBEC) [6], apolipoprotein E (ApoE) [7], cholesterol 7-alpha hydroxylase [8], lecithin:cholesterol acyltransferase (LCAT) [9], hepatic lipase [10], and many other genes to the liver. These studies have established adenovirus as a reliable vector for hepatic overexpression, and helped elucidate the roles of several important genes in lipid metabolism.

While adenovirus has proven to be a very powerful gene delivery tool, it is not without limitations. The most important of these is the potent immunogenicity of the virus. The adenoviral capsid produces a strong cytotoxic immune response that leads to the selective killing of the transduced cells [11, 12]. Consequently, expression of the transgene generally peaks between 3 and 7 days post infection and then dramatically declines [13]. In the best cases, lower level expression can be observed out to 6 weeks or slightly longer. These inflammatory side effects limit the practicality of using adenovirus for long-term studies. Adenovirus has been used successfully to study the impact of several genes on the development of [14–21] and in a few examples regression of atherosclerosis [22–24]. However, extremely delicate timing is required to balance peak transgene expression with the window for detecting quantifiable differences in lesions. Great care must also be taken to ensure that the vectors are of sufficient purity to make appropriate comparisons. Since adenoviral vectors elicit such a potent inflammatory response, they would in principle not be the ideal choice for studies of atherosclerosis—which is categorically an inflammatory process. The validity of this approach is questionable in situations when the expressed transgene is expected to modulate inflammatory or immune responses. Investigators should carefully consider these pitfalls when deciding on adenovirus as a gene delivery vehicle. Adenoviral vectors would still be useful in situations where the transgene is large, only robust short-term expression is required, and a high degree of inflammation can be tolerated.

1.3 Adeno-associated Virus

AAV vectors are markedly less inflammatory than adenovirus [25], and generally avoid the problems of liver toxicity and transient expression. Even though as much as 80 % of the human population may be seropositive for some form of AAV, these viruses are not known to cause any pathology in humans [26]. This has generated a great deal of interest in the use of recombinant AAV as vectors for human gene therapy. Several key proof of concept studies have been successfully demonstrated in mice. The first involved delivery of Factor IX [27–30], resulting in the sustained phenotypic correction of hemophilia B [31]. AAV encoding Factor IX and several other therapeutic gene products are currently in human clinical trials. This is an exciting time for human gene therapy, and AAV has clearly taken the spotlight.

AAVs are rapidly gaining popularity as gene delivery tools to answer important questions in basic science research as well. In recent years, AAV vectors have been successfully applied to the study of lipoprotein metabolism and atherosclerosis. In an early example, phospholipid transfer protein (PLTP) was found to play an unexpected pro-atherogenic role in ApoE−/− mice [32]. Soon after, overexpression of LDLR by AAV serotypes 7 and 8 was shown to rescue the lipid abnormalities, as well as to completely prevent atherosclerosis in LDLR-deficient mice [33]. In a similar study, overexpression of ApoE by AAV7 and -8 was able to prevent atherosclerosis in ApoE−/− mice, affording therapeutic protection for a full year after administration [34]. A comparison of AAV2 pseudotyped with capsids from serotypes 1, 2, and 5 showed that AAV1 and -5 were most effective at producing persistent expression of ApoA-I in knockout mice [35]. Aside from showing the strength and utility of recombinant AAV, other studies have also answered important biological questions not feasible with conventional genetic manipulation. For example, the famed ApoA-I Milano variant was found to be no better than wild type ApoA-I in reducing the progression of atherosclerosis in LDLR-deficient mice [36]. Such a head-to-head comparison of these two proteins would be nearly impossible using traditional mouse genetics. In another recent study, AAV transduction of CETP improved the rate of macrophage reverse cholesterol transport in mice [37], calling into question whether CETP inhibition is a desirable therapeutic goal in humans. These studies have clearly demonstrated the power and versatility of AAV as a gene delivery tool. It is expected that AAV will become the preferred platform for somatic overexpression studies in the years to come, and will be used to answer many important questions related to lipoprotein metabolism and atherosclerosis.

AAVs are small single-stranded DNA viruses that require a helper virus such as adenovirus for replication. Initially discovered as contaminants in adenovirus preparations [38], AAV is one of the smallest viruses in existence with a non-enveloped capsid 22 nm in diameter (Fig. 1). The icosahedral capsid is a spherical shell of 60 subunits that encases a single-stranded DNA genome of only 4.7 kb. Because it is so small, AAV has evolved to be remarkably efficient in the use of its genes. The AAV genome contains only two genes, called Rep and Cap. The Rep gene gives rise to four different viral replicase proteins—Rep78, Rep68, Rep52, and Rep40. The Rep proteins are required not only for viral replication but also for transcription, encapsidation, integration, and rescue of the virus from the latent state. The Cap gene encodes three different capsid proteins known as VP1, VP2, and VP3. In addition to providing the capsid of the virus, these proteins are important for replication and packaging of AAV. The AAV genome is flanked on either side by 145 base pair inverted terminal repeats (ITRs). The ITRs contain Rep binding elements and terminal resolution sites that are required for second-strand synthesis of the DNA [39].

AAVs have been traditionally classified by serotype, and 11 of these have been described thus far. In addition, hundreds of different AAV sequences have recently been isolated, corresponding to multiple novel clades of AAV which may or may not fall within previously known serotypes [40]. The capsid serotype is the key determinant of the tropism of AAV for different tissues. After binding to receptors on the cell surface, the virus is rapidly internalized by clathrin-mediated endocytosis. Once inside the cell, AAV is trafficked to the nucleus where the capsid is uncoated, releasing the viral DNA. At this point, the single-stranded AAV genome must undergo second-strand synthesis before gene expression can occur. In the presence of the Rep gene products, the AAV DNA can integrate at a specific site on the long arm of human chromosome 19 [41]. When the Rep and Cap genes are missing, as in recombinant AAV vectors, the viral DNA will remain episomal [42].

The ITRs are the only DNA sequences required in *cis* for packaging into the virus. Recombinant AAV can be created by removing the viral Rep and Cap genes and replacing them with an

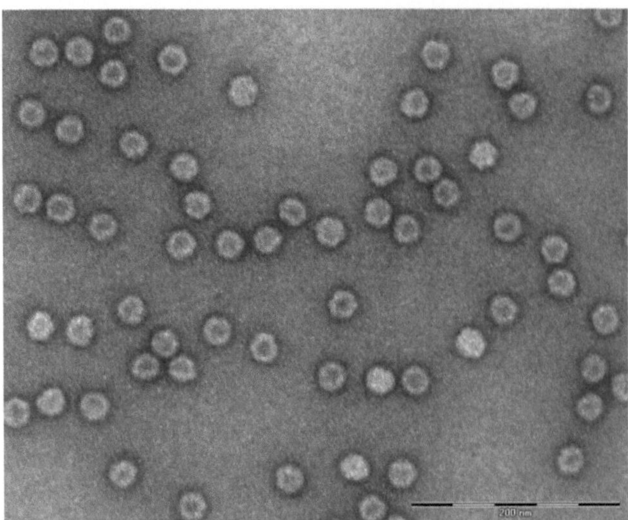

Fig. 1 Adeno-associated virus particles—electron micrograph of recombinant AAV2/8 particles from a vector preparation after negative staining with uranyl acetate

expression cassette containing the transgene of interest. The Rep and Cap genes, as well as the required adenoviral helper genes (E1A, E1B, E4, E2A, and VAl), can be easily provided in trans by co-transfection into a suitable host cell line. This strategy can be used to produce AAV preparations free of contaminating adenovirus. The purified virus contains only single-stranded DNA consisting of the expression cassette flanked by the AAV ITRs. The major advantage of this method is the ability to pseudotype viral vectors with capsids from different AAVs. This affords the investigator great flexibility in selecting the appropriate serotype for the tissue of interest.

AAV serotype 2 is the most widely studied AAV serotype, and many in vitro and in vivo studies have been performed based on this vector [43]. Despite the impressive longevity of transgene expression obtained with AAV2, its application has been limited because of the low level of transgene expression. Blocks at the level of vector entry and post-entry processing contribute to these inefficiencies [44]. In addition, the prevalence of preexisting antibodies against AAV2 in human populations (30 % are serum positive in some populations) could also hinder the application of AAV2 as a gene transfer vector in humans [45]. Progress in overcoming these barriers has been made through the development of vectors based on other serotypes that enter the cell via receptors distinct from those that recognize AAV2 and that are not affected by the presence of preexisting anti-AAV2 antibodies.

The ability to pseudotype AAV vectors with capsids from different AAV serotypes has greatly facilitated the targeting of particular tissues [46]. A variety of other AAV serotypes including AAV1, -3, -4, -5, and -6 have been isolated and used in vectors for gene delivery [47]. The recently isolated serotypes 7, 8, and 9 appear to have superb transduction ability and potential as vectors for gene therapy [48]. AAV serotype 8 has been found to have a particularly strong tropism for the liver [49], making this a preferred choice for hepatic overexpression. AAV9 is also very efficient at targeting the liver, but is perhaps the best serotype thus far for targeting heart and skeletal muscle. Recently, a head-to-head comparison was made for AAV serotypes 1–9 in their ability to transduce different tissues in mice by tail vein injection using a luciferase reporter gene [50]. It was found that AAV2, -3, -4, and -5 gave low overall expression, AAV1, -6, and -8 gave moderate expression, and AAV7 and -9 gave highest expression. All serotypes were capable of transducing liver to some degree, with 7 and 9 being the most effective, followed next by AAV8. Serotype 9 was the most efficient at transducing heart, consistent with many other reports in the literature [51, 52]. AAV4 showed the highest expression of all the serotypes in the lung and kidney. It should be noted that the promoter, transgene product and route of administration also have a substantial impact on expression, so caution is advised in extrapolating these results to other scenarios. AAV8 has been the most widely and successfully used serotype for studies of lipoprotein metabolism in mice. For the sake of simplicity, we will focus on the use of AAV8 for in vivo transduction of liver. It should be noted that the methods for viral production and purification are applicable to the other AAV serotypes as well.

This chapter will describe methodology for the use of AAV vectors for studies of lipoprotein metabolism or atherosclerosis. We will first discuss the rationale for choosing an AAV vector for somatic overexpression. The next step is cloning the transgene of interest into an AAV *cis* plasmid. This AAV *cis* plasmid is then used to deliver the expression cassette to HEK 293 cells for packaging into AAV vector using the triple transfection method. After transfection, the virus is then purified, titered, and assessed for purity. The purified virus may then be tested in vitro to verify infectivity and transgene expression. Finally, the viral vector is injected into mice to investigate the effects of the transgene on lipoprotein metabolism. Transgene expression and phenotypic changes are monitored over the subsequent weeks or months following infection. The basic theory of the method as well as important considerations before starting are provided in narrative form before each step-by-step protocol. Additional tips to assist the investigator in troubleshooting are provided in Subheading 4.

2 Materials

2.1 Transfection

1. Dulbecco's Modified Eagle's Medium (DMEM).
2. Dulbecco's Phosphate-Buffered Saline Solution (DPBS).
3. Fetal bovine serum (FBS).
4. Sterile tissue culture plates (150 mm).
5. Sterile, disposable pipettes.
6. Penicillin/streptomycin (P/S).
7. 2.5 M $CaCl_2$.
8. AAV *trans* plasmid [53, 54] (*see* **Note 1**).
9. ΔF6 adenoviral helper plasmid (*see* **Note 1**).
10. AAV *cis* plasmid.
11. 2× HBS (HEPES-buffered saline): 280 mM NaCl, 50 mM HEPES, 75 mM sodium phosphate, pH 7.05–7.1.
12. Sterile 125-ml receiver.
13. 37 °C Water bath.
14. 37 °C, 5 % CO_2, incubator.
15. Sterile 50-ml conical tubes.
16. Sterile aspirating pipettes.
17. Sterile 500-ml centrifuge bottles.
18. Resuspension buffer: 50 mM Tris–HCl, 200 mM NaCl, 2 mM $MgCl_2$.

2.2 Purification

1. Sterile 50-ml conical tubes.
2. Resuspension buffer: 50 mM Tris–HCl, 200 mM NaCl, 2 mM $MgCl_2$.
3. Benzonase (EMSCO 1682–1.01697.0002).
4. 10 % (W/V) Octyl-β-d-glucopyranoside (Sigma 08001).
5. Tabletop centrifuge.
6. Beckman ultracentrifuge.
7. Heavy (H) cesium chloride solution: 1.5 g/ml CsCl solution.
8. Light (L) cesium chloride solution: 1.3 g/ml CsCl solution.
9. 1.4 g/ml CsCl solution.
10. SW28 centrifuge tubes.
11. Tygon-silicone tubing, 2 ft long, presterilized, fitted with 2 1/16th in. male luers.
12. Sterile 96-well round bottom plates.
13. Sterile 1.5-ml Eppendorf tubes.
14. 18 Gauge, 1″ needle.

15. 16 Gauge needle.
16. Refractometer.
17. 70.1 Ti quick-seal tubes, 13 ml.
18. 3-ml syringe; 10-ml syringe.
19. Portable heat-sealer device for 70.1 Ti quick-seal tubes.
20. Dulbecco's Phosphate-Buffered Saline (DPBS).
21. PBS/NaCl (1× DPS containing 35 mM NaCl).
22. Amicon 100 kDa MWCO filtration device, 15 ml (Cat# UFC9 100 08).
23. Sterile 100 % glycerol.
24. 70 % Ethanol.
25. Masterflex pump.

2.3 Genome Copy Number Titration

1. DNase digestion buffer: 13 mM Tris–HCl, pH 7.5, 5 mM $MgCl_2$, 0.12 mM $CaCl_2$.
2. *Cis* plasmid.
3. Forward PCR primer (10 μM).
4. Reverse PCR primer (10 μM).
5. Fluorescent probe (10 μM).
6. 96-well optical reaction plate.
7. Optical plate sealer sheet.
8. Distilled water DNAse/RNAse free.
9. DNase I.
10. qPCR Mastermix (ABI).
11. ABI Prism Sequence Detector or equivalent real-time PCR system.
12. 37 °C water bath.
13. rAAV control sample (previously titrated rAAV sample).
14. Restriction enzyme with buffer.
15. QIAquick PCR Purification Kit (Qiagen, Valencia, CA) (Cat # 28104).
16. Aerosol barrier pipette tips.
17. Spectrophotometer.

2.4 Endotoxin Assay

1. Heat block, with 96-well plate adaptor.
2. LAL reagent water, 100 ml (Lonza, Walkersville, MD) (Cat # W50-100).
3. Sterile glass tubes (Lonza).
4. 96-well plate (sterile, polystyrene, flat-bottom).

5. Stop solution (25 % v/v glacial acetic acid in H_2O).
6. LAL kit, QCL-1000 (Lonza) (Cat # 50-648u).
7. Positive control sample (endotoxin concentration determined previously).
8. 96-well absorbance plate reader.
9. Biosafety containment hood.

2.5 Capsid Purity Assay

1. 10 mM Tris–HCl pH 8.0.
2. Vertical gel apparatus (Xcell SureLock, Invitrogen, Carlsbad, CA, EI0020).
3. Pre-cast, 10-Well Polyacrylamide 4–12 % Bis-Tris Gel (Invitrogen #NP0321BOX).
4. NuPAGE LDS sample buffer (4×, Invitrogen NP0007).
5. NuPAGE sample reducing agent (10×, Invitrogen NP0004).
6. NuPAGE MOPS Running buffer (20×, Invitrogen NP0002).
7. NuPAGE antioxidant (Invitrogen NP0005).
8. BenchMark Protein Ladder (Invitrogen #10747-012).
9. Methanol, reagent grade.
10. Acetic acid, reagent grade.
11. Sypro Ruby Protein Gel Stain (Invitrogen #S12000).
12. Positive control-purified AAV sample.
13. Gel imaging system with UV-transillumination (302 nm).

2.6 Transduction of HepG2 Cells

1. Minimum essential media (MEM).
2. 10 % Fetal bovine serum (FBS).
3. Antibiotic/antimycotic mix.
4. NuPAGE LDS sample buffer (4×, Invitrogen NP0007).
5. NuPAGE sample reducing agent (10×, Invitrogen NP0004).
6. NuPAGE MOPS running buffer (20×, Invitrogen NP0002).
7. NuPAGE antioxidant (Invitrogen NP0005).

2.7 In Vivo Liver Transduction

1. Purified and titered AAV vector.
2. Sterile PBS.
3. 1-ml syringe with 26 G needle.
4. Mice.

3 Methods

3.1 Choosing the Vector

AAV is the preferred vector when seeking stable, long-term expression. AAV is far less immunogenic than adenovirus, and avoids the problems of liver toxicity and antibodies to the viral capsid. Because of the time needed for second-strand DNA synthesis, there is a significant lag before transgene expression. Practically speaking, stable expression can normally be obtained by 2 weeks post infection, and should persist out to a year or more. This makes AAV the ideal vector for long-term experiments such as atherosclerosis studies. Transgene size is another important consideration when choosing AAV. The 4.7 kb packaging capacity of the virus places strict limitations on size. The entire expression cassette should not exceed 4.9 kb from ITR to ITR inclusive. Exceeding this limit drastically affects viral yield. Considering the size of the ITRs, the promoter, and polyadenylation signal, this typically prevents AAV from accepting transgenes larger than roughly 2.6 kb in size. Self-complementary AAVs, which are recombinant self-annealing vectors designed to circumvent second-strand synthesis, would be limited to 1.3 kb or less.

Tissue-specific expression of the transgene depends primarily on three factors: (1) the serotype of the virus, (2) the promoter driving the transgene, and (3) the route of administration. The tropism of AAV for different tissues is entirely dependent upon the capsid serotype. All commonly used AAV serotypes can efficiently transduce liver, although AAV7, -8, and -9 have been found to be the most effective [50, 53]. AAV8 has been shown to transduce greater than 80 % of the hepatocytes in the liver at a dose of 1×10^{12} genome copies (GC) per mouse when given via portal vein injection [33]. Since AAV8 and all other AAV serotypes are not restricted to liver in terms of transduction ability, it is necessary to include a tissue-specific promoter in the expression cassette. Choosing a strong promoter that is specifically liver-expressed will minimize expression from other tissues. In our experience, the human thyroid-binding globulin (TBG) promoter [53, 55] offers a good compromise between tissue specificity and high-level expression. Tissue-specific expression also depends upon the accessibility of the virus to the tissue. Injections that place the virus in close proximity to the targeted tissue will generally be most effective in obtaining high-level expression. For example, the highest levels of expression in the liver are obtained by portal vein injection. For most studies, however, less invasive procedures are preferred. Fortunately, the liver has a particularly high affinity for AAV in general, and can easily be targeted using systemic administrations. Tail vein injection is a reliable and established method for delivering AAV to the liver. In our experience, intraperitoneal injection of AAV8 also gives excellent and reproducible transduction. Studies using AAV to target extrahepatic tissues may be hindered by this remarkable ability of the liver to clear virus from the circulation.

3.2 Generating the Expression Cassette

The most critical component of the expression cassette is the promoter driving transgene expression. The TBG promoter is an

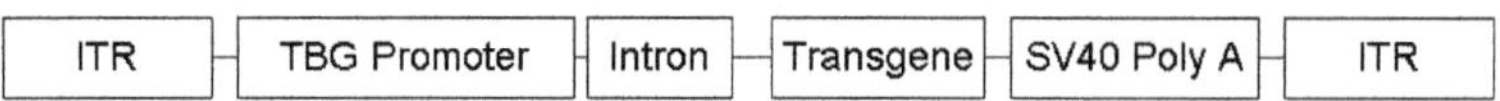

Fig. 2 AAV expression cassette. Typical arrangement of an expression cassette for making recombinant AAV vectors. The inverted terminal repeats (ITRs) on either side are the only wild type viral sequence required in *cis* for vector production. Expression is driven by a strong promoter, in this case thyroxine-binding globulin which gives high-level expression that is restricted to the liver. An intron is included 5′ of the transgene to improve splicing and nuclear export. Downstream of the transgene, a strong poly A signal (in this case from SV40) ensures message stability and translatability

excellent choice for liver-restricted expression. Other key features of the expression cassette include an intron upstream of the transgene to improve nuclear export linked to RNA splicing, as well as a strong polyadenylation signal such as that from SV40 or bovine growth hormone (BGH) for transcription termination and mRNA polyadenylation. These elements are critical to ensure the stability and translatability of the message. As an example, Fig. 2 shows an illustration of an expression cassette used for preparing AAV vectors.

Since the ITRs are the only DNA sequences required for packaging into AAV, preparing the virus is fairly straightforward. First, the transgene of interest is cloned into a *cis* plasmid including the promoter, intron, and polyadenylation signal, all flanked by the AAV ITRs. This is facilitated by using a vector with a multi-cloning site with common restriction enzymes as shown below in Fig. 3.

The transgene may be obtained by PCR or by screening a cDNA library. However, in practice, it is often more convenient and cost-effective to purchase a commercially available clone for this purpose. Good molecular biological practice dictates that the clone should be digested and sequenced for verification purposes before cloning into the AAV plasmid.

3.3 AAV Cis Plasmid Quality Control

Due to the impact of starting materials on the final vector preparation, it is critical that a thorough and systemic plasmid identity assay be completed prior to the initiation of vector production. It is recommended that all plasmid DNA master stocks used for the production of vector undergo complete sequence analysis as well as thorough restriction enzyme analysis to assess the purity and integrity of the material. Plasmid DNA preparations subsequently amplified from master stocks may also undergo full sequence analysis or less rigorous testing by restriction enzyme analysis. For the restriction enzyme analysis, the presence of each functional DNA element is targeted by restriction enzyme digest and the correct restriction pattern is verified with the plasmid map and the seed DNA preparation. For AAV *cis* plasmids, these elements include the 5′ and 3′ ITR, transgene, promoter, polyadenylation sequence, and any other regulatory element. For AAV *trans* plasmids, the AAV *rep* gene and the correct *cap* gene are verified using

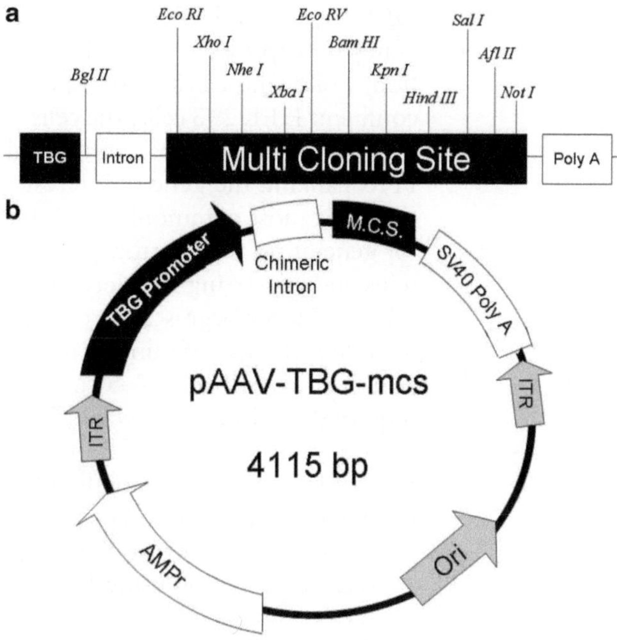

Fig. 3 AAV *cis* plasmids. A typical AAV *cis* plasmid used to deliver the transgene and expression cassette to the cells for packaging into recombinant AAV. (**a**) The multiple cloning site includes common restriction sites to allow easy cloning of transgenes into the expression cassette. (**b**) This plasmid, pAAV-TBG-mcs, is commonly used to produce AAV2/8 targeting the liver. It consists of the expression cassette on a plasmid backbone that allows for selection (ampicillin resistance in this case) and propagation (*ori* origin of replication) in bacteria

serotype-specific restriction enzymes. Due to the presence of repeated elements in the AAV *cis* plasmid, in particular, it is recommended that the plasmid be maintained in a *rec* minus cell line (e.g., stbl2 or SURE) and each preparation fully characterized prior to use. Full plasmid characterization incorporates the following: concentration, determination of $A_{260/280}$ ratio, detection of endotoxin, and plasmid identity.

When the plasmid has been fully characterized and is determined to be of high quality, it can be used for production of recombinant AAV vectors. The following sections provide detailed procedures for producing and purifying AAV, characterizing copy number and purity of purified AAV, and verifying its transduction and transgene expression in vitro. The chapter concludes with strategies for in vivo studies using AAV vectors to target the liver.

3.4 Production of Recombinant AAV by Transient Transfection Using HEK 293 Cells

Recombinant AAV vectors are routinely produced by the triple transfection method [56] using three plasmids: an AAV *cis* plasmid containing the AAV ITRs and encoding a gene of interest, an AAV *trans* plasmid encoding the AAV *rep* and *cap* genes, and an adenovirus helper plasmid pΔF6 which provides helper functions from

Ad E2A, E4, and VA RNA genes. The adenovirus E1A and E1B genes are expressed in the HEK 293 cells used for vector production. Following calcium phosphate-mediated transfection of subconfluent HEK293 cells, the cells are harvested and lysed to release the AAV virions encapsidating single-stranded genomes with AAV ITRs flanking the gene of interest. The transient transfection procedure is most commonly used due to its simplicity and flexibility for generating AAV vectors pseudotyped with different capsid proteins and expressing a variety of genes; however, alternative methods suitable for large-scale production of recombinant AAV vectors, including the use of human, mammalian, or insect cell lines, have also been described [57–59]. Note that adeno-associated viruses require that Biosafety Level 2 (BSL2) practices be followed, and all work should be performed in approved areas. The investigator is advised to consult the Biosafety Committee at their institution for specific requirements and permission to use AAV.

A detailed procedure for production of recombinant AAV by transient transfection using HEK 293 cells is provided below.

3.4.1 Preparation of HEK 293 Cells for Transfection

1. Seed 40 × 150 mm plates 1 day prior to transfection to achieve 80–85 % confluency the following day. HEK 293 cells are cultured in DMEM supplemented with 5 % FBS, 1 % penicillin/ streptomycin.
2. On the day of transfection, check cells for optimal confluency and change the media 2 h before transfection by removing the media and replacing with 20 ml fresh media without disturbing the monolayer.

3.4.2 Preparation of Transfection Cocktail

1. For transfection, add the following to a sterile 125-ml receiver (note: this is approximately 50 μg total DNA per plate with 1:1:1 molar ratios of each plasmid DNA):

   Sterile water to 50 ml

   5.2 ml 2.5 M $CaCl_2$

   1,040 μg Ad helper plasmid

   520 μg AAV *cis* plasmid

   520 μg AAV *trans* plasmid
2. Add 12.5 ml 2× HBS to 4 × 50 ml conical tubes.
3. Add 12.5 ml of solution from Subheading 3.4.2, **step 1**, above to each of the 4 × 50 ml conical tubes from Subheading 3.4.2, **step 2**, (each containing 12.5 ml 2× HBS). Add dropwise while vortexing. This is the transfection cocktail. Incubate the cocktail at room temperature for 5 min.

3.4.3 Transfection of HEK 293 Cells

1. Add 2.5 ml of the transfection cocktail to each plate. Gently swirl the plate to evenly distribute the cocktail over the entire plate.

Place tissue culture plates in a 37 °C, 5 % CO_2 incubator.

16 h post transfection, aspirate the media from the plates and replace with fresh media.

3.4.4 Harvesting of Transfected Cells

Harvest cells approximately 72 h post transfection.

1. Remove tissue culture plates from the incubator and place them in the biosafety cabinet. Aspirate ~10 ml of the media from each plate. Use a cell scraper to scrape off all the cells from each plate. Using a pipette, transfer the cell suspension from each plate to a 500-ml sterile centrifuge bottle.
2. Place each centrifuge bottle into the super-speed centrifuge and pellet the cells by spinning at 4,000 rpm for 15 min at 4 °C. Remove the bottle from the centrifuge and return to the biosafety cabinet. Using a sterile aspirating pipette, remove the supernatant from the bottles.

3.4.5 Storage of Cell Pellet/Suspension

1. For cesium chloride column purification, resuspend the cell pellet in approx. 18 ml of resuspension buffer for a final volume of 20 ml. Transfer the resuspended material to a sterile 50-ml conical tube, then place it at −80 °C until processing.

3.5 Purification of Recombinant AAV Vectors by Cesium Chloride Gradient Centrifugation

A variety of methods have been developed for the purification of AAV vectors including cesium chloride gradient centrifugation, iodixanol gradient centrifugation, affinity column, and liquid chromatography-based methods [60]. Purification by cesium chloride gradient centrifugation is a useful method for routine-scale AAV production, suitable for the purification of all known AAV serotypes, and utilizes equipment commonly found in biology laboratories. In addition, CsCl purification enables the separation of full particles from empty capsids lacking AAV genomes and, for that reason, may be included as a necessary step in the production of clinical grade vector [61]. The generic nature of the method allows comparisons to be made between different AAV serotype vectors prepared using the same method, an important consideration for studies utilizing more than one serotype vector. The major disadvantage of the method is its limited scalability. However, it is included here due to its simplicity and general utility for the production of research-grade material. Methods for production of scalable quantities of clinical grade AAV vectors are described elsewhere [62].

The following procedure describes generation of the cell lysate and purification of recombinant AAV vectors by cesium chloride gradient centrifugation.

3.5.1 Generation of Cell Lysate

1. Thaw cell suspension for 10 min at 37°C.

Subject the cells to two more freeze–thaw cycles by alternating incubation in a dry ice/ethanol bath with incubation in a 37 °C water bath.

2. Add 100 μl Benzonase (250 U/prep) and invert gently. Incubate the samples at 37 °C for 20 min; invert the tube every 5 min.
   Centrifuge in a Sorvall at 8 K for 15 min at 4 °C.
3. Add 4.57 ml 5 M NaCl to a final concentration of 1 M NaCl. Add 1.5 ml of 10 % octyl-β-d-glucopyranoside for a final concentration of 0.5 % and mix gently by inversion.

3.5.2 Preparation of First Cesium Chloride Gradient

1. For each vector preparation, prepare step gradients consisting of 7.5 ml heavy (H) cesium chloride solution and 15 ml light (L) cesium chloride solution in Beckman SW-28 tubes. Dispense 15 ml of l-CsCl solution into the bottom of a SW28 centrifuge tube, then carefully underlay 7.5 ml of H-CsCl solution.
2. Overlay 20 ml of the lysate onto each 2-tier CsCl gradient. Without disturbing the gradient, place the above SW28 tubes into the rotor buckets and balance them. Place the buckets onto the SW28 rotor and spin at 25,000 rpm (82,705 × *g*) for 18–20 h at 15 °C.

3.5.3 Fraction Collection After the First Centrifugation

1. Carefully remove the centrifuge tubes from the buckets without disturbing the gradient. Secure the first tube on a tube holder.
2. Take a presterilized 2 ft length of 1.6 mm tygon-silicone tubing fitted with 2 1/16th in. male luers and insert 18 G 1″ needles into the luers.
3. Pierce the tube at a right angle as close to the bottom of the tube as possible with one of the 18 G 1″ needles with the bevel facing up and clamp the tubing into the easy load rollers of the masterflex pump.
   Gently increase the speed of the pump to 1 ml/min. Collect the first 4.5 ml into a 15-ml Falcon tube and then start to collect 250 μl fractions into a 96-well round bottom plate. Collect 48 fractions.
4. Dispense the remainder of the gradient into a beaker containing a 20 % bleach solution; rinse the tubing with PBS and collect the rinse in the beaker. Repeat collecting the fractions for the second tube. Discard the needle/tubing assembly after use.

3.5.4 Determination of Refractive Index (RI)

1. Using a multichannel pipettor, transfer 10 μl of each fraction (of the 48 collected) to a fresh plate and leave the remainder of the fractions in the biosafety cabinet.
2. Take 5 μl of each fraction from the new plate and read the RI using a refractometer. The fractions corresponding to those containing AAV should have an RI that falls within the range

of **1.3740–1.3660**. Pool the fractions that are within range and discard the fractions that are out of range. Repeat this procedure for the second gradient.

3.5.5 Preparation of Second Cesium Chloride Gradient

1. The total pooled volume from each gradient should be 5–6 ml. Pool the two gradient harvests in a 50-ml conical tube and bring the final volume to 13 ml with 1.4 g/ml solution of CsCl. Mix with a pipette.
2. Using a 10-ml syringe and 18 G needle, add the pooled first gradient harvest to a 13-ml sealable centrifuge tube. The solution should be added to the line on the neck of the tube without introducing bubbles. Seal the tube using a portable heat-sealer device. Place the tube in a 70.1Ti rotor with the appropriate balance tube. Insert the rotor caps and centrifuge at 60 K for 20 h at 15 °C in a Beckman ultracentrifuge.

3.5.6 Fraction Collection After the Second Centrifugation

1. Following the second centrifugation, a band may be visible; however, the fractions are still collected as described in Subheadings 3.5.3 and 3.5.4 above, with the exception of pooling fractions corresponding to RI within the **1.3750–1.3650** range.

3.5.7 Desalting and Buffer Exchange

1. Aliquot 50 ml PBS/NaCl (sterile PBS with additional 35 mM NaCl) into a 50-ml Falcon tube. Add the pooled fractions from Subheading 3.5.6, **step 1**, above, into 15 ml PBS/NaCl. Mix gently and load onto a pre-wet Amicon 100 kDa MWCO filtration device.
2. Centrifuge in a benchtop Sorvall centrifuge at 2–4 K for 2 min to determine the flow rate of the sample. Repeat additional short spins to reduce the volume of the retentate to approx.1.8 ml (keeping the sample above the level of the membrane). The flowthrough should be discarded after each centrifugation.
3. Add an additional 13 ml PBS/NaCl, mix by pipetting with the retentate remaining in the device, and centrifuge again as above. Continue this process until all of the 50 ml PBS/NaCl (in Subheading 3.5.7, **step 1**) has been added to the device.
4. Rinse the membrane with the final retentate by repeatedly pipetting against the membrane surface. Recover the retentate into a suitably sized sterile centrifuge tube initially using a 1 ml Eppendorf pipette tip and ending with a 200 μl Eppendorf pipette tip in order to reach the remaining retentate at the bottom of the device.
   1. Rinse the membrane 2× with 200 μl PBS/NaCl and add to the tube containing the retentate above.

2. Determine the exact volume of material and add glycerol to 5 % final concentration.

5. Aliquot the purified preparation into appropriate volumes, keeping aside several small-volume samples (e.g., 4×50 μl samples) for subsequent quality control. Freeze aliquots immediately at −80 °C (*see* **Note 2**).

3.6 Characterization of Purified AAV

Appropriate characterization of the resulting purified AAV vector is critical as the outcome of any research project can be dramatically influenced by the quality and quantity of viral vector used. Unlike clinical grade vector, for which purified preparations must meet a set of predetermined specifications, there are no established release criteria which must be met for the use of research-grade vector. The set of criteria which are considered most appropriate by the researcher and institution may differ according to the ultimate use of the vector (i.e., a more stringent set of criteria may be placed upon vectors to be used in large animal studies compared to vector generated for proof of concept studies in small animal models). At a minimum, starting materials and purified vector should be assayed for quantity and purity. Additional assessments of potency, identity, and safety may be recommended for large animal studies or for other uses with specific considerations. The most basic quality control of research-grade AAV vector requires AAV genome titer, detection of endotoxin contamination, and an assessment of protein contamination. Procedures for these assays are detailed below.

3.7 AAV Genome Copy Number Titration

Several methods are available for determining the physical genome titer of a purified AAV vector preparation, the most commonly used of which are dot blot analysis and quantitative PCR. Quantitative PCR utilizes TaqMan (Applied Biosystems, Foster City, CA) reagents and machines to determine the genome copy (GC), and number of AAV vector lots as a measure of AAV particles with full genome content. This term GC is synonymous with vector genomes (vg) and DNase-resistant particles (DRP). The accuracy and reliability of the quantitative PCR GC titration is critical for AAV characterization and affects the outcome of other downstream assays. A number of standards, validation samples (both viral and plasmid), and controls (for background and DNA contamination) are recommended for use in this assay and careful attention to the parameters of the standard curves generated for reactions as well as in the preparation of standards, controls, and samples using the highest quality materials is essential.

A detailed protocol for AAV genome titration is provided below.

3.7.1 Cis Plasmid Standard Preparation

1. Linearize the *cis* plasmid used to generate the AAV with a restriction enzyme that cuts outside of the intended PCR target.
2. Purify the linearized plasmid using the QIAquick PCR Purification Kit.
   Determine the concentration of the linearized plasmid by spectrophotometry and convert the concentration readout to grams per liter (g/l).
3. Calculate the formula weight (F.W.) of the standard plasmid:

$$\text{F.W.} = \text{Plasmid size}\left(\text{in base pairs}\right) \times 662\ \text{g} / \text{m}$$

4. Calculate the molar concentration (M) of the linearized plasmid:

$$\text{M} = \text{mol} / \text{l} = \left(\text{mass}\ \left(\text{in grams}\right) / \text{F.W.}\right),$$

5. Determine copy number per microliter of the linearized plasmid based on molar concentration (1 M is equivalent to about 6.0221415×10^{23} copies/l).
6. Make the first dilution to a final concentration of 1×10^{10} copies per 20 μl (5×10^{8} copies/μl). Use the $C_1 V_1 = C_2 V_2$ relationship, where V_1 is the unknown *volume*, C_1 is the stock linearized plasmid *concentration* (copies/μl), V_2 is the final volume (100 μl), and C_2 is the final concentration (5×10^{8} copies/μl). Add V_1 volumes of stock linearized plasmid to enough nuclease-free water to make a 100 μl solution.
7. Carry out a serial tenfold dilution of 1×10^{10} copies/μl of the linearized standard in nuclease-free water. Use the tubes containing 1×10^{8} copies/μl through 10 copies/μl dilutions as standards. Aliquot and store at −20 °C.

3.7.2 Digestion of Un-encapsidated DNA

1. Prepare an experimental plan for the position of sample, controls, and DNA standards in duplicate in a 96-well format.
2. Place 5 μl of rAAV preparation and controls (PBS and rAAV control) in 1.5-ml Eppendorf tubes. Add 45 μl of DNAse digestion buffer and 10 U of DNAse I per tube. Incubate for 30 min at 37 °C. Then store tubes on ice.

3.7.3 Sample Preparation

In duplicate, prepare a serial dilution of extract sample and controls: Dilutions 1 and 2: 90 μl H_2O + 10 μl sample (tenfold dilution).

Dilutions 3–7: 40 μl H_2O + 10 μl sample (fivefold dilution).

3.7.4 PCR Preparation

1. Real-time PCR primers and probes should be designed using application-specific software (Applied Biosystems) and diluted in water to 10 μM.

2. Prepare enough mix for two reactions of each DNA sample, a standard curve, two control samples (PBS+rAAV control), plus two reactions for the no template control (H_2O).

The sample volumes per reaction are:

25 μl TaqMan Universal PCR Mix (2×)

1 μl Forward primer (10 μM)

1 μl Reverse primer (10 μM)

0.5 μl Fluorescent probe (10 μM)

2.5 μl Nuclease-free water

20 μl sample DNA (or H_2O for control)

3. Place the optical cover sheet over the plate and centrifuge briefly, then place in the real-time machine.

Cycle as follows:

1. Hold at 95 °C 10 min
2. Alternate between 95 °C for 15 s and

   60 °C for 1 min for 40 cycles.

3.7.5 Data Analysis

1. Refer to the appropriate instrument user guide for instructions on how to analyze the data. The general process for analyzing the data involves the following procedures: (a) View the amplification plots and set the baseline and threshold values; during early PCR cycles, the background signal in all wells is used to determine the baseline fluorescence. The threshold should be placed in the region of the exponential phase. (b) View the C_t (threshold cycle) values: the replicates should be tight, within 0.5 C_t.
2. Plot the *Ct*s of the standard plasmid (*Y* axis) versus the log of the initial quantity (*X* axis) to generate a standard curve. The standard curve should be linear over the entire range of where the unknowns are expected to fall. Standard curve specifications should be correlation coefficient ≥0.99, efficiency of amplification 95–105 %. The *Ct*s of negative controls (PBS and H_2O) should be above 35.

3.7.6 Determination of Titer

1. Determine the number of copies per PCR reaction for each dilution of test article using the standard curve. This feature is automated in most real-time PCR software packages. For those performing the calculations manually, determine the number of copies of each sample by applying the equation of the regression line (plasmid range):

$$\text{Copy number} = e^{[(C_t - b)/}$$

2. According to the dilution of the sample and volume loaded, calculate the number of copies per ml (vg/ml). NB: Since real-time PCR only targets one of the two strands (plus or minus) packaged within AAV capsids, a multiplication factor of 2 is also required.
3. Carry out the average of the titers obtained for each dilution to arrive at the final titer. For the rAAV control check that the titer obtained is included in the interval of data previously obtained.

3.8 Assessment of Purity

A comprehensive assessment of purity for AAV vector preparations could include detection of contaminants introduced as a result of the production process such as residual host cell DNA and protein or helper plasmid DNA, as well as endotoxin and mycoplasma. While careful assessment of these impurities is an integral part of the quality control of clinical grade material [57], critical assays for research material are determination of protein purity and endotoxin contamination which can impact on the immunogenic properties of the final product. A number of kits for the detection of endotoxin are commercially available, including the Limulus Amebocyte Lysate (LAL) assay (QCL-1000, Bio Whittaker) which can be used to detect and quantitatively determine the gram-negative bacterial endotoxin level in plasmid or vector preparations. The protein purity of a vector preparation can be assessed following SDS-polyacrylamide gel electrophoresis and densitometry analysis. In this way, bands corresponding to the viral structural proteins, VP1, VP2, and VP3, may be visualized and their size and relative intensity assessed relative to other contaminating proteins. It is important to note that the purification method can impact significantly on the purity of a vector preparation. As indicated previously, while cesium chloride gradient ultracentrifugation is a useful method for the purification of a variety of AAV serotype vectors and has particular utility for the generic separation of full (genome containing) and empty particles, the level of purity achievable using this method is less than that for column-chromatography methods which have been developed for a limited number of AAV serotypes. An appropriate release criteria for a cesium chloride gradient purified vector preparation is 85 % or greater viral structural proteins as depicted in Fig. 5.

3.9 Endotoxin Analysis

3.9.1 Sample Preparation (Biosafety Containment Hood)

1. Spin down the vector sample tubes and place them in a rack until they will be used for serial dilutions. With LAL reagent water in a sterile hood, prepare four tenfold serial dilutions of each pipet ranging from 1:10 to 1:10,000 in glass tubes.
2. Pipet 50 μl of each sample and the positive control sample into a 96-well plate with the dilutions placed in descending order from most to least concentrated.

3. Place the heat block in the hood, make sure the block is equilibrated to 37 °C, and then place the 96-well plate in it.

3.9.2 Kit Reagents Dilution and Standards Formulation

1. Reconstitute the lyophilized endotoxin (standard) with 1 ml of LAL reagent water. Vortex for 15 min using a foam vortex adaptor.
2. Reconstitute the chromogenic substrate by adding 6.5 ml of LAL reagent water into the vial. Prewarm the substrate to 37 °C by placing it in an incubator.
3. Reconstitute the LAL reagent vial with 3 ml of LAL reagent water.
4. Set up 7 glass tubes for serial dilution of the standard and 1 tube for the blank. Label the seven tubes to reflect their EU concentration, e.g., 1, 0.8, 0.6, 0.5, 0.3, 0.2, 0.1 EU/ml.
5. Pipet the required volume of water into each tube. Then pipet the required amount of endotoxin standard into EU tubes. Perform dilutions as necessary to achieve the standard curve.
6. In duplicate, pipet 50 μl of the standards into the 96-well plate resting in the heat block. Be sure to vortex the glass tubes vigorously before taking the aliquot. Heat the plate for at least 5 min before performing the reaction.

3.9.3 Performing the Reaction

1. At time = 16 min, add 50 μl of LAL to the first column of the plate. Begin timing as LAL is added to the first column. Mix by pipetting. At a consistent rate, add LAL to the columns across the plate and mix. At time = 6 min, add 100 μl of substrate as above. At time = 0 min, add 50 μl of stop solution as above.

3.9.4 OD Measurement and Data Interpretation

1. Take off the lid, place the plate in a 96-well absorbance reader and read absorbance at 405 and 600 nm. (The substrate absorbs at 405 nm; 600 nm is a background subtraction.)
2. Graph the standard curve (405–600 nm vs. EU/ml) and determine the linear range. The absorbance of test sample and controls falling within the linear range should be interpolated and multiplied by the dilution to determine the endotoxin concentration.

3.10 Capsid Purity Assessment by SDS-PAGE

3.10.1 Electrophoresis Apparatus Assembly and Gel Loading

1. Combine 1×10^{10} test or control AAV particles with NuPAGE LDS sample buffer/NuPAGE reducing agent and bring the final volume to 10–15 μl with 10 mM Tris pH 8.0.
2. Heat at 80 °C for 15 min.
3. Add 5 μl of Benchmark ladder to separate tubes and add the reducing agent/sample buffer—do not heat.

4. Remove the adhesive strip from the lower buffer contact of the NuPAGE gel and assemble the gel into the Xcell SureLock system (vertical gel apparatus). If only running one gel use a buffer dam for the other side of the apparatus.
5. Place 200 ml 1× MOPS running buffer containing 500 μl antioxidant in the upper chamber and carefully remove the comb so as not to disturb the wells. Check for leaks and then place 600 ml 1× MOPS running buffer in the lower chamber.
6. Load the samples and marker and run the gel at ~150 V for approximately 1 h or until the dye front reaches the bottom of the gel.

3.10.2 Gel Staining

1. Fix the gel in 100 ml of the fixing solution (50 % methanol, 7 % acetic acid) solution with gentle agitation for 2 × 30 min.
2. Decant the fix solution and replace it with 50 ml SYPRO Ruby stain.
3. Wrap the staining container in aluminum foil to protect the stain from light and allow the gel to stain overnight.
4. Remove the SYPRO Ruby stain from container and replace with wash solution (10 % methanol, 7 % acetic acid).
5. Wash the foil-covered gel for 30 min with gentle agitation followed by two 5 min washes in Milli-Q water.

3.10.3 Gel Documentation and Densitometric Analysis

1. View the stained gel using a gel imaging system with 302 nm UV-transillumination. Capture an image of the gel.
2. Using the densitometry features of the gel imaging system, obtain a chromatogram measuring the background and intensity of all bands for each gel lane according to the software manual.
3. Integrate the areas under the peaks of the chromatogram. Determine the intensity of VP1, VP2, VP3, and contaminating bands as a percentage of total area under all peaks.
4. Calculate the sample purity as the combined intensity of VP1, VP2, and VP3 as a percentage of total intensity of all peaks as shown in Fig. 4.

3.10.4 Anticipated Results

Using the methods described above, it is possible to generate research-grade AAV vectors in adequate quantities and with sufficient titers for use in a variety of preclinical models. AAV2/8-based vectors (so named for the AAV2 serotype ITR genome pseudotyped with AAV8 capsid protein) as well as other novel AAV serotype vectors suitable for liver-directed gene deliver such as AAV2/7 and AAV2/9 can be produced with average yields of greater than 1×10^{13} genome copies, with undetectable levels of

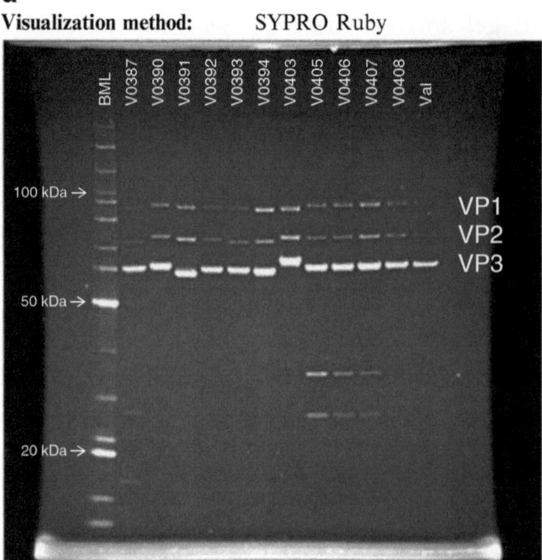

b

Lane	Vector Serotype	Lot #	Vector Load	VP1 (% total lane intensity)	VP2	VP3	VP Ratio	Purity (%)	Result
1									
2	Benchmark Ladder	Ladder							
3	AAV2	V0387	1.00E+10	2.70	3.74	81.58	1:1.4:30.2	88.01	Pass
4	AAV2/8	V0390	1.00E+10	5.45	6.45	85.38	1:1.2:15.7	97.28	Pass
5	AAV2/7	V0391	1.00E+10	6.59	10.95	77.77	1:1.7:11.8	95.31	Pass
6	AAV2/Rh64R1	V0392	1.00E+10	3.81	5.48	89.59	1:1.4:23.5	98.88	Pass
7	AAV2/hu37	V0393	1.00E+10	4.46	6.50	87.92	1:1.5:19.7	98.87	Pass
8	AAV2/1	V0394	1.00E+10	9.05	5.74	84.80	1.6:1:14.8	99.59	Pass
9	AAV2/rh32.33	V0403	1.00E+10	8.56	10.12	79.28	1:1.2:9.3	97.96	Pass
10	AAV2/8	V0405	1.00E+10	4.88	4.18	64.62	1.2:1:15.5	73.68	Fail
11	AAV2/8	V0406	1.00E+10	6.49	5.72	74.54	1.1:1:13	86.75	Pass
12	AAV2/8	V0407	1.00E+10	8.56	8.33	74.02	1:1:8.9	90.91	Pass
13	AAV2/8	V0408	1.00E+10	5.58	6.63	82.91	1:1.2:14.8	95.13	Pass
Val	Validation	V0081		3.14	4.73	86.34	1:1.5:27.5	94.21	Pass
15									

Fig. 4 SDS-PAGE and densitometry of AAV vector preparations. (**a**) AAV proteins were separated by SDS-PAGE and then detected by staining with Sypro Ruby stain (Invitrogen). Densitometry was then performed on G Box chemi HR16 imager (Syngene). *Lanes* are indicated in the table below and the position of the AAV virion proteins VP1, VP2, and VP3 are indicated. (**b**) Densitometric analysis of *lane 2*. In the majority of cases AAV virion proteins comprise >85 % of the detected proteins

endotoxin contamination and with a purity of at least 85 % virion protein. Additional assays for in vitro infectivity or in vivo transduction capacity are recommended for further characterization of AAV vector preparations. Particularly for vectors destined for use in large animal models, an assessment of in vivo transduction efficiency by detection of gene expression in a small animal model is recommended.

3.11 Verification of Transduction and Transgene Expression In Vitro

It is often advisable to test the virus in vitro before proceeding on to animal studies. In cases where a liver-specific promoter is used (such as TBG), it is necessary to use a liver-derived cell line. In cases where a ubiquitously active promoter is used such as CMV and beta-actin, any cell line that can express and/or secrete the protein of interest would be acceptable. Since it is our preference to restrict expression to the liver, the following steps will work well for infection of HepG2 cells, a human hepatoma line. It should be noted that these conditions may not be optimal for other cell lines.

3.12 Transduction of HepG2 Cells

3.12.1 Growth of HepG2 Cells

HepG2 cells are grown in minimum essential media (MEM) containing 10 % fetal bovine serum (FBS), along with antibiotic/antimycotic mix at 37 °C. Cells are typically seeded at 1:6 to ensure rapid growth.

3.12.2 Seeding Plates with HepG2 Cells (Day 1)

HepG2 cells at 90 % confluence are seeded at a 1:6 ratio in 12-well sterile tissue culture plates.

3.12.3 Infection with AAV Vector (Day 2)

Cells should be treated with the following doses of virus: 2×10^{11} GC, 4×10^{11} GC, and 8×10^{11} GC. Also include a set of cells treated with the same doses of a control virus, as well as cells treated with media alone. The virus should be diluted in MEM to yield a final volume of 250 μl per well.

1. Wash cells two times with PBS.
2. Add 250 μl of the virus mixture to each well.
3. Incubate the cells at 37 °C for 5 h. Manually rotating the plates to mix the media will help to ensure equal infection. This is typically done every half an hour.
4. After the incubation, bring the volume in the wells up to 1 ml with growth media.
5. Incubate the cells at 37 °C for 72 h. If the transgene encodes a secreted protein that will be assayed in the media, the media can be changed to serum-free MEM at 48 h post infection and allowed to incubate overnight at 37 °C prior to harvest.

3.12.4 Harvesting Transduced Cells and Media (Day 5)

1. Collect the media from the transfected cells (if protein of interest is secreted) and transfer to a 1.5-ml tube. If protein is not secreted, then the media can be removed by aspiration and Subheading 3.11.4, **step 2**, can be skipped.
2. Briefly centrifuge media at 1,000×*g* for 2 min at 4 °C to remove dead cells and debris. Save the supernatant and store at −20 °C.
3. Wash the cells 3× with ice-cold PBS.

4. Add 1 ml of LDS sample buffer to the wells (without reducing agent). Scrape the wells using a small cell scraper, and pass the crude lysate through a syringe with an 18 G needle to fragment the genomic DNA. Transfer the lysate to a 1.5-ml tube and store at −20 °C.

3.12.5 Western Blotting

1. For the media, combine 13 μl of media with 5 μl of 4× LDS sample buffer and 2 μl of 10× reducing agent in a 1.5-ml tube.
2. For the cell lysates, add 2 μl of 10× reducing agent to 18 μl of lysate in a 1.5-ml tube.
3. Denature proteins by boiling tubes for 5 min. Let cool to room temperature and load on a precast NuPAGE gel as described in Subheading 3.9.1.
4. After the gel has run, electrophoretically transfer the proteins to a nitrocellulose membrane. Block the membrane and immunoblot with an antibody to the protein of interest as required.

3.12.6 Expected Results

It is expected that marked overexpression of the transgene's protein product will be detected in cells transduced with the recombinant AAV. The transduction should be dose dependent; however, some toxicity may be expected at the highest dose. Overexpression will be most noticeable when the antibody does not detect the endogenous human protein made by the HepG2 cells, or if the transgene is not normally expressed in this cell line.

3.12.7 Alternative Means of Verifying Expression

1. Detecting the protein by Western Blotting is the preferred method for verifying expression. This is an established method that gives information about the relative abundance of the protein, as well as its molecular mass. ELISAs are more quantitative, and can easily substitute for a Western blot if such an assay is readily available. In this situation it may be necessary to harvest the cells in a more mild lysis buffer that does not contain bromophenol blue or other dyes.
2. In cases where a measurable activity exists, a biochemical assay is preferred. This gives valuable information that can verify the transgene product is catalytically active.
3. In cases where no antibody or activity assay is available, measuring mRNA is helpful but by no means definitive. As with any expression vector, the transgene may produce message that is either not translated or encodes a nonfunctional protein. In addition, there are important caveats to measuring mRNA from viral transgenes (*see* **Note 3**).

3.13 In Vivo Studies Using AAV Vectors to Target the Liver

Designing an animal study with AAV involves several important considerations: controls, mice, dose, time, measuring transgene expression, and assessing biological effects. These aspects will be discussed individually in the following subsections. It should be noted that most of these conditions may be varied depending upon the model and the hypothesis being tested.

3.14 In Vivo Liver Transduction

1. Controls
Each experiment should include a group of mice infected with a control virus. This is important to account for the effects of viral infection. Although AAV is far less immunogenic than adenovirus, some low level of inflammation or toxicity may still be present. This will primarily depend upon the purity of the vector that is used. There are several options for control viruses. The simplest of these is an AAV that contains stuffer DNA in lieu of a transgene, or "null virus." AAVs that express irrelevant nonmammalian proteins such as β-galactosidase (LacZ) or green fluorescent protein (GFP) are commonly used as well. AAVs expressing reporter genes such as these can be useful in determining transduction efficiency, as well as cell type specificity. In our hands these control viruses (AAV2/8-LacZ, AAV2/8-null, AAV2/8-GFP) do not have any significant impact on lipid parameters. The control virus should be given at the same time, by the same route, and in the same amount (GC/mouse) as the virus for the gene being investigated.
2. Mice
As a starting point, it is advisable to use a group of age- and sex-matched wild type mice. The choice of strain is somewhat arbitrary, but ideally would be the same as that used for subsequent experiments. When looking for changes in plasma lipids, we find that an "n" of 6 mice per group is a good starting point. Pre-bleeding the mice to obtain plasma for baseline lipid measurements is an essential part of a successful experiment. If working with genetically modified mice, or those that are not well matched for sex and age, it is advisable to presort according to the analyte of interest (i.e., total cholesterol levels). In these situations, a larger number of mice per group are recommended.
3. Delivering Virus
Each mouse is given an injection of 1×10^{12} total GC of the AAV. This is our standard dose of AAV and has been found to give reproducible transduction and expression. The virus should be thawed immediately before use, and only the exact number of aliquots needed should be taken. The AAV should be diluted in sterile PBS or saline to a final volume of 200–300 μl. It is important to account for the void volume of the syringe in the calculation. In addition, preparing 15–20 % extra virus solution ensures that the last dose can be loaded accurately. After careful mixing by inversion, the viruses are warmed to room temperature and injected. Systemic administration of

AAV has traditionally been accomplished by tail vein injection. However, we have had good success with intraperitoneal (IP) injection (*see* **Note 4**). If giving an IP injection, extreme care should be taken to avoid puncturing the gut or liver. Inadvertently delivering a subcutaneous injection will drastically reduce expression.

4. Monitoring Transgene Expression
   Second-strand synthesis of the DNA is thought to be the rate-limiting step for in vivo transduction of liver. Several days may be required for complete conversion to dsDNA. It is generally expected that stable expression of the transgene will be obtained 10–14 days postinfection. At this point, the mice may be bled to measure plasma analytes, or even transgene expression in the case of secreted proteins. Bleeding every week to 2 weeks will generate a clear picture of expression over time. Typically, expression of the transgene will reach optimal levels between 2 and 6 weeks. The level of expression may drop off, but should persist out to a year or more. Much of this depends upon the expression cassette and the transgene itself.

5. Determining Optimal Dose
   In many instances it is necessary to vary the dose of AAV depending upon the nature of the transgene and its biological activity. This is particularly true of proteins with enzymatic activity, but also applies to other transgenes as well. The high level of expression obtained with 1×10^{12} GC carries the risk of introducing artifactual results. In contrast to transgenic mouse models, somatic overexpression has the distinct advantage of dosability. Transduction of liver with recombinant AAV2 has been found to give a linear dose response from 3.7×10^{9} to 3.0×10^{11} vector genomes (vg) per mouse, with saturation reached at 1.8×10^{12} vg [63]. In our experience, AAV8 can be readily titrated over a range from 1×10^{12} GC down to 1×10^{10} GC/mouse. Transduction efficiency may be more variable at lower doses, but in practice this works quite well. The optimal dose will depend greatly on the biological activity of the gene product. In many cases, very small doses of virus are sufficient. Figure 5 shows a dose response of AAV2/8 encoding human CETP.

6. Time Courses
   Once the optimal dose has been determined, performing a time course is also worthwhile. This is most easily accomplished when plasma or sera can be assayed either for transgene expression or the presence of an analyte or activity of interest. Tracking these measurements over time will yield valuable information about the stability of expression and the biological activity of the gene product. It should be noted that although AAV is far less immunogenic than Adenovirus, neutralizing antibodies

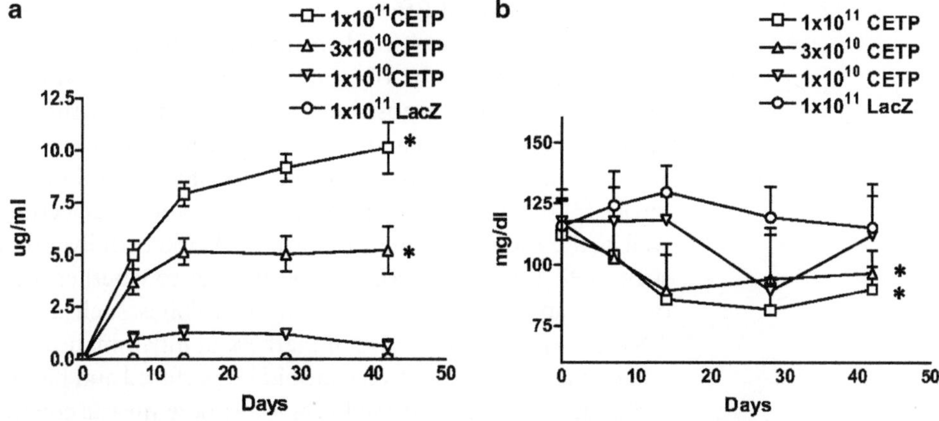

Fig. 5 Dose-titration of AAV8 encoding human CETP in LDLR–/– ApoBEC–/– mice. LDLR/Apobec-1–DKO mice were given three dosages of human CETP (hCETP), adeno-associated virus serotype 8 (AAV8), or LacZ via intraperitoneal injection ($n = 5$/each group). (**a**) The change of plasma hCETP mass level of each time point determined by ELISA (day 0, 7, 14, 28, 42). (**b**) The change of plasma HDL expression level of each time point determined on a Cobas Fara. The differences between each CETP group and 1×10^{11} LacZ group were evaluated. $^{*}P < 0.05$ versus 1×10^{11} LacZ group. Reproduced with permission from *Circulation* [37] (*Tanigawa* et al. 2007;116:1267–1273)

can still be generated over time. This is most likely to be a problem with non-murine transgenes, and should be taken into consideration when designing experiments lasting more than 4–6 weeks. Relying only on protein or mRNA measurements yields an incomplete picture of expression.

7. Detecting the Transgene

Detecting the virally expressed transgene is an important control in any somatic overexpression experiment. The method of detection is usually predicated by the resources available. In cases where an antibody exists, detecting the transgene by Western blotting or ELISA adds credence to any phenotypic observations. In cases where this is not possible, the investigator may consider tagging the viral construct for easy detection. This should only be done when the tag does not interfere with the biological activity of the gene product, and the approach is consistent with the hypothesis being tested. In practice, this is often too big of a risk considering the time and resources required for these studies. Measurement of mRNA levels by real-time RT-PCR can be used as a surrogate marker for protein expression. While convenient, this cannot confirm that the protein is being produced or at what level. Real-time RT-PCR is particularly useful in determining quantitatively how much a gene has been overexpressed relative to the endogenous transcript. However, certain precautions should be taken in measuring virally overexpressed mRNA levels (*see* **Note 3**).

8. Assessing Biological Effects

If the transgene is expected to influence lipoprotein metabolism, plasma should be obtained from the mice before infection, at various times post infection, and finally at the time of sacrifice. It is best to plan to obtain suitable amounts of plasma at each time point to assay the analytes of interest, i.e., cholesterol, HDL cholesterol, phospholipids, and triglycerides, as well as the transgene when applicable. Markers of liver damage such as ALT can give a rough indication as to whether the virus or the transgene product is toxic. Cytokines such as IL6 or TNF-alpha can serve as useful markers of inflammation. At the end of the experiment, livers should be perfused and harvested. Weighing the mice and the livers is a simple measurement that can offer valuable clues about the function of the gene being interrogated. Liver should be saved as well for the specific purpose of RNA isolation. Preserving and recording these samples will pay great dividends when the data is analyzed. Having extra tissue or plasma on hand allows for easy testing of new hypotheses in the future. As with all experiments in science, careful and methodical planning is the way to ensure success.

9. Effects of Mouse Strain and Gender on Transduction Efficiency

Although AAV8 is very efficient in targeting mouse hepatocytes, the efficiency of the process has been described to be dependent on genetic background, sex, and estrous cycle stage for female mice. Indeed, for certain AAV serotypes, dramatic differences between the levels of transgene expression have been observed between different mouse strains. In our experience, immunodeficient animals such as $RAG^{-/-}$ or NCR nude mice are highly permissive for AAV transduction. C57Bl/6 mice give higher levels of gene transfer in hepatic tissue in comparison to mice with a Balb/c genetic background. The mechanisms underlying these differences are likely due to the hosts' immunology, the biology of vector-cell interaction on an entry or post-entry level, or the efficiency of AAV genome conversion to a transcriptionally active and stable form.

Several reports have indicated the influence of the gender on AAV transduction efficiency in vivo [64–67]. In order to achieve equivalent levels of transgene expression, higher doses of vector may be required in female compared to male mice. Elegant experiments by Davidoff et al. have shown that castration abrogates this effect and testosterone injection in female mice can rescue their lower permissivity for vector transduction. These data suggest a clear endocrine impact on vector transduction [65]. Supportive of this are the findings that gene transfer efficiency may be influenced by the stage of the estrous cycle of females [64]. Others have shown that the route of injection and the AAV vector serotype can overcome this gender differential in terms of vector transduction [54, 66]. It is important to be aware that strain and sex-specific differences in AAV transduction efficiency do exist. It is advisable to take these factors into consideration when designing experiments involving mice of different sex or genetic background.

4 Notes

1. AAV trans plasmids, Ad helper plasmid, vector maps, and sequences are available from the Penn Vector Core upon request (University of Pennsylvania, Suite 2000 Translational Research Labs (TRL), 125S. 31st Street, Philadelphia, PA, 19104–3403, Phone: 215-898-0226).
2. After titering, the virus should be aliquotted and stored immediately at –80 °C. Based on a typical animal experiment, we find it convenient to aliquot the virus in tubes containing no more than 3×10^{12} GC. Although this is mostly a matter of preference, having more aliquots will prevent wasting of valuable virus. In practical terms, this usually corresponds to roughly 100 μl aliquots of the untitered virus. Once the AAV is thawed it should be stable for several weeks at 4 °C, although refreezing at –80 °C is recommended. Subjecting AAV to more than one freeze–thaw cycle may adversely affect transduction efficiency and should be avoided.
3. Real-time RT-PCR is a simple and convenient way to measure transgene message levels. It should be noted that transgene DNA exists episomally either as a single species or as concatemers, and is not integrated into the genomic DNA. This transgene DNA is often co-isolated using most RNA isolation protocols. As such, it is critically important to treat the isolated RNA with DNase before proceeding to reverse transcription and amplification. Including a control reaction lacking the reverse transcriptase will show that the DNase treatment was effective in removing the contaminating DNA.
4. Although AAV may also be delivered by intravenous injection, we find that intraperitoneal injection is equally effective in terms of expression, and easily executed with minimal hands-on training. This is our routine method for delivering AAV to the liver.

Acknowledgments

The authors would like to thank Anne Douar at Genethon for her contributions to the genome titer assay method. We would also like to acknowledge Peter Bell who kindly provided the electron micrograph image of AAV shown in Fig. 1.

References

1. Rux JJ, Burnett RM (2004) Adenovirus structure. Hum Gene Ther 15:1167–1176
2. Stratford-Perricaudet LD, Levrero M, Chasse JF, Perricaudet M, Briand P (1990) Evaluation of the transfer and expression in mice of an enzyme-encoding gene using a human adenovirus vector. Hum Gene Ther 1:241–256
3. Herz J, Gerard RD (1993) Adenovirus-mediated transfer of low density lipoprotein receptor gene acutely accelerates cholesterol clearance in normal mice. Proc Natl Acad Sci USA 90:2812–2816
4. Ishibashi S, Brown MS, Goldstein JL, Gerard RD, Hammer RE, Herz J (1993) Hypercholesterolemia in low density lipoprotein receptor knockout mice and its reversal by adenovirus-mediated gene delivery. J Clin Invest 92:883–893
5. Kopfler WP, Willard M, Betz T, Willard JE, Gerard RD, Meidell RS (1994) Adenovirus-mediated transfer of a gene encoding human apolipoprotein A-I into normal mice increases circulating high-density lipoprotein cholesterol. Circulation 90:1319–1327
6. Hughes SD, Rouy D, Navaratnam N, Scott J, Rubin EM (1996) Gene transfer of cytidine deaminase apoBEC-1 lowers lipoprotein(a) in transgenic mice and induces apolipoprotein B editing in rabbits. Hum Gene Ther 7:39–49
7. Stevenson SC, Marshall-Neff J, Teng B, Lee CB, Roy S, McClelland A (1995) Phenotypic correction of hypercholesterolemia in apoE-deficient mice by adenovirus-mediated in vivo gene transfer. Arterioscler Thromb Vasc Biol 15:479–484
8. Spady DK, Cuthbert JA, Willard MN, Meidell RS (1995) Adenovirus-mediated transfer of a gene encoding cholesterol 7 alpha-hydroxylase into hamsters increases hepatic enzyme activity and reduces plasma total and low density lipoprotein cholesterol. J Clin Invest 96:700–709
9. Seguret-Mace S, Latta-Mahieu M, Castro G, Luc G, Fruchart JC, Rubin E, Denefle P, Duverger N (1996) Potential gene therapy for lecithin-cholesterol acyltransferase (LCAT)-deficient and hypoalphalipoproteinemic patients with adenovirus-mediated transfer of human LCAT gene. Circulation 94:2177–2184
10. Applebaum-Bowden D, Kobayashi J, Kashyap VS, Brown DR, Berard A, Meyn S, Parrott C, Maeda N, Shamburek R, Brewer HB Jr, Santamarina-Fojo S (1996) Hepatic lipase gene therapy in hepatic lipase-deficient mice. Adenovirus-mediated replacement of a lipolytic enzyme to the vascular endothelium. J Clin Invest 97:799–805
11. Yang Y, Jooss KU, Su Q, Ertl HC, Wilson JM (1996) Immune responses to viral antigens versus transgene product in the elimination of recombinant adenovirus-infected hepatocytes in vivo. Gene Ther 3:137–144
12. Jooss K, Ertl HC, Wilson JM (1998) Cytotoxic T-lymphocyte target proteins and their major histocompatibility complex class I restriction in response to adenovirus vectors delivered to mouse liver. J Virol 72:2945–2954
13. Kass-Eisler A, Falck-Pedersen E, Elfenbein DH, Alvira M, Buttrick PM, Leinwand LA (1994) The impact of developmental stage, route of administration and the immune system on adenovirus-mediated gene transfer. Gene Ther 1:395–402
14. Kozarsky KF, Donahee MH, Glick JM, Krieger M, Rader DJ (2000) Gene transfer and hepatic overexpression of the HDL receptor SR-BI reduces atherosclerosis in the cholesterol-fed LDL receptor-deficient mouse. Arterioscler Thromb Vasc Biol 20:721–727
15. Tsukamoto K, Tangirala RK, Chun S, Usher D, Pure E, Rader DJ (2000) Hepatic expression of apolipoprotein E inhibits progression of atherosclerosis without reducing cholesterol levels in LDL receptor-deficient mice. Mol Ther 1:189–194
16. Quarck R, De Geest B, Stengel D, Mertens A, Lox M, Theilmeier G, Michiels C, Raes M, Bult H, Collen D, Van Veldhoven P, Ninio E, Holvoet P (2001) Adenovirus-mediated gene transfer of human platelet-activating factor-acetylhydrolase prevents injury-induced neointima formation and reduces spontaneous atherosclerosis in apolipoprotein E-deficient mice. Circulation 103:2495–2500
17. Juan SH, Lee TS, Tseng KW, Liou JY, Shyue SK, Wu KK, Chau LY (2001) Adenovirus-mediated heme oxygenase-1 gene transfer inhibits the development of atherosclerosis in apolipoprotein E-deficient mice. Circulation 104:1519–1525
18. Belalcazar LM, Merched A, Carr B, Oka K, Chen KH, Pastore L, Beaudet A, Chan L (2003) Long-term stable expression of human apolipoprotein A-I mediated by helper-dependent adenovirus gene transfer inhibits atherosclerosis progression and remodels atherosclerotic plaques in a mouse model of familial hypercholesterolemia. Circulation 107:2726–2732
19. Dong J, Liu J, Lou B, Li Z, Ye X, Wu M, Jiang XC (2006) Adenovirus-mediated overexpression of sphingomyelin synthases 1 and 2 increases the atherogenic potential in mice. J Lipid Res 47:1307–1314

20. Ng CJ, Bourquard N, Hama SY, Shih D, Grijalva VR, Navab M, Fogelman AM, Reddy ST (2007) Adenovirus-mediated expression of human paraoxonase 3 protects against the progression of atherosclerosis in apolipoprotein E-deficient mice. Arterioscler Thromb Vasc Biol 27:1368–1374
21. Hu Q, Zhang XJ, Zhang C, Zhao YX, He H, Liu CX, Feng JB, Jiang H, Yang FL, Zhang CX, Zhang Y (2008) Peroxisome proliferator-activated receptor-gamma1 gene therapy attenuates atherosclerosis and stabilizes plaques in apolipoprotein E-deficient mice. Hum Gene Ther 19:287–299
22. Tangirala RK, Tsukamoto K, Chun SH, Usher D, Pure E, Rader DJ (1999) Regression of atherosclerosis induced by liver-directed gene transfer of apolipoprotein A-I in mice. Circulation 100:1816–1822
23. Tsukamoto K, Tangirala R, Chun SH, Pure E, Rader DJ (1999) Rapid regression of atherosclerosis induced by liver-directed gene transfer of ApoE in ApoE-deficient mice. Arterioscler Thromb Vasc Biol 19:2162–2170
24. Harris JD, Graham IR, Schepelmann S, Stannard AK, Roberts ML, Hodges BL, Hill V, Amalfitano A, Hassall DG, Owen JS, Dickson G (2002) Acute regression of advanced and retardation of early aortic atheroma in immunocompetent apolipoprotein-E (apoE) deficient mice by administration of a second generation [E1(−), E3(−), polymerase(−)] adenovirus vector expressing human apoE. Hum Mol Genet 11:43–58
25. Zaiss AK, Liu Q, Bowen GP, Wong NC, Bartlett JS, Muruve DA (2002) Differential activation of innate immune responses by adenovirus and adeno-associated virus vectors. J Virol 76:4580–4590
26. Goncalves MA (2005) Adeno-associated virus: from defective virus to effective vector. Virol J 2:43
27. Koeberl DD, Alexander IE, Halbert CL, Russell DW, Miller AD (1997) Persistent expression of human clotting factor IX from mouse liver after intravenous injection of adeno-associated virus vectors. Proc Natl Acad Sci USA 94:1426–1431
28. Herzog RW, Hagstrom JN, Kung SH, Tai SJ, Wilson JM, Fisher KJ, High KA (1997) Stable gene transfer and expression of human blood coagulation factor IX after intramuscular injection of recombinant adeno-associated virus. Proc Natl Acad Sci USA 94:5804–5809
29. Snyder RO, Miao CH, Patijn GA, Spratt SK, Danos O, Nagy D, Gown AM, Winther B, Meuse L, Cohen LK, Thompson AR, Kay MA (1997) Persistent and therapeutic concentrations of human factor IX in mice after hepatic gene transfer of recombinant AAV vectors. Nat Genet 16:270–276
30. Nakai H, Herzog RW, Hagstrom JN, Walter J, Kung SH, Yang EY, Tai SJ, Iwaki Y, Kurtzman GJ, Fisher KJ, Colosi P, Couto LB, High KA (1998) Adeno-associated viral vector-mediated gene transfer of human blood coagulation factor IX into mouse liver. Blood 91:4600–4607
31. Wang L, Takabe K, Bidlingmaier SM, Ill CR, Verma IM (1999) Sustained correction of bleeding disorder in hemophilia B mice by gene therapy. Proc Natl Acad Sci USA 96:3906–3910
32. Yang XP, Yan D, Qiao C, Liu RJ, Chen JG, Li J, Schneider M, Lagrost L, Xiao X, Jiang XC (2003) Increased atherosclerotic lesions in apoE mice with plasma phospholipid transfer protein overexpression. Arterioscler Thromb Vasc Biol 23:1601–1607
33. Lebherz C, Gao G, Louboutin JP, Millar J, Rader D, Wilson JM (2004) Gene therapy with novel adeno-associated virus vectors substantially diminishes atherosclerosis in a murine model of familial hypercholesterolemia. J Gene Med 6:663–672
34. Kitajima K, Marchadier DH, Miller GC, Gao GP, Wilson JM, Rader DJ (2006) Complete prevention of atherosclerosis in apoE-deficient mice by hepatic human apoE gene transfer with adeno-associated virus serotypes 7 and 8. Arterioscler Thromb Vasc Biol 26:1852–1857
35. Kitajima K, Marchadier DH, Burstein H, Rader DJ (2006) Persistent liver expression of murine apoA-l using vectors based on adeno-associated viral vectors serotypes 5 and 1. Atherosclerosis 186:65–73
36. Lebherz C, Sanmiguel J, Wilson JM, Rader DJ (2007) Gene transfer of wild-type apoA-I and apoA-I milano reduce atherosclerosis to a similar extent. Cardiovasc Diabetol 6:15
37. Tanigawa H, Billheimer JT, Tohyama J, Zhang Y, Rothblat G, Rader DJ (2007) Expression of cholesteryl ester transfer protein in mice promotes macrophage reverse cholesterol transport. Circulation 116:1267–1273
38. Atchison RW, Casto BC, Hammon WM (1965) Adenovirus-associated defective virus particles. Science 149:754–756
39. Dai J, Rabie AB (2007) The use of recombinant adeno-associated virus for skeletal gene therapy. Orthod Craniofac Res 10:1–14
40. Gao G, Alvira MR, Somanathan S, Lu Y, Vandenberghe LH, Rux JJ, Calcedo R, Sanmiguel J, Abbas Z, Wilson JM (2003) Adeno-associated viruses undergo substantial evolution in primates during natural infections. Proc Natl Acad Sci USA 100:6081–6086

41. Kotin RM, Siniscalco M, Samulski RJ, Zhu XD, Hunter L, Laughlin CA, McLaughlin S, Muzyczka N, Rocchi M, Berns KI (1990) Site-specific integration by adeno-associated virus. Proc Natl Acad Sci USA 87:2211–2215
42. Schnepp BC, Jensen RL, Chen CL, Johnson PR, Clark KR (2005) Characterization of adeno-associated virus genomes isolated from human tissues. J Virol 79:14793–14803
43. Kotin RM (1994) Prospects for the use of adeno-associated virus as a vector for human gene therapy. Hum Gene Ther 5:793–801
44. Sanlioglu S, Monick MM, Luleci G, Hunninghake GW, Engelhardt JF (2001) Rate limiting steps of AAV transduction and implications for human gene therapy. Curr Gene Ther 1:137–147
45. Chirmule N, Truneh A, Haecker SE, Tazelaar J, Gao G, Raper SE, Hughes JV, Wilson JM (1999) Repeated administration of adenoviral vectors in lungs of human CD4 transgenic mice treated with a nondepleting CD4 antibody. J Immunol 163:448–455
46. Rabinowitz JE, Rolling F, Li C, Conrath H, Xiao W, Xiao X, Samulski RJ (2002) Cross-packaging of a single adeno-associated virus (AAV) type 2 vector genome into multiple AAV serotypes enables transduction with broad specificity. J Virol 76:791–801
47. Kay MA, Glorioso JC, Naldini L (2001) Viral vectors for gene therapy: the art of turning infectious agents into vehicles of therapeutics. Nat Med 7:33–40
48. Gao G, Vandenberghe LH, Wilson JM (2005) New recombinant serotypes of AAV vectors. Curr Gene Ther 5:285–297
49. Gao G, Lu Y, Calcedo R, Grant RL, Bell P, Wang L, Figueredo J, Lock M, Wilson JM (2006) Biology of AAV serotype vectors in liver-directed gene transfer to nonhuman primates. Mol Ther 13:77–87
50. Zincarelli C, Soltys S, Rengo G, Rabinowitz JE (2008) Analysis of AAV serotypes 1–9 mediated gene expression and tropism in mice after systemic injection. Mol Ther 16:1073–1080
51. Inagaki K, Fuess S, Storm TA, Gibson GA, McTiernan CF, Kay MA, Nakai H (2006) Robust systemic transduction with AAV9 vectors in mice: efficient global cardiac gene transfer superior to that of AAV8. Mol Ther 14:45–53
52. Pacak CA, Mah CS, Thattaliyath BD, Conlon TJ, Lewis MA, Cloutier DE, Zolotukhin I, Tarantal AF, Byrne BJ (2006) Recombinant adeno-associated virus serotype 9 leads to preferential cardiac transduction in vivo. Circ Res 99:e3–e9
53. Gao GP, Alvira MR, Wang L, Calcedo R, Johnston J, Wilson JM (2002) Novel adeno-associated viruses from rhesus monkeys as vectors for human gene therapy. Proc Natl Acad Sci USA 99:11854–11859
54. Hildinger M, Auricchio A, Gao G, Wang L, Chirmule N, Wilson JM (2001) Hybrid vectors based on adeno-associated virus serotypes 2 and 5 for muscle-directed gene transfer. J Virol 75:6199–6203
55. Hayashi Y, Mori Y, Janssen OE, Sunthornthepvarakul T, Weiss RE, Takeda K, Weinberg M, Seo H, Bell GI, Refetoff S (1993) Human thyroxine-binding globulin gene: complete sequence and transcriptional regulation. Mol Endocrinol 7:1049–1060
56. Xiao X, Li J, Samulski RJ (1998) Production of high-titer recombinant adeno-associated virus vectors in the absence of helper adenovirus. J Virol 72:2224–2232
57. Wright JF (2008) Manufacturing and characterizing AAV-based vectors for use in clinical studies. Gene Ther 15:840–848
58. Cecchini S, Negrete A, Kotin RM (2008) Toward exascale production of recombinant adeno-associated virus for gene transfer applications. Gene Ther 15:823–830
59. Van Vliet KM, Blouin V, Brument N, Agbandje-McKenna M, Snyder RO (2008) The role of the adeno-associated virus capsid in gene transfer. Methods Mol Biol 437:51–91
60. Grieger JC, Choi VW, Samulski RJ (2006) Production and characterization of adeno-associated viral vectors. Nat Protoc 1:1412–1428
61. Qu G, Bahr-Davidson J, Prado J, Tai A, Cataniag F, McDonnell J, Zhou J, Hauck B, Luna J, Sommer JM, Smith P, Zhou S, Colosi P, High KA, Pierce GF, Wright JF (2007) Separation of adeno-associated virus type 2 empty particles from genome containing vectors by anion-exchange column chromatography. J Virol Methods 140:183–192
62. Grieger JC, Samulski RJ (2005) Adeno-associated virus as a gene therapy vector: vector development, production and clinical applications. Adv Biochem Eng Biotechnol 99:119–145
63. Nakai H, Thomas CE, Storm TA, Fuess S, Powell S, Wright JF, Kay MA (2002) A limited number of transducible hepatocytes restricts a wide-range linear vector dose response in recombinant adeno-associated virus-mediated liver transduction. J Virol 76:11343–11349
64. Dodge JC, Clarke J, Passini MA, Song A, O'Riordan CR, Cheng SH, Stewart GR (2005) 497. Sex and estrous cycle stage influence the

efficiency of AAV-mediated gene transfer in the rodent brain. Mol Ther 11:S192–S193

65. Davidoff AM, Ng CY, Zhou J, Spence Y, Nathwani AC (2003) Sex significantly influences transduction of murine liver by recombinant adeno-associated viral vectors through an androgen-dependent pathway. Blood 102:480–488
66. Berraondo P, Crettaz J, Ochoa L, Paneda A, Prieto J, Troconiz IF, Gonzalez-Aseguinolaza G (2006) Intrahepatic injection of recombinant adeno-associated virus serotype 2 overcomes gender-related differences in liver transduction. Hum Gene Ther 17:601–610
67. Voutetakis A, Zheng C, Wang J, Goldsmith CM, Afione S, Chiorini JA, Wenk ML, Vallant M, Irwin RD, Baum BJ (2007) Gender differences in serotype 2 adeno-associated virus biodistribution after administration to rodent salivary glands. Hum Gene Ther 18:1109–1118

Chapter 14

Modulation of Lipoprotein Metabolism by Antisense Technology: Preclinical Drug Discovery Methodology

Rosanne M. Crooke and Mark J. Graham

Abstract

Antisense oligonucleotides (ASOs) are a new class of specific therapeutic agents that alter the intermediary metabolism of mRNA, resulting in the suppression of disease-associated gene products. ASOs exert their pharmacological effects after hybridizing, via Watson-Crick base pairing, to a specific target RNA. If appropriately designed, this event results in the recruitment of RNase H, the degradation of targeted mRNA or pre-mRNA, and subsequent inhibition of the synthesis of a specific protein. A key advantage of the technology is the ability to selectively inhibit targets that cannot be modulated by traditional therapeutics such as structural proteins, transcription factors, and, of topical interest, lipoproteins. In this chapter, we will first provide an overview of antisense technology, then more specifically describe the status of lipoprotein-related genes that have been studied using the antisense platform, and finally, outline the general methodology required to design and evaluate the in vitro and in vivo efficacy of those drugs.

Key words Antisense technology, 2′-O-2-methoxyethyl (2′MOE) antisense oligonucleotides (ASOs), RNase H, Lipoprotein metabolism, Apolipoprotein(a), Apolipoprotein B-100, Lipoprotein(a), Apolipoprotein C-III, Very low density lipoprotein cholesterol (VLDL-C), Low density lipoprotein cholesterol (LDL-C), High density lipoprotein cholesterol (HDL-C), Triglycerides (TG), Acyl-coenzyme A:cholesterol acyltransferase 2 (ACAT2), Stearoyl-CoA desaturase-1 (SCD-1), Acetyl-coenzyme A carboxylases 1 and 2 (ACC1, ACC2), Diacylglycerol acyltransferase-2 (DGAT2)

1 Introduction

Unlike traditional small-molecule therapeutic agents that interfere with or modulate the function of proteins, antisense oligonucleotides (ASOs) are single-stranded, chemically modified DNA-like drugs that bind with high affinity to RNA targets through Watson-Crick base pairing, ultimately resulting in the suppression of disease-associated gene products [1, 2]. ASO-mediated reduction of specific targeted proteins may be accomplished through several potential pathways. A well-defined terminating mechanism, and one that is used by the majority of antisense drugs in the clinic, utilizes RNase H-mediated cleavage of an RNA–DNA duplex [3–7].

Lita A. Freeman (ed.), *Lipoproteins and Cardiovascular Disease: Methods and Protocols*, Methods in Molecular Biology, vol. 1027, DOI 10.1007/978-1-60327-369-5_14,

The most advanced ASOs, which are chimeric molecules generally 20 nucleotides in length, contain methoxyethyl (MOE) groups in the 2′ position of the ribose moieties in the termini or "wings" (positions 1–5 and 15–20) and DNA-like 2′ deoxyphosphorothioate nucleotides in the center or "gapped" regions [1, 8–10]. These "2′MOE gapmers" (1) are 5- to 15-fold more potent than first-generation drugs (containing only uniform deoxyphosphorothioate linkages) due to enhanced affinity for their RNA targets, (2) support RNase H, and (3) display improved pharmacokinetic properties (greater stability, longer half-lives) permitting weekly or possibly quarterly dosing in the clinic. They also have improved therapeutic indices because of a reduced propensity to induce pro-inflammatory effects [1, 8].

Antisense drugs directed to inflammatory, metabolic, and cardiovascular disease targets have been successfully evaluated in vitro, in vivo, and in humans over the past several years [8–15]. They are water soluble and may be administered in saline without special formulation via subcutaneous (s.c.) injection and other routes of delivery, including intravenous, topical, aerosol, enema, intravitreal, intraventricular/intrathecal, and oral administration [16–20]. Additionally, these drugs are well tolerated in animals and humans. Key safety issues have been identified and meaningful progress has been made towards understanding potential adverse events [21–23]. Since antisense therapeutics are not substrates for the cytochrome P450 system, they can be used safely when co-administered with traditional therapeutics with different mechanisms of action [24].

Pharmacokinetic properties of ASOs have been extensively characterized in multiple species and, most importantly, in man [16–20, 25]. At parenterally administered therapeutic doses, the liver, kidney, bone marrow, adipose tissue, spleen, and lymph nodes accumulate significant amounts of drug. Conversely, ASOs distribute poorly to intestine, skeletal muscle, heart, lung, reproductive organs, pancreas, and brain. As the 2′MOE ASOs are resistant to exonuclease-mediated degradation, tissue half-lives are prolonged and may range from 10 to 30 days. Generally, these drugs are cleared from tissue by a slow endonucleolytic metabolic process, resulting in the urinary excretion of lower molecular weight metabolites (8–12mers).

2 Antisense Inhibitors to Lipoproteins

Lipoproteins are complex macromolecules that consist of a central core of nonpolar lipid components (triglycerides and cholesteryl esters) and a surface monolayer of apolipoproteins [apoA-I, A-II, A-IV, B-48, B-100, C-I, C-II, C-III, D, E2, E3, E4, M, apo(a)] and polar lipids (primarily phospholipids) [26, 27]. The apolipoproteins are the lipid-binding and structural elements of lipoproteins.

They also act as cofactors in various enzymatic processes regulating their own metabolism or function as ligands in receptor-lipoprotein interactions. In general, the lipoprotein classes vary in density, electrophoretic mobility, percentage of surface components, and core lipids. The liver is a major site of synthesis for the majority of the apolipoproteins, and intestine can also synthesize many apolipoproteins, especially postprandially.

The dysregulation of the synthesis and metabolism of certain lipoproteins and their core components are well-recognized risk factors for the development of coronary heart disease (CHD), which is still the leading cause of death in the United States and all industrialized nations [28–31]. For example, data from several large interventional clinical trials have demonstrated that elevated levels of cholesterol in the apoB-containing lipoproteins, such as very-low-density lipoprotein cholesterol (VLDL-C) and low-density lipoprotein cholesterol (LDL-C), are implicated in the development of atherosclerosis and CHD [32–37]. ApoB-100 is also thought to be directly atherogenic through direct binding interactions with the vascular subendothelium [38]. Elevated levels of lipoprotein(a) (Lp(a)), a lipoprotein particle with apoB-100 covalently linked to a second apolipoprotein—apolipoprotein(a)—via a disulfide bond, represent another independent risk factor for the development of CHD, peripheral, and cerebral vascular disease [39–41]. Increased levels of apolipoprotein C-III (apoC-III) [42–44] are believed to contribute to the development of hypertriglyceridemia or elevated triglycerides (TG) which are also recognized by the National Cholesterol Education Program (NCEP) Adult Treatment Panel (ATP) as contributing to the risk of CHD [45–47]. Conversely, clinical trial data suggest that low serum levels of apoA-I-containing high-density lipoprotein cholesterol (HDL-C) can inversely predict CHD risk [48, 49]. More recently, it has been suggested the apolipoprotein B/A-I ratio is even more predictive of cardiovascular risk than that of LDL-C and HDL-C [50, 51].

Antisense technology has proven to be a valuable tool, not only for the development of novel therapeutic agents that specifically target many of the risk factors that predispose individuals to cardiovascular disease but also as a rapid means for the identification and validation of gene targets (i.e., functional genomics) that may be involved in the complex pathways regulating lipoprotein metabolism. A number of genes that affect the production or composition of lipoprotein particles and are principally expressed within the liver have been successfully inhibited both in vitro and in vivo and will be discussed in further detail.

The most advanced of these lipoprotein antisense inhibitors, and exemplar of the strength of the technology, is a 2′MOE gapmer targeted to human apoB-100 (Mipomersen-ISIS 301012). Mipomersen has been extensively studied in the clinic, both in healthy volunteers and in hypercholesterolemic subjects [12, 52–55]. In Phase 2 trials in polygenic and homozygous and heterozygous familial hypercholesterolemic (HoFH; HeFH) subjects, mipomersen, as monotherapy, produced linear, significant, prolonged, and dose-dependent reductions in apoB, LDL-C, all atherogenic lipids, and triglycerides. Co-administration of mipomersen with statins and ezetimibe produced additive benefit. The drug was well tolerated whether administered as monotherapy or in combination with other lipid-lowering agents. On January 29, 2013 the FDA approved mipomersen (KYNAMRO™) as an adjunct to lipid-lowering medications and diet to reduce LDL-C, apoB, total cholesterol and non-HDL-C in patients with HoFH.

3 Antisense Technology Preclinical Drug Discovery Process

Irrespective of whether an ASO-targeted gene is destined solely for functional genomic/target validation studies or will advance to clinical candidacy, the fundamental steps involved in preclinical drug discovery are the same [56–63] (Fig. 1). While ASOs of varying lengths, chemical modifications, and structures can be designed [62, 63], our focus will be on 2′MOE-gapmers whose terminating mechanism relies on the RNase H enzyme described in the Introduction.

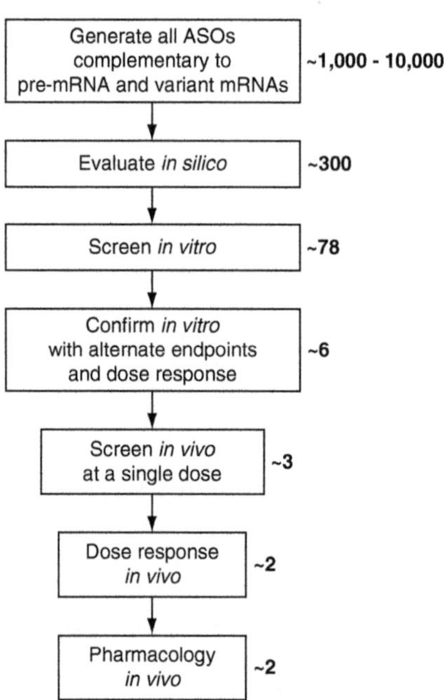

Fig. 1 Typical screening process to identify antisense oligonucleotide (ASO) drugs for targeted genes. This screening paradigm significantly reduces the number of possible candidates, ultimately producing several optimized leads that can be evaluated in vivo for pharmacological effects

3.1 Identification of Lipoprotein-Related Targets

At Isis, the Cardiovascular Group has successfully modulated lipoproteins and lipid pathways with ASOs [12, 13, 39, 64–69] by exploiting the fact that the liver, a key site of cholesterol, fatty acid/triglyceride, and apolipoprotein biosynthesis, also represents a major organ of ASO distribution. Previous studies in our laboratories have localized 2′MOE ASOs within hepatocytes as well as non-parenchymal cells, using a variety of detection techniques [25, 70–72]. Importantly, antisense drugs distributed in vivo without the need for special delivery vehicles such as polycations or liposomes.

3.2 Evaluation of ASOs In Silico

Antisense technology is uniquely suited to exploit the wealth of human and animal genomic sequencing data [56, 57, 59, 61, 63]. This process begins by designing a series of ASOs that can hybridize to different sites on the mRNA, pre-mRNA, or noncoding regions of RNA of interest. Common sources of information for known primary and processed RNA variants include public sequence databases such as EMBL (European Molecular Biology Laboratory) and GenBank. Expressed sequence tags (ESTs) and complementary DNAs (cDNAs) are generally aligned against the genomic sequence using the bioinformatics program, Basic Local Alignment Search Tool, or BLAST [73].

Theoretically, any region within an RNA sequence can be targeted and full-length gene sequences are not necessary to obtain effective antisense inhibitors. Although thousands of sites may be targeted by ASOs, various computational methods and algorithms are used to restrict this number. For example, some sequence motifs containing CpGs that are known to be immunostimulatory in mouse and humans [21, 22] are eliminated, as are some pyrimidine-rich motifs that may produce hepatotoxicity. Strings of Cs and Gs that might lead to aptameric structures are typically avoided as well [59].

Careful evaluation of gene sequence homology may also reduce the number of potentially cross-reactive ASO binding sites. This entails a GenBank BLAST alignment of an ASO sequence to avoid targeting of unrelated genes with conserved nucleotide motifs. In general, ASOs containing a minimum of 3 base mismatches to another unrelated gene sequence are inactive against such targets. Off-target mRNA suppression can be further assessed using RNA isolated from cells treated with the ASO of interest by performing real-time PCR (RT-PCR) with alternative gene-specific probes. The presence of single nucleotide polymorphisms (SNPs) is also another important concern as a target advances towards clinical candidacy as even a single mismatch within an ASO binding site would likely reduce the potency of the drug in human subpopulations possessing such a mutation.

Finally, another key consideration for ASO design relates to whether or not one would desire an ASO that works across multiple species. Because RNA sequences and secondary/

tertiary structures may often differ between species, thus potentially reducing ASO hybridization efficiency [59], it is not always possible to select a site that is conserved between humans and, say, mouse, rat, and nonhuman primates, the latter representing three commonly used species for toxicological and pharmacological evaluations. However, if sufficient gene sequence homology exists, pan-species ASO can be developed. This would provide obvious cost and time savings as the same drug to be used in the clinic could be employed for pre-IND (Investigational New Drug) safety and efficacy studies in rodents and nonhuman primates.

After careful consideration of the factors listed above, generally 78 ASOs are designed against each target. This strategy usually yields a handful of lead compounds that will be studied in subsequent pharmacological screening. A clear exception to this is during the process of identifying a human clinical candidate. In that case, several hundred ASOs will be evaluated (to be discussed in more detail in later sections).

3.3 Evaluation of ASOs In Vitro

The next steps in determining an optimal ASO are to screen multiple compounds in relevant cell lines or primary hepatocytes in vitro, first at a single dose (Fig. 2a) and subsequently in 4 to 6 point dose–response studies (Fig. 2b). Obviously, cultured cells must express the target of interest. Primary hepatocytes have proven to be a very reliable cell type, and are particularly useful for studying lipoprotein targets [12]. Endothelial cells, adipocytes, and other immortalized cell lines such as A549, Hep3B, and HepG2 cells have also been used at this stage [59, 61].

Although 2′MOE ASOs demonstrate potent pharmacological activity in animal models and in the clinic when administered in buffered saline solutions, screening in tissue culture involves complexation of these drugs with cationic lipids/transfecting agents such as lipofectin or lipofectamine [61, 74]. In a typical single-dose screening experiment, 150 nM of the ASOs will be formulated with 6 μg/ml lipofectin, with cells being exposed to drugs for 24–48 h. Another commonly used method for transfecting cells grown in suspension is electroporation, where an externally applied electric current causes a temporary increase in plasma membrane permeability. Regardless of the means to facilitate cellular uptake, control ASOs containing nucleotide mismatches or those directed to unrelated targets are also evaluated in the same experiments to demonstrate the specificity of target mRNA reduction.

The primary endpoint for ASO activity is suppression of the targeted mRNA. For high-throughput analysis, RT-PCR is traditionally used to identify the most active compounds [75, 76]. One important consideration when using RT-PCR is that ASOs which hybridize within the amplicon region of the primer probe site can often be competitive inhibitors during the PCR amplification step,

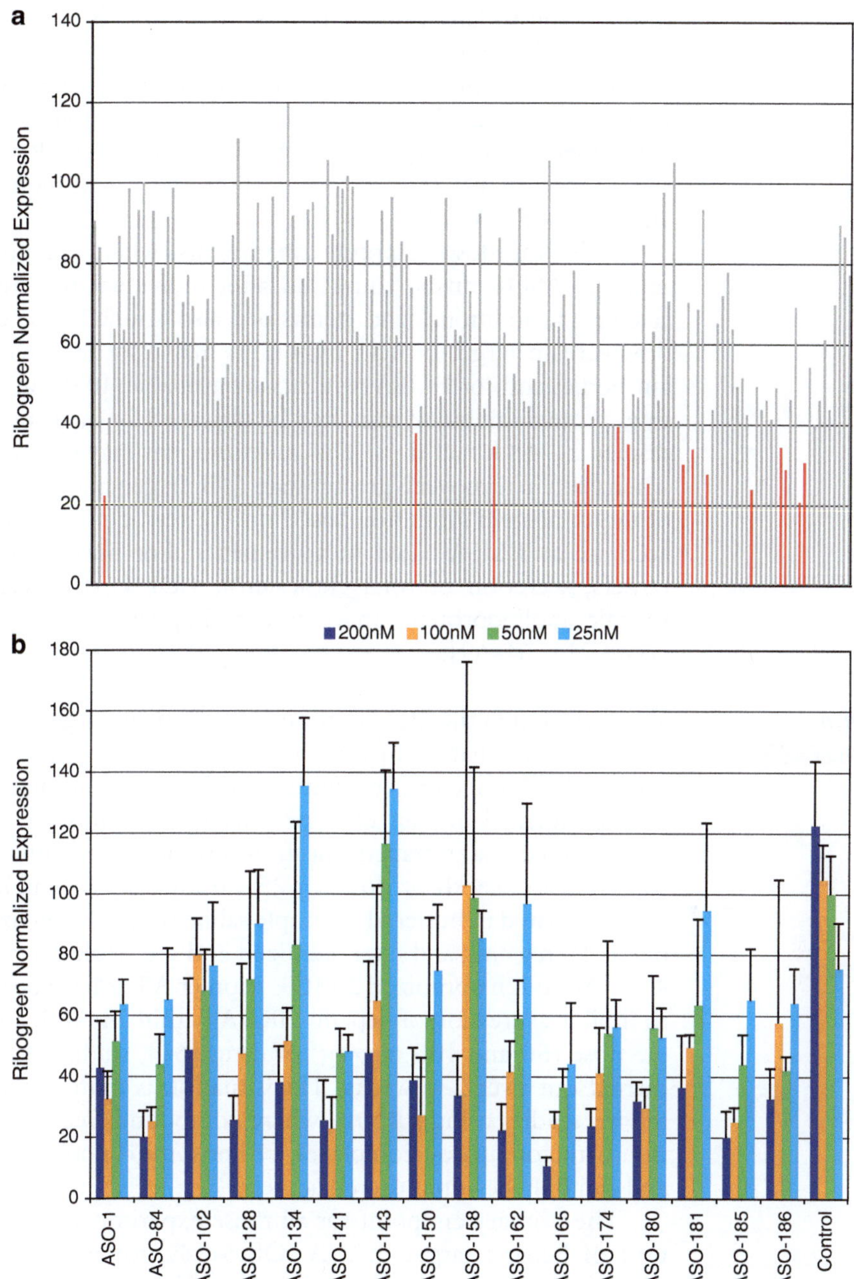

Fig. 2 Results from an initial screen of 2′MOE chimeric antisense inhibitors designed to inhibit human CRP. (**a**) 185 ASOs are tested in Hep3B cells treated with each compound at 150 nM for 24 h prior to mRNA isolation. A mismatch control ASO is shown on the far *left*. After Ribogreen normalization, data were expressed as a percentage of the untreated controls. Bars highlighted in "*red*" were selected for further characterization based upon their enhanced potency. (**b**) Confirmatory 4-point dose response performed with 16 of the most active compounds selected from the initial screen in Hep3B cells treated 24 h at doses of 200, 100, 50, and 25 nM. Each treatment was performed in triplicate wells with standard deviation indicated for each treatment concentration. A mismatch control ASO is shown on the far *right*

thus leading to false-positive results. Therefore, it is important to either avoid designing ASOs within the amplicon or, alternatively, retest mRNA expression using a second probe set that hybridizes to a different region of the transcript. Northern blots may also be performed to confirm results obtained using RT-PCR, although these types of analyses have lower throughput and require larger amounts of RNA isolate [59, 77].

Following dose–response experiments, where the potency of multiple ASOs can be compared after individual IC_{50} values are calculated, additional end points are also studied. For example, downstream target protein inhibition may also be determined if antibodies and/or ELISA reagents are available. In the in vitro setting, protein inhibition can often be observed for many targets that display 24–48 h half-lives. This measurement may be problematic when protein turnover is greater than 72 h. In general, quantification of target protein inhibition based on Western blots may not be as accurate as ELISA-based measurements. For lipoprotein-related targets, secretion of total cholesterol, LDL-C, triglycerides, or specific apolipoproteins may serve as important surrogate endpoints [12, 64–69].

3.4 In Vivo Evaluation of ASO Leads

The final step in the identification of nonhuman targets consists of the in vivo evaluation of three or four of the most potent ASOs previously identified in cell culture. At Isis, these studies are typically performed in lean mice or rats. Compounds are generally administered via intraperitoneal, intravenous, or subcutaneous routes, twice weekly at doses of 25–50 mg/kg. Dosing solutions are formulated in buffered or simple saline from a lyophilized powder at 10 mg/ml, and filter-sterilized and are stable for months at 4 °C. Although optimized 2′MOE gapmer ASOs have been shown to inhibit expression of hepatic mRNA within 24–48 h after a single dose (50 and 100 mg/kg) of drug [61], studies are typically carried out over 2–6 weeks. This situation is quite distinct from in vivo studies using siRNAs, where compounds require formulation with conjugates, complexes, or liposome/lipoplexes [78, 79] and can only be administered acutely.

The primary endpoint for all these experiments is the reduction of hepatic target mRNA. Obviously, for non-lipoprotein-related targets, pharmacological activity has also been demonstrated in multiple tissues including the kidney, adipose, bone, lymphoid tissues, and inflammatory cells [25, 61]. ASOs with reductions in mRNA >70 %, compared to saline-treated control animals, are considered to be potent leads. At study termination, total RNA is extracted from whole liver using Qiagen RNeasy isolation kits and subjected to RT-PCR analysis. Values are generally normalized to glyceraldehyde-3-phosphate dehydrogenase and/or Ribogreen levels [12]. Probes are typically dual-labeled with 5′FAM (a 6-carboxyfluorescein reporter) and 3′

TAMRA, a 5(6)-carboxytetra-methyl-rhodamine quencher. After approximately 40 amplification cycles, absolute values are obtained using SDS analysis software from Applied Biosystems.

Non-target-related toxicities may also be identified at this step through evaluation of secondary endpoints such as body and organ weights and serum levels of transaminases, urea, and creatinine. The most common and well-characterized toxicological response in mice following ASO administration is a pro-inflammatory reaction mediated by the direct activation of cells of the innate immune system that results in lymphoproliferation and lymphohistiocytic infiltrates producing increases in serum transaminases [8, 21]. Rat kidneys are particularly sensitive to ASOs (independent of sequence), which may manifest as a low-grade proteinuria. While these well-characterized phenomena are not observed in nonhuman primates, or in man [21, 23], given that rodents are generally required for more advanced toxicological evaluations of human clinical candidates, the exclusion of ASOs that are highly pro-inflammatory or potentially nephrotoxic at this stage is sensible.

Following this initial single-dose study, two or more of the most potent ASOs are generally advanced to the next level, where the pharmacological/phenotypic effects of the drugs are evaluated as a function of dose and drug concentration in more sophisticated dyslipidemia models. Although rodents carry most of their plasma cholesterol in the form of HDL-C and are, therefore, resistant to the development of atherosclerosis, genetic manipulation (transgenic/knockout mice) or feeding mice various high-fat or high-cholesterol diets obviate this limitation [12, 64–69, 80, 81]. Two widely used models include the *ldlr*- and *apoe*-deficient mouse strains, which develop significant hyperlipidemia and extensive atherosclerotic plaque burden within 12–16 weeks on high-fat/cholesterol diets [80, 81]. *Ldlr*-deficient transgenic mice fed a hyperlipidemic diet and co-expressing human apoB-100, apo(a), or cholesterol ester transfer protein (CETP) also display accelerated vascular disease [82–84]. In many of these models, apolipoprotein mRNA and protein levels, particle composition, and lipid levels have been significantly modulated by specific ASOs [12, 13, 39, 64–69]. Amelioration of atherosclerotic plaque progression has been successfully demonstrated using the murine apoB and ACAT2 antisense inhibitors, as well as mipomersen, in mice containing the human apoB transgene [64, 65, 85].

Studies performed using the apoB ASO highlight the variety of preclinical models and end points that can be evaluated. For example, the in vivo administration of an apoB antisense inhibitor has demonstrated rapid and sustained reduction in hepatic apoB mRNA/protein and plasma apoB, LDL-C, and total cholesterol in a dose- and time-dependent fashion in lean and high-fat-fed C57BL/6 mice. These results have also been confirmed in high-fat-fed *ldlr*, *apoe*-deficient, and human apoB-transgenic mice, as well as hamsters, rats, rabbits, and monkeys [12, 64]. In addition to pronounced effects on lipids, reduction of apoB produced a significant reduction of atherosclerosis in three murine models. Pharmacological reduction of apoB has been well tolerated and no safety issues have been identified to date. Unlike small-molecule MTP inhibitors [86, 87], hepatic and intestinal steatosis were not produced in mouse and monkey, likely due to favorable secondary changes in the transcription of key hepatic genes involved in lipogenesis and fatty acid oxidation [12]. Positive pharmacological data in multiple hyperlipidemic animal models and the successful completion of 6-month and 1-year toxicological evaluations in mice and monkeys provided supportive evidence to advance the human apoB drug into the clinic.

4 Identification of Human Drug Candidates

Once an ASO directed against a novel target has demonstrated robust pharmacology in several animal models and there is compelling evidence, from either animal knock-in/out models or human data, that validates the importance of the gene in man, then the careful identification and evaluation of a human clinical candidate becomes critically important. At Isis, this process routinely involves multiple steps (Fig. 3):

1. >500 ASOs are designed to multiple mRNA and pre-mRNA sites (coding, noncoding sequences, etc.) and tested at single doses in a variety of immortalized cell lines such as HepG2, Hep3B, and A549 cells or primary human/monkey hepatocytes.

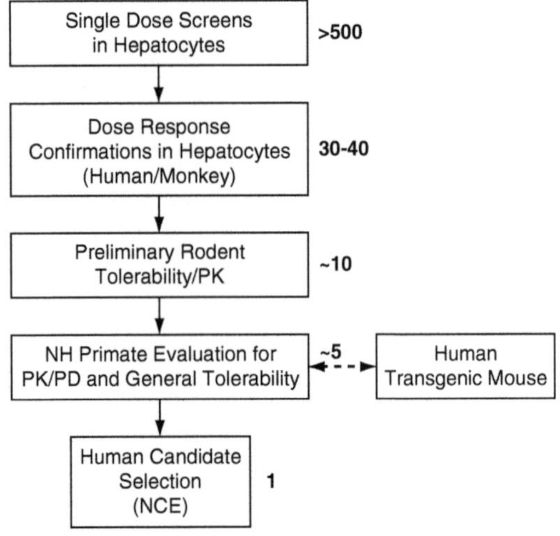

Fig. 3 Flow scheme for identification of human ASO drug candidates

At this point, a decision can be made (or not) to ensure that the sites are conserved between human and nonhuman primates.

2. Approximately 30 or 40 of the most potent ASOs are then screened in at least three independent 5-point dose response experiments where their IC_{50} values are cross-compared to narrow the choices to roughly ten final candidates. Target mRNA inhibition is generally confirmed using two to three probe sets and if possible, other secondary endpoints (e.g., cellular protein levels, lipoprotein secretion rates) are evaluated as well.
3. Tolerability and pharmacokinetics of the most potent leads (approximately 10–20 drugs) are determined in rodents.
4. Compounds which advance through these initial steps are elevated to second-tier status and are subjected to more extensive characterizations that include PK/PD (pharmacokinetic/pharmacodynamic) and tolerability studies in nonhuman primates administered increasing doses of the ASOs for up to 3 months, in vitro pro-inflammatory testing, serum protein binding assays, and finally, analysis for SNPs within ASO binding sites. PK/PD experiments in hyperlipidemic monkeys or transgenic mice expressing the human gene of interest might also be performed. This extensive screening paradigm was successfully used to identify the human apoB drug, mipomersen.

5 Conclusion

The identification and evaluation of small-molecule therapeutics is a time-consuming and expensive proposition. In general, this process may take 7–10 years, require 20–40 FTEs per year, and cost \$40–100 million [57, 61]. Because of the unambiguous ability to rapidly design, synthesize, and test numerous compounds in vitro and in vivo, the predictable chemical class-related toxicological and pharmacokinetic properties of ASOs [56, 57], an antisense drug may be advanced from bench to clinic in just 2–5 years, at a cost of \$5–10 million using 3–5 FTEs. With the ever-expanding list of potential targets involved in lipid and lipoprotein metabolism being identified by genome-wide association studies [88–91], antisense technology may indeed prove the most facile method to rapidly advance novel drugs for the treatment of dyslipidemias.

Acknowledgments

The authors would like to thank Drs. Stanley T. Crooke and Sue Freier for their careful review of the manuscript, Tracy Reigle for her excellent help with figure production, and finally, Pamela Black for her expertise in producing the final document.

References

1. Crooke ST (2004) Progress in antisense technology. Annu Rev Med 55:61–95
2. Crooke ST (1999) Molecular mechanism of action of antisense drugs. Biochim Biophys Acta 1489:31–44
3. Crooke ST (1998) Molecular mechanisms of antisense drugs: RNaseH. Antisense Nucleic Acid Drug Dev 8:133–134
4. Wu H, Lima WF, Zhang H, Fan A, Sun H, Crooke ST (2004) Determination of the role of the human RNAse H1 in the pharmacology of DNA-like antisense drugs. J Biol Chem 279:17181–17189
5. Lima W, Rose JB, Nichols JG, Wu H, Migawa MT, Wyrzykiewicz TK, Vasques G, Swayze EE, Crooke ST (2007) The positional influence of the helical geometry of the heteroduplex substrate on human RNase H1 catalysis. Mol Pharmacol 71:73–82
6. Lima W, Rose JB, Nichols JG, Wu H, Migawa MT, Wyrzykiewicz TK, Siwkowski AM, Crooke ST (2007) Human RNaseH1 discriminates between subtle variations in the structure of the heteroduplex substrate. Mol Pharmacol 71:83–91
7. Lima W, Wu H, Crooke ST (2008) The RNase H mechanism. In: Crooke ST (ed) Antisense drug technology. Principles, strategies and applications. CRC, Boca Raton, FL, pp 47–74
8. Henry SP, Geary RS, Yu R, Levin AA (2001) Drug properties of second generation antisense oligonucleotides: how do they measure up to their predecessors? Curr Opin Invest Drugs 2:1444–1449
9. Dean NM, Butler M, Monia BP, Manoharan M (2001) Pharmacology of 2′-O-(2-methoxyethyl) modified antisense oligonucleotides. In: Crooke ST (ed) Antisense technology: principles, strategies and applications. Marcel Dekker, Inc., New York, NY, pp 319–338
10. Zellweger T, Miyake H, Cooper S, Conklin BS, Monia BP, Gleave ME (2001) Antitumor activity of antisense clusterin oligonucleotides is improved in vitro and in vivo by incorporation of 2′-O-(2-methoxy)ethyl chemistry. J Pharmacol Exp Ther 298:934–940
11. Kastelein JJP, Wedel MW, Baker BB, Su J, Bradley JD, Yu RZ, Chuang E, Graham MJ, Crooke RM (2006) Potent reduction of apolipoprotein B and low density lipoprotein cholesterol by short term administration of an antisense inhibitor of apolipoprotein B. Circulation 114:1729–1735
12. Crooke RM, Graham MJ, Lemonidis KM, Whipple CP, Koo S, Perera RJ (2005) An apolipoprotein B antisense oligonucleotide lowers LDL cholesterol in hyperlipidemic mice without causing hepatic steatosis. J Lipid Res 46:872–884
13. Crooke RM (2005) Antisense oligonucleotides as therapeutics for hyperlipidaemias. Expert Opin Biol Ther 5:907–917
14. Gleave ME, Monia BP (2005) Antisense therapy for cancer. Nature 5:468–479
15. Sewell KL, Geary RS, Baker BF, Glover JM, Mant TG, Yu RZ, Tami JA, Dorr FA (2002) Phase I trial of ISIS 104838, a 2′ methoxyethyl modified antisense oligonucleotide targeting tumor necrosis factor-alpha. J Pharmacol Exp Ther 303:1334–1343
16. Geary RS, Khatsenko O, Bunker K, Crooke R, Moore M, Burckin T, Troung L, Sasmor H, Levin AA (2001) Absolute bioavailability of 2′-O-2-(methoxyethyl)-modified antisense oligonucleotides following intraduodenal instillation in rats. J Pharmacol Exp Ther 296:898–904
17. Crosby JR, Guha M, Tung D, Miller DA, Bender B, Condon TP, York-Defalco C, Geary RS, Monia BP, Karras JG, Gregory SA (2007) Inhaled CD86 antisense oligonucleotide suppresses pulmonary inflammation and airway hyper-responsiveness in allergic mice. J Pharmacol Exp Ther 321:938–946
18. Miner PB Jr, Geary RS, Matson J, Chuan E, Xia SM, Baker BF, Wedel MK (2006) Bioavailability and therapeutic activity of alicaforsen (ISIS 2302) administered as a rectal retention enema to subjects with active ulcerative colitis. Aliment Pharmacol Ther 23:1427–1434
19. Zhang RW, Iyer RP, Yu D, Tan WT, Zhang XZ, Lu ZH, Zhao H, Agrawal S (1996) Pharmacokinetics and tissue disposition of chimeric oligodeoxynucleoside phosphorothioates in rats after intravenous administration. J Pharmacol Exp Ther 278:971–979
20. Hardee GE, Tillman LG, Geary RS (2008) Routes and formulations for delivery of antisense oligonucleotides. In: Crooke ST (ed) Antisense drug technology. Principles, strategies and applications. CRC, Boca Raton, FL, pp 217–236
21. Henry SP, Kim T-W, Kramer-Strickland K, Zanardi TA, Fey RA, Levin AA (2008) Toxicologic properties of phosphorothioate 2′-methoxyethyl chimeric antisense inhibitors in animals and man. In: Crooke ST (ed) Antisense drug technology. Principles, strategies and applications. CRC, Boca Raton, FL, pp 327–363
22. Henry SP, Stecker K, Brooks D, Monteith D, Conklin B, Bennett CF (2000) Chemically modified oligonucleotides exhibit decreased immune stimulation in mice. J Pharmacol Exp Ther 292:468–479

23. Kwoh JT (2008) An overview of the clinical safety experience of first and second generation oligonucleotides. In: Crooke ST (ed) Antisense drug technology. Principles, strategies and applications. CRC, Boca Raton, FL, pp 365–399
24. Geary RS, Bradley JD, Watanabe T, Kwon Y, Wedel M, van Lier JJ, VanVliet AA (2006) Lack of pharmacokinetic interaction for ISIS 113715, a 2′-O-methoxyethyl modified antisense oligonucleotide targeting protein tyrosine phosphatase 1B messenger RNA, with oral antidiabetic compounds metformin, glipizide or rosiglitazone. Clin Pharmacokinet 45:789–801
25. Geary RS, Yu RZ, Siwkowski A, Levin AA (2008) Pharmacokinetic/pharmacodynamic properties of phosphorothioate 2′-*O*-(2-methoxyethyl)-modified antisense oligonucleotides in animals and man. In: Crooke ST (ed) Antisense drug technology. Principles, strategies and applications. CRC, Boca Raton, FL, pp 236–305
26. Kane JP, Havel RJ (2001) Introduction: structure and metabolism of plasma lipoproteins. In: Scriver CR, Beaudet AL, Sly WS, Valle D (eds) The metabolic and molecular bases of inherited disease, 8th edn. McGraw-Hill, New York, pp 2705–2716
27. Mahley RW, Bersott TP (2006) Drug therapy for hypercholesterolemia and dyslipidemia. In: Brunton LL (ed) Goodman and Gilman's the pharmacological basis of therapeutics, 11th edn. McGraw-Hill, New York, pp 933–966
28. Lloyd-Jones DM, Leip EP, Larson MG, D'Agostino RB, Beiser A, Wilson PWF, Wolf PA, Levy D (2006) Prediction of lifetime risk for cardiovascular disease by risk factor burden at 50 years of age. Circulation 113:791–798
29. Epstein JA, Rader DJ, Parmacek MS (2002) Perspective: cardiovascular disease in the post-genomic era-lessons learned and challenges ahead. Endocrinology 143:2045–2050
30. Davis RA, Hui TY (2001) 2000 George Lyman Duff memorial lecture. Atherosclerosis is a liver disease of the heart. Arterioscler Thromb Vasc Biol 21:887–898
31. Nabel EG (2003) Cardiovascular disease. N Engl J Med 349:60–72
32. Expert Panel on Detection, Evaluation, and Treatment of High Blood Cholesterol in Adults (2001) Executive summary of the Third Report of the National Cholesterol Education Program (NCEP) Expert panel on detection, evaluation and treatment of high blood cholesterol in adults (Adult Treatment Panel III). JAMA 285:2486–2497
33. Grundy SM, Cleeman JI, Bairey CN, Brewer HB, Clark LT, Hunninhake D, Pasternak RC, Smith SC, Sone NJ, Coordinating Committee of the National Cholesterol Education Program (2004) Implications of recent clinical trials for the National Cholesterol Education Program Adult Treatment Panel III Guidelines. Circulation 110:227–239
34. Knopp RH (1999) Drug treatment of lipid disorders. N Engl J Med 341:498–511
35. (1998) LIPID Study Group: prevention of cardiovascular events and death with pravastatin in patients with coronary heart disease and a broad range of initial cholesterol levels. N Engl J Med 339:1339–1357
36. Downs GR, Clearfield M, Weiss S (1998) Primary prevention of acute coronary events with lovastatin in men and women with average cholesterol levels: results of the AFCAPS/TEXCAPS (Air Forces/Texas Coronary Atherosclerosis Prevention Study). JAMA 279:1615–1622
37. Heart Protection Study Collaborative Group (2002) MRC/BHF heart protection study of cholesterol lowering with simvastatin in 20,536 high risk individuals: a randomized placebo controlled trial. Lancet 360:7–22
38. Olofsson S-O, Boren J (2005) Apolipoprotein B: a clinically important apolipoprotein which assembles atherogenic lipoproteins and promotes the development of atherosclerosis. J Intern Med 258:395–410
39. Merki E, Graham MJ, Mullick AE, Miller ER, Crooke RM, Pitas RE, Witztum JL, Tsimikas S (2008) Antisense oligonucleotide directed to human apolipoprotein-B-100 reduces lipoprotein(a) levels and oxidized phospholipids on human apolipoprotein B-100 particles in lipoprotein(a) transgenic mice. Circulation 118:743–753
40. Tsimikas S, Brilakis ES, Miller ER, McConnell JP, Kornman KS, Witztum JL, Berger PB (2005) Oxidized phospholipids, Lp(a) lipoprotein, and coronary artery disease. N Engl J Med 353:46–57
41. Annurad E, Boffa MB, Koschinsky ML (2006) Lipoprotein(a): a unique risk factor for cardiovascular disease. Clin Lab Med 26:751–772
42. Bobik A (2008) Apolipoprotein CIII and atherosclerosis: beyond effects on lipid metabolism. Circulation 118:702–704
43. Kawakami A, Osaka M, Tani M, Azuma H, Sacks FM, Shimokado K, Yoshida M (2008) Apolipoprotein CIII links hyperlipidemia with vascular cell dysfunction. Circulation 118:731–742

44. Scheffer PG, Teerlink T, Dekker JM, Bos G, Dijpels G, Diamant M, Kostense PJ, Stehouwer CD, Heine RJ (2008) Increased plasma apolipoprotein C-III concentration independently predicts cardiovascular mortality: the Hoorn Study. Clin Chem 54:1325–1330
45. Giorda CB, Avogaro A, Maggini M, Lombardo F, Mannucci E, Turco S, Alegiani SS, Raschetti R, Velussi M, Ferrannini E, DAI Study Group (2008) Recurrence of cardiovascular events in patients with type 2 diabetes: epidemiology and risk factors. Diabetes Care 31(11):2154–2159
46. Mitka M (2004) Guidelines: new lows for LDL target levels. JAMA 292:911–913
47. Rubins HB (2000) Triglycerides and coronary heart disease: implications of recent clinical trials. J Cardiovasc Risk 7:339–345
48. Watts GF, Barrett PH, Chan DC (2008) HDL metabolism in context: looking on the bright side. Curr Opin Lipidol 19:395–404
49. deGoma EM, deGoma RL, Rader DJ (2008) Beyond high-density lipoprotein cholesterol levels. Evaluating high-density lipoprotein function as influenced by novel therapeutic approaches. J Am Coll Cardiol 51:2199–2211
50. McQueen MJ, Hawken S, Wang X, Ounpuu S, Sniderman A, Probstfield J, Steyn K, Sanderson JE, Hasani M, Volkova E, Kazmi K, Yusuf S, INTERHEART Study Investigators (2008) Lipids, lipoproteins, and apolipoproteins as risk markers of myocardial infarction in 52 countries (the INTERHEART study): a case controlled study. Lancet 372:224–233
51. Romero-Corral A, Somers VK, Lopez-Jimenez F (2008) Apolipoproteins: promising biomarkers for cardiovascular risk prediction. J Intern Med 264:1–3
52. El Harchaoui K, Akdim F, Stroes ES, Trip MD, Kastelein JJ (2008) Current and future pharmacologic options for the management of patients unable to achieve low-density lipoprotein-cholesterol goals with statins. Am J Cardiovasc Drugs 8:233–242
53. Athyros VG, Kakafika AI, Tziomalos K, Karagiannis A, Mihkailidis DP (2008) Antisense technology for the prevention or treatment of cardiovascular disease: the next blockbuster? Expert Opin Investig Drugs 17:969–972
54. Ito MK (2007) ISIS 301012 gene therapy for hypercholesterolemia: sense, antisense, or nonsense? Ann Pharmacother 41:1669–1678
55. Burnett JR (2006) Drug evaluation: ISIS 31012, an antisense oligonucleotide for the treatment of hypercholesterolemia. Curr Opin Mol Ther 8:461–467
56. Crooke ST, Vickers T, Lima W, Hongjiang W (2008) Mechanisms of antisense drug action, an introduction. In: Crooke ST (ed) Antisense drug technology. Principles, strategies and applications. CRC, Boca Raton, FL, pp 3–46
57. Bennett CF (2002) Efficiency of antisense oligonucleotide drug discovery. Antisense Nucleic Acid Drug Dev 12:215–224
58. Dean NM, Bennett CF (2003) Antisense oligonucleotide-based therapeutics for cancer. Oncogene 22:9087–9096
59. Freier SM, Watt AT (2008) Basic principles of antisense drug discovery. In: Crooke ST (ed) Antisense drug technology. Principles, strategies and applications. CRC, Boca Raton, FL, pp 117–141
60. Gleave ME, Monia BP (2005) Antisense therapy for cancer. Nat Rev Cancer 5:468–479
61. Bennett CF (2008) Pharmacological properties of 2′-*O*-methoxyethyl-modified oligonucleotides. In: Crooke ST (ed) Antisense drug technology. Principles, strategies and applications. CRC, Boca Raton, FL, pp 273–303
62. Swayze EE, Bhat B (2008) The medicinal chemistry of oligonucleotides. In: Crooke ST (ed) Antisense drug technology. Principles, strategies and applications. CRC, Boca Raton, FL, pp 143–182
63. Monia BP, Yu RZ, Lima W, Siwkowski A (2008) Optimization of second-generation antisense drugs: going beyond generation 2.0. In: Crooke ST (ed) Antisense drug technology. Principles, strategies and applications. CRC, Boca Raton, FL, pp 487–506
64. Crooke R, Baker B, Wedel M (2008) Cardiovascular therapeutic applications. In: Crooke ST (ed) Antisense drug technology. Principles, strategies and applications. CRC, Boca Raton, FL, pp 601–639
65. Bell TA, Brown JM, Graham MJ, Lemonidis KM, Crooke RM, Rudel LL (2006) Liver-specific inhibition of acyl-coenzyme A:cholesterol acyltransferase 2 with antisense oligonucleotides limits atherosclerosis development in apolipoprotein B100-only low-density lipoprotein receptor−/− mice. Arterioscler Thromb Vasc Biol 26:1814–1820
66. Brown JM, Bell TA, Alger HM, Sawyer JK, Smith TL, Kelley K, Shah R, Wilson MD, Davis MA, Lee RG, Graham MJ, Crooke RM (2008) Targeted depletion of hepatic ACAT2-driven cholesterol esterification reveals a non-biliary route for fecal neutral sterol loss. J Biol Chem 283:10522–10534
67. Graham MJ, Lemonidis KM, Whipple CP, Subramaniam A, Monia BP, Crooke ST, Crooke RM (2007) Antisense inhibition of

proprotein convertase subtilisin/kexin type 9 reduces serum LDL in hyperlipidemic mice. J Lipid Res 48:763–767

68. Zhang YL, Hernandez-Ono A, Weisberg S, Conlon D, Graham MJ, Crooke RM, Huang LS, Ginsberg HN (2006) Aberrant hepatic expression of PPAR γ2 stimulates hepatic lipogenesis in a mouse model of obesity, insulin resistance, dyslipidemia and hepatic steatosis. J Biol Chem 281:37603–37615
69. Imai Y, Varela GM, Jackson MB, Graham MJ, Crooke RM, Ahima RS (2007) Reduction of hepatosteatosis and lipid levels by an adipose differentiation-related protein antisense oligonucleotide. Gastroenterology 132:1947–1954
70. Geary RS, Yu RZ, Watanabe T, Henry SP, Hardee GE, Chappell A, Matson J, Sasmor H, Cummins L, Levin AA (2003) Pharmacokinetics of a tumor necrosis factor alpha phosphorothioate 2′-O-(2-methoxyethyl) modified antisense oligonucleotide: comparison across species. Drug Metab Dispos 31:1419–1428
71. Geary RS, Watanabe TA, Truong L, Freier SA, Lesnik EA, Sioufi NB, Sasmor H, Manoharan M, Levin AA (2001) Pharmacokinetic properties of 2′-O-(2-methoxyethyl)-modified oligonucleotide analogs in rats. J Pharmacol Exp Ther 296:890–897
72. Yu R, Geary RS, Monteith DK, Matson J, Truong L, Fitchett J, Levin AA (2004) Tissue disposition of 2′-O-(2-methoxyethyl)-modified oligonucleotides in monkeys. J Pharm Sci 93:48–59
73. Altschul SF, Gish W, Miller W, Myers EW, Lipman DJ (1990) Basic local alignment search tool. J Mol Biol 215:403–410
74. Dalby B, Cates S, Harris A, Ohki EC, Tilkins ML, Price PJ, Ciccarone VC (2004) Advanced transfection with Lipofectamine 2000 reagent: primary neurons, siRNA, and high-throughput applications. Methods 33:95–103
75. Nolan T, Hands RE, Bustin SA (2006) Quantification of mRNA using real-time RT-PCR. Nat Protoc 1:1559–1582
76. Bustin SA (2000) Absolute quantification of mRNA using real-time reverse transcription polymerase chain reaction assays. J Mol Endocrinol 25:169–193
77. Alwine JC, Kemp DJ, Stark GR (1977) Method for detection of specific RNAs in agarose gels by transfer to diazobenzyloxymethyl paper and hybridization with DNA probes. Proc Natl Acad Sci U S A 12:5350–5354
78. deFougerolles AR, Maraganore JM (2008) Discovery and development of RNAi therapeutics. In: Crooke ST (ed) Antisense drug technology. Principles, strategies and applications. CRC, Boca Raton, FL, pp 465–484
79. Soutschek J, Akinc A, Bramlage B, Charisse K, Constien R, Donoghue M, Elbashir S, Geick A, Hadwiger P, Harborth J, John M, Kesavan V, Lavine G, Pandey RK, Racie T, Rajeev KG, Rohl I, Toudjarska I, Wang G, Wuschko S, Bumcrot D, Koteliansky V, Manoharan M, Vornlocher HP (2004) Therapeutic silencing of an endogenous gene by systemic administration of modified siRNAs. Nature 432:173–178
80. Krause BR, Princen HMG (1998) Lack of predictability of classical animal models for hypolipidemic activity: a good time for mice? Atherosclerosis 140:15–24
81. Zadelaar S, Kleeman R, Verschuren L, deVries-Van der Weij J, van der Hoorn J, Princen HM, Kooistra T (2007) Mouse models for atherosclerosis and pharmaceutical modifiers. Arterioscler Thromb Vasc Biol 27:1706–1721
82. Grass DS, Saini U, Felkner RH, Wallace RE, Lao WJ, Young SG, Swanson ME (1995) Transgenic mice expressing both human apolipoprotein B and human CETP have a similar lipoprotein cholesterol distribution similar to that of normolipidemic humans. J Lipid Res 36:1082–1091
83. Naik SU, Wang X, Da Silva JS, Jaye M, Macphee CH, Reilly MP, Billheimer JT, Rothblat GH, Rader DJ (2006) Pharmacological activation of liver X receptors promotes reverse cholesterol transport in vivo. Circulation 113:90–97
84. Arai T, Wang N, Bezouevski M, Welch C, Tall AR (1999) Decreased atherosclerosis in heterozygous low density lipoprotein receptor-deficient mice expressing the scavenger receptor B1 transgene. J Biol Chem 274:2366–2371
85. Mullick AE, Graham MJ, Fu W, York-DeFalco C, Crooke RM (2008) Antisense inhibition of apolipoprotein B ameliorated diet-induced hypercholesterolemia and reduced atherosclerosis in LDL receptor-deficient mice. Arteriosclerosis, Thrombosis and Vascular Biology 2008 Annual Conference, Atlanta, GA, p 345, 16–18 April
86. Chandler CE, Wilder DE, Pettini JL, Savoy YE, Petras SF, Chang G, Vincent J, Harwood HJ Jr (2003) CP-346086: an MTP inhibitor that lowers plasma cholesterol and triglycerides in experimental animals and humans. J Lipid Res 44:1887–1901
87. Cuchel M, Bloedon LT, Szapary PO, Kolansky DM, Wolfe ML, Sarlis A, Millar JS, Ikewaki K, Siegelman ES, Gregg RE, Rader DJ (2007) Inhibition of microsomal triglyceride transfer protein in familial hypercholesterolemia. N Engl J Med 356:148–156

88. Willer CJ, Sanna S, Jackson AU, Scuteri A, Bonnycastle LL et al (2008) Newly identified loci that influence lipid concentrations and risk of coronary artery disease. Nat Genet 40:161–169
89. Breslow JL (2000) Genetics of lipoprotein abnormalities associated with coronary heart disease susceptibility. Annu Rev Genet 34:233–254
90. Lunetta KL, D'Agostino RB, Karasik D, Benjamin EJ, Gui CY, Govindaraju R, Kiel DP, Kelly-Hayes M, Massaro JM, Pencina MJ, Seshadri S, Murabito JM (2007) Genetic correlates of longevity and selected age-related phenotypes: a genome-wide association study in the Framingham Study. BMC Med Genet 19:8(Suppl 1):S13
91. Holleboom AG, Vergeer M, Hovingh GK, Kastelein JJP, Kuivenhoven JA (2008) The value of HDL genetics. Curr Opin Lipidol 19:385–394

Part IV

Special Topics

Chapter 15

Chromatin Immunoprecipitation

Grant D. Barish and Rajenda K. Tangirala

Abstract

Recent studies have elucidated molecular mechanisms underlying the transcriptional control of metabolism in complex metabolic disorders such as metabolic syndrome and atherosclerosis. Chromatin immunoprecipitation (ChIP) is an important technique to study protein–DNA interactions in vivo. Chemical cross-linking of DNA and its associated proteins, followed by chromatin shearing, immunoprecipitation of a protein of interest, DNA isolation, and PCR interrogation, can identify specific interactions between protein and DNA or sites of histone epigenetic alteration. Transcription factors and epigenetic modifications are key determinants of transcription. Accordingly, ChIP experiments can provide powerful mechanistic insights to understand gene expression.

Key words Chromatin, Immunoprecipitation, ChIP, Transcription, Protein–DNA interaction, Cardiovascular disease

1 Introduction

Work over the past two decades has redefined atherosclerosis as a chronic inflammatory disease driven by metabolic dysregulation. Transcription underlies much of metabolism and inflammation. Accordingly, methods to understand transcriptional control are gaining importance in cardiovascular research. Chromatin immunoprecipitation (ChIP) is a technique to examine interactions between proteins and DNA, in order to determine the occupancy of a protein along a specified region of the genome. When properly performed and interpreted, ChIP can provide powerful mechanistic insights to understand gene expression.

In general terms, chromatin immunoprecipitation begins by covalently linking protein to DNA using chemical cross-linking with formaldehyde. Cell nuclei are subsequently isolated, and the chromatin, now containing fixed protein–DNA complexes, is sheared with sonication into DNA fragments (generally to average fragment sizes of 200–1,000 base pairs). Sheared chromatin is then incubated with antibodies directed against a protein of interest,

Lita A. Freeman (ed.), *Lipoproteins and Cardiovascular Disease: Methods and Protocols*, Methods in Molecular Biology, vol. 1027, DOI 10.1007/978-1-60327-369-5_15,

and immune complexes of antibody and chromatin are precipitated using protein A or G beads. The precipitated DNA is then isolated and purified. Finally, the precipitated DNA is interrogated using quantitative PCR. The abundance of a protein along a particular region of the genome is reflected by its enrichment of DNA. An overview is shown in Fig. 1.

Many ChIP protocols have been developed by academic labs as well as commercial vendors. These techniques are generally similar but differ somewhat in their lysis, shearing, and wash buffers or DNA isolation methods following immunoprecipitation. We utilize a protocol modified from that described by Nelson et al. [1]. It is a relatively straight-forward method with fewer steps than most protocols, making it particularly approachable for beginners. With this and all ChIP protocols, however, proper controls and data interpretation are critical. Special emphasis on these areas is provided below, along with a detailed ChIP protocol.

2 Materials

2.1 Equipment

1. Diagenode Bioruptor (recommended) or other sonicator.
2. Refrigerated microcentrifuge.
3. Tube rotator in cold (4 °C) room.
4. Thermal microcentrifuge tube mixer.
5. DNA gel electrophoresis box.
6. Quantitative PCR system.

2.2 Reagents

1. Ultrapure Formaldehyde (Polysciences 16 % Ultrapure EM Grade).
2. 1 M Glycine in ddH_2O (37.6 g glycine brought up to 500 ml in ddH_2O).
3. Phosphate-buffered saline.
4. Cell Lysis/Wash Buffer: 150 mM NaCl, 50 mM Tris–HCl (pH 7.5), 5 mM EDTA, 0.5 % NP-40, 1.0 % Triton X-100 (4.383 g NaCl, 25 ml of 100 mM EDTA pH 8.0, 25 ml of 1 M Tris–HCl pH 7.5, 25 ml of 10 % NP-40, and 50 ml of 10 % Triton X-100 brought up to 500 ml in ddH_2O).
5. Roche Complete Mini Protease Inhibitor Tablets (EDTA-free).
6. Shearing Buffer: 1 % SDS, 10 mM EDTA, 50 mM Tris–HCl pH 8.0.
7. Dilution Buffer: 0.01 % SDS, 1.1 % Triton X-100, 1.2 mM EDTA, 16.7 mM Tris–HCl, pH 8.1, 167 mM NaCl.
8. "ChIP grade" antibody to protein of interest and species-specific control antibody.
9. Eppendorf LoBind 1.5 ml microfuge tubes.

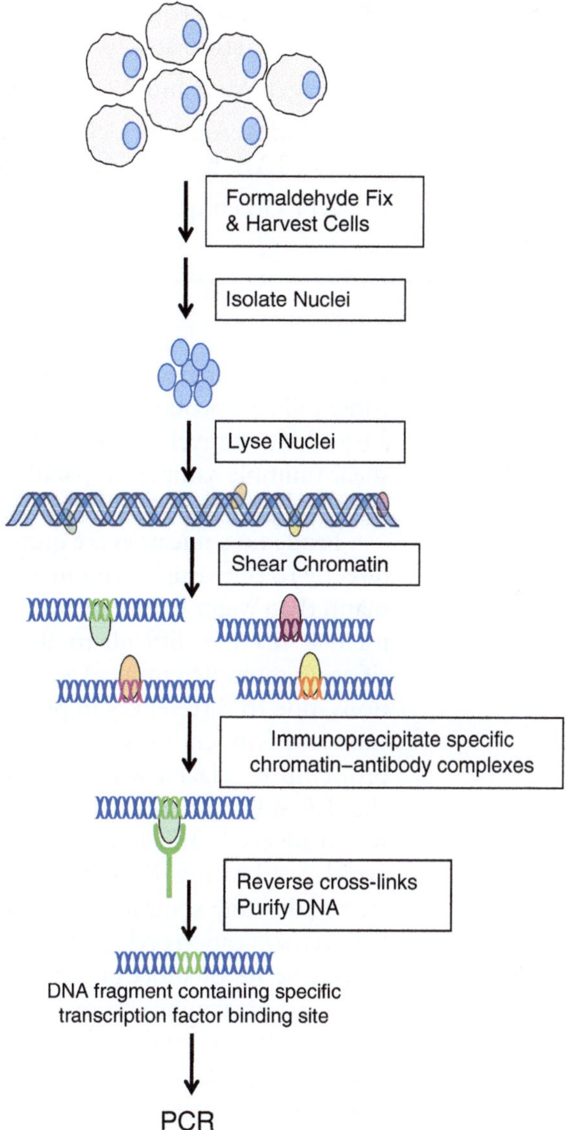

Fig. 1 Overview of Chromatin Immunoprecipitation. Cells are fixed to covalently cross-link protein to DNA. Next, the cells are lysed and their nuclei are isolated. The nuclei are lysed and chromatin is sheared into small fragments. Immunoprecipitation is performed using antibody to a specific DNA-bound protein of interest. Precipitated DNA is then isolated and purified. PCR is then performed to interrogate whether the protein of interest binds to a specific genomic region

10. Protein A or G agarose beads blocked with salmon sperm DNA (Millipore).
11. Chelex 100 (Bio-Rad): prepare a 10 % solution in ddH_2O using 1 g Chelex resin in a 50 ml conical tube brought up to 10 ml with ddH_2O.
12. Boiling Rack (USA Scientific).

13. 20 mg/ml Proteinase K in ddH_2O (dissolve 20 mg in 1 ml ddH_2O, store at −20 °C).
14. Qiagen MinElute PCR Purification Kit.
15. Colorless Agarose Gel Loading Buffer (20 g sucrose plus 10 ml 50× TAE, brought up to a volume of 50 ml).
16. Sybr Green quantitative PCR buffer.

3 Methods

3.1 Before the Experiment

3.1.1 Optimize Shearing with Your Sonicator

Every sonicator has a different efficiency for shearing chromatin. Most utilize a probe-tip to deliver sonic pulses to a single sample at a time. Alternatively, water bath sonicators have been developed to shear multiple samples in parallel. There are advantages and disadvantages to either form of sonication.

Probe tip sonicators are quite powerful, delivering sonic energy directly to the sample, and may be more efficient in shearing chromatin than water bath systems. However, use of a probe tip sonicator may make it difficult to ensure that each experimental sample (if using multiple samples) is subjected to the same shearing conditions, due to differences in probe placement within the microfuge tube and temperatures during sonication. It is important that the probe tip be placed within a few millimeters from the bottom of the 1.5 ml microfuge tube when sonicating to avoid foaming, which adversely affects shearing. Also, special care must be taken to avoid overheating during sonication by placing the sample on ice and dividing sonication into multiple rounds, cooling the sample between sonication cycles. Chromatin shearing occurs more efficiently when a series of short pulses are used rather than one continuous pulse. To determine optimal shearing times, prepare samples containing 10–20 million cells per milliliter shearing buffer (*see* Subheading 3.2.1, below) in 1.5 ml microfuge tubes. Set the power output of the probe to 50 %, and sonicate samples using ten 1 s long pulses for two, four, six, or eight rounds, pausing to ice samples for 2 min between each round of sonication. Process the sheared chromatin (*see* Subheadings 3.2.4 and 3.2.5) and determine DNA fragment sizes using gel electrophoresis.

Water bath sonicators allow multiple samples to be sheared simultaneously in closed tubes, and some feature integrated cooling systems to maintain samples at cold temperatures during sonication. We use a Diagenode Bioruptor, which can sonicate multiple samples simultaneously in 1.5 ml microfuge tubes or 15 ml conicals and does not produce foaming. Using the high power settings set to 30 s on/30 s off, shear samples for one to six 5 min cycles (total time = 5–30 min, sonication time = 2.5–15 min). Process the sheared chromatin (*see* Subheadings 3.2.4 and 3.2.5) and determine DNA fragment sizes using gel electrophoresis.

Optimal fragment sizes depend upon your experimental needs. Fragment sizes of 500–1,000 base pairs are generally sufficient if the goal is to determine whether or not a protein binds somewhere along the regulatory regions of a particular gene. If greater spatial resolution is required, shearing down to fragment sizes of 200–300 base pairs may be needed. Use of more stringent shearing necessitates that the primer set used to interrogate the ChIP sample amplifies DNA in very close proximity to the binding site of your protein of interest. For poorly characterized genes, use of less stringent shearing to first ascertain whether a protein binds somewhere along your gene of interest may be the best initial approach.

Once you have determined optimal shearing conditions for your sonicator, be sure to keep conditions consistent for your actual experiment. Many factors beyond the sonicator and its settings affect shearing efficiency. These include formaldehyde concentration and cross-linking time, cell type, lysis buffers, and cell concentrations. Higher formaldehyde concentrations and/or longer cross-linking times impair shearing but may be necessary for weakly bound or indirectly DNA bound proteins. Nuclei from different cell types vary in their sonication efficiencies. Additionally, viscosity impacts shearing efficiency, which is influenced by buffer conditions and cell concentrations. In light of these variables, sonication parameters may need to be re-optimized if experimental conditions are changed.

3.1.2 Design Primers

Commercially available or online primer design programs should be used to design primers to interrogate DNA regions bound by your protein(s) of interest. Quantitative PCR primers made for assessment of gene expression are generally not usable for ChIP analysis. Many of these are either designed to span intron–exon boundaries and/or are designed to amplify exon regions far removed from the promoter or other gene regulatory regions. Therefore, custom primers will generally need to be made. If little or no information is known about the interactions between your protein and gene of interest, several primer sets may need to be designed to investigate interactions along multiple stretches of the DNA (such as several predicted binding sites for a transcription factor; *see* **Note 1**). We use ABI Primer Express to design primers with amplicon sizes of ~50 base pairs. Primer sets for each experiment should include:

1. Positive control primer set(s): Design primers flanking a known DNA binding site (if available) for your factor of interest.
2. Negative control primer set(s): Design primers that amplify DNA regions distant from the known or suspected binding site(s) of a factor of interest. For example, design primers that amplify DNA 1,000–2,000 base pairs upstream or downstream from a transcription factor's known/suspected binding site.

3. Housekeeping gene primer set(s): Design primers to the promoter region of a housekeeping gene, such as GAPDH or actin. Assuming that your factor of interest does not regulate the housekeeping gene, this can be an additional or alternative negative control primer set. Since housekeeping genes are actively transcribed, this primer set can also be utilized as a positive control for your chromatin immunoprecipitation if positive control immunoprecipitations are set up with antibodies to RNA Polymerase II or acetylated histones.
4. Region of interest primer set(s): Design primers that flank or closely approximate the stretch of DNA you wish to interrogate for occupancy by your protein(s) of interest.

Primers may be tested by amplifying purified genomic DNA prior to using them to interrogate chromatin immunoprecipitation samples.

3.1.3 Obtain Antibodies

Antibodies are required and sometimes limiting reagents for the performance of ChIP assays. Antibodies for ChIP not only must be suitable for immunoprecipitation, but they must also recognize their target following formaldehyde cross-linking, which can mask epitopes. Generally, polyclonal antibodies are more likely to work for ChIP, as with immunoprecipitation in general, although some monoclonal antibodies are also quite effective and have high specificity. Several commercial vendors now indicate whether their antibodies are usable for ChIP. These “ChIP” qualified designations can be helpful, although such labeling can be deceiving. Many vendors base their labeling simply on customer feedback rather than in-house testing. Moreover, in many cases antibody specificity is not confirmed using negative control materials, such as cells devoid of the factor of interest (knockout or knockdown reagents). When possible, we recommend that knockout or knockdown cells be used in experiments as additional negative controls to ensure antibody specificity.

In addition to antibodies against the protein(s) of interest, control antibodies should be obtained. Species-specific pre-immune antibodies should be acquired for negative control chromatin immunoprecipitations and are available from several commercial vendors. Positive control antibodies should also be obtained and may include antibodies to RNA Polymerase II, or alternative antibodies to histone proteins, such as histone H3 or acetylated histone H3.

3.2 Procedure for Chromatin Immunoprecipitation

3.2.1 Cross-Link and Harvest Cells

1. Choose the appropriate cell type for fixation (*see* **Note 2**). We generally utilize 20 million cells (adherent or in suspension) as starting material, which is sufficient for ~10 chromatin immunoprecipitations. Add 16 % formaldehyde (0.67–9.33 ml of tissue culture media) directly into the cell culture media to achieve a final concentration of 1 % formaldehyde and incubate at room temperature for 10 min (*see* **Notes 3** and **4**).

2. Add 1.44 ml of 1 M glycine per 10 ml of medium (final concentration: 125 mM glycine) to quench the formaldehyde. Incubate at room temperature for 5 min.
3. Spin to collect cells in suspension and remove media, or for adherent cells, just aspirate media. Wash twice with cold PBS, and collect cells into microfuge tubes (use cell lifter or scraper for adherent cells). Spin cells at 2,000 × *g* × 5 min in a 4 °C refrigerated microfuge. Carefully aspirate off residual PBS. Cells may be frozen at −80 °C for later use or utilized immediately.

3.2.2 Lyse Cells and Shear Chromatin

Perform all steps below on ice or at 4 °C.

1. Add 1 ml of cold Cell Lysis/Wash Buffer containing protease inhibitors (dissolve one Roche Complete Protease Inhibitor EDTA-free mini tablet in 10 ml Cell Lysis/Wash Buffer) to the cell pellet. Pipette up and down several times to generate a homogenous cell slurry. The cell slurry may be further aspirated and pushed through a 1 ml insulin syringe to assist in breaking apart cell membranes. Centrifuge the cell slurry at 12,000 × *g* × 1 min and aspirate the supernatant.
2. Repeat **step 1**, above, one additional time.
3. Add 1 ml of cold Shearing Buffer containing protease inhibitors to the nuclear pellet isolated from **step 2**, above. Pipette up and down several times to resuspend the nuclei.
4. If using a Diagenode Bioruptor, transfer the 1 ml of resuspended nuclei into a single 15 ml conical tube or into four 1.5 ml microfuge tubes, each with 250 μl. If using a different sonicator, transfer the nuclei in shearing buffer as needed for your sonicator (*see* Subheading 3.1.1).
5. Shear chromatin, keeping the sample cold (at 4 °C) at all times. Shearing conditions must be empirically determined depending upon your cell type as well as the fixation conditions used (longer cross-linking times and/or higher formaldehyde concentrations result in less efficient shearing; *see* Subheading 3.1.1).

3.2.3 Immunoprecipitation

1. Centrifuge the sheared chromatin at 12,000 × *g* × 10 min at 4 °C. Transfer the supernatant to a fresh 15-ml conical tube and determine its volume.
2. Pipette 1/12th of sheared chromatin to a fresh 1.5 ml microcentrifuge tube labeled "input" and store at −20 °C. This will be used to determine the quantity of input chromatin used for the immunoprecipitations and to assess chromatin shearing. Input DNA isolation is described in Subheading 3.2.4 (*see* **Note 5**).
3. To the remaining chromatin in the 15-ml conical tube, add nine volumes of Dilution Buffer containing protease inhibitors and mix the sample gently by inversion.

4. Aliquot the diluted chromatin in volumes of 10× input (from **step 2**, above) into 1.5 ml microfuge tubes, so each tube will contain the same total amount of chromatin as the input sample (**step 2**, above).
5. Add antibody to each sample of diluted chromatin. We recommend using experimental duplicates or triplicates for each immunoprecipitation. Saturating amounts of antibody should be used for immunoprecipitations, to ensure that the precipitated DNA reflects the abundance of a specified protein and is not limited by inadequate amounts of antibody to fully precipitate the protein bound DNA. Generally, 1–3 μg of affinity-purified antibody is sufficient for each immunoprecipitation.
6. Incubate the microtubes of diluted chromatin and antibody on a rotating tube rack or platform at 4 °C overnight.
7. The following day, prepare protein A-agarose (for immunoprecipitations with rabbit polyclonal antibodies) or protein G-agarose (for immunoprecipitations with mouse monoclonal antibodies) beads which have been pre-blocked with salmon sperm DNA. Aliquot a volume of 20 μl of beads per immunoprecipitation, plus an additional 25 % to a microfuge tube (for pipetting error), into a single tube. Most agarose beads are supplied as a 50 % slurry, so if 20 immunoprecipitations are planned, aliquot 40 μl of 50 % slurry per immunoprecipitation × 20 immunoprecipitations = 800 μl, plus an additional 200 μl = 1,000 μl.
8. Spin the agarose beads down at 1,500 × g × 1 min, aspirate the supernatant, and add 1 ml of Cell Lysis/Wash Buffer. Resuspend the beads by inverting the tube several times, spin, and aspirate the supernatant. Repeat for a total of three bead washes.
9. After the final wash, add a volume of Cell Lysis/Wash buffer equal to 50 % of the volume of the agarose slurry aliquotted in **step** 7, above. Mix the agarose beads plus Cell Lysis/Wash buffer into a slurry by repeated inversion or gentle pipetting with a wide-bore pipette tip.
10. Aliquot 40 μl of washed protein A or protein G-agarose bead slurry into fresh 1.5 ml Eppendorf LoBind tubes using wide-bore pipette tips (40 μl bead slurry/ChIP). Place the tubes on ice.
11. Centrifuge the microtubes from **step 6** containing immune complexes (chromatin bound to antibodies) at 12,000 × g for 10 min at 4 °C using a refrigerated microfuge. Move the top 90 % of the supernatant, carefully avoiding the bottom of the tubes, to the LoBind microfuge tubes containing protein A or G-agarose beads (from **step 10**, above) (*see* **Note 6**).
12. Incubate the LoBind tubes containing chromatin–antibody complexes and protein A- or G-agarose beads for 1 h at 4 °C using a rotating tube rack or platform.

13. Centrifuge the samples from **step 12** at 1,500 × *g* × 1 min at 4 °C and carefully aspirate the supernatant, avoiding the beads (*see* **Note 7**). Wash the beads by adding 1 ml of cold Cell Lysis/Wash Buffer. Place the samples back on the rotating tube rack or platform for 1 min, spin down and repeat for a total of six washes per sample. During the final wash, be sure to aspirate all liquid from the tube, leaving only the beads.

3.2.4 Input DNA Isolation

1. Take the sheared "input" chromatin sample from Subheading 3.2.3, **step 2** and add 3× its volume of pure ethanol.
2. Vortex briefly and put the chromatin–ethanol mixture on dry ice or at −80 °C × 15 min.
3. Spin the chilled chromatin–ethanol mixture at 12,000 × *g* × 10 min at 4 °C.
4. Decant the supernatant, using care to avoid losing the precipitated DNA pellet.
5. Air-dry the precipitated DNA. To facilitate drying, the sample may be placed in a SpeedVac.

3.2.5 DNA De-Crosslinking and Purification

1. Pipette 100 μl of 10 % Chelex 100 slurry into each washed ChIP sample (from Subheading 3.2.3, **step 13**) or input DNA pellet (from Subheading 3.2.4, **step 5**) using wide-bore pipette tips. The Chelex slurry settles quickly and should be vortexed immediately prior to adding to each sample.
2. Vortex each sample.
3. Place samples in a boiler rack and boil for 10 min.
4. Centrifuge the samples at 12,000 × *g* × 1 min at 4 °C.
5. Add 1 μl of proteinase K to each sample, vortex, and place in a thermal mixer at 55 °C for 30 min at a medium mixing speed.
6. Place samples back in a boiler rack and boil for an additional 10 min.
7. Centrifuge the samples at 12,000 × *g* × 1 min at 4 °C.
8. Collect the top 80 μl of supernatant from each sample and move to new 1.5 ml microcentrifuge tubes. Use care to avoid transferring beads.
9. Add an additional 120 μl of MilliQ H_2O to each sample, vortex, and centrifuge the samples at 12,000 × *g* × 1 min.
10. Collect the top 120 μl of supernatant from each sample and add this to the supernatants from Subheading 3.2.5, **step 8**, yielding a total volume of 200 μl per sample. Again, use care to avoid transferring Chelex resin or agarose beads.
11. Use the Qiagen MinElute PCR Purification Kit per manufacturer's instructions to further purify the precipitated DNA

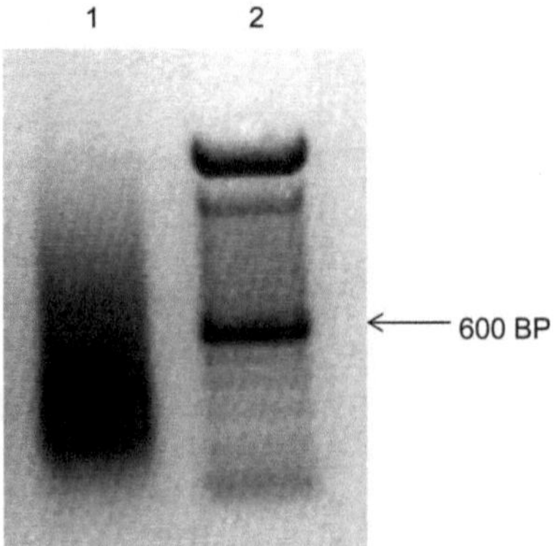

Fig. 2 Assessment of Chromatin Shearing. 1 % agarose gel in TAE buffer stained with ethidium bromide. *Lane 1*—sheared chromatin following a 15 min cycle of 30 s ON/30 s OFF using a Diagenode Bioruptor, *Lane 2*—100 BP DNA ladder (Invitrogen). Most fragments in *Lane 1* are 200–600 base pairs

samples and input chromatin. Elute both the precipitated DNA samples and input chromatin samples into microfuge tubes using 20 μl of kit supplied elution buffer (10 mM Tris–Cl pH 8.5).

3.2.6 Assessment of Shearing

Combine 5 μl of MinElute column-purified input chromatin with 2 μl of colorless agarose gel loading buffer and load sample into a 1 % agarose TAE gel. Also load a 100 base pair DNA ladder. Run gel for ~45 min at 100 V and photograph gel to determine the size of the sheared chromatin fragments. An example is shown in Fig. 2.

3.3 Quantitative PCR Analysis

3.3.1 Quantitative PCR

1. Quantitative PCR should be used to assess the enrichment of specific regions of DNA in your immunoprecipitated samples and input chromatin(s). Extensive discussion of quantitative PCR methodology is beyond the scope of this protocol but is extensively discussed in other reviews [2–4]. It can be performed using either delta Ct or relative standard curve methodologies, the latter of which is our preference. Use species-specific genomic DNA, beginning with a concentration of 3.0 ng/μl, and serially dilute it by fivefold to generate five standards with concentrations of 3.0, 0.6, 0.12, 0.024, 0.0048 ng/μl. These genomic DNA dilutions will be used to generate a 5-point standard curve, from which the relative quantities of DNA in your ChIP samples and input chromatin will be determined.
2. Dilute the ChIP sample MinElute column-purified DNA samples (Subheading 3.2.5, **step 11**) with 180 μl of MilliQ H_2O, bringing up their volumes from 20 to 200 μl.
3. Dilute 10 μl (half) of the input DNA(s) (Subheading 3.2.5, **step 11**) with 190 μl of MilliQ H_2O.

4. Set up quantitative PCR reactions. We use a 384 well qPCR format, with 10 μl reactions consisting of 5 μl SYBR GreenER master mix, 0.1 μl primer pair (10 μM each), 0.9 μl H_2O, and 4 μl ChIP or input sample diluted as described in Subheading 3.3.1, **steps 2** and **3**, respectively. Plates are run using an ABI Prism 7900HT and analyzed with SDS 2.3 software.

3.3.2 Data Analysis

Quantitative PCR data for chromatin immunoprecipitations can be presented in a variety of ways. Indeed, there is little consensus as to the best method for data normalization, which is an under-appreciated component of the ChIP technique. This topic is extensively reviewed elsewhere [5, 6]. Data may be presented with no normalization (showing absolute qPCR signals) or following normalization with any one of a variety of techniques including background subtraction (qPCR signal from the negative control ChIP samples are subtracted from qPCR signals from the other ChIP samples), fold enrichment (qPCR signals from ChIP samples are divided by the qPCR signals from the negative control ChIPs), percentage of input (qPCR signals from ChIP samples are divided by qPCR signals from the input chromatin), normalization to control sequences (qPCR ChIP signals from the DNA region of interest are divided by qPCR ChIP signals from a positive control DNA region), normalization to nucleosome density (qPCR signals from ChIP samples are divided by qPCR signals from ChIPs performed with antibody directed against an invariant histone, such as histone H3). Ideally, any of these methods would yield similar results, but in practice they may not. The fold enrichment technique, which is commonly used in the literature, is particularly fraught with potential for errors. Indeed, the trace amounts of background DNA found in negative control ChIP samples may yield qPCR signals only after a very high number of amplification cycles ($Ct > 35$), producing nonlinear results that may spuriously affect the quotient of sample ChIP/negative control ChIP, as used to express fold enrichment.

We recommend using the percentage of input technique or normalizing qPCR signals with relative input chromatin concentrations, graphing both negative control ChIP and experimental ChIP samples (*see* Subheading 3.3.3, below). Statistical analysis to compare the quantities of precipitated DNA, a reflection of protein occupancy, can be performed using Student's *t*-tests or Anova as appropriate.

3.3.3 Example

Wild type C57 mouse bone marrow differentiated macrophages were plated on 15-cm tissue culture dishes with 2.0×10^7 cells/plate. The following day, LPS or control solution was added to the tissue culture media for 2 h. Cells were then fixed, harvested, and sheared to generate chromatin fragments of 500–1,000 base pairs in length as per the protocol described. Immunoprecipitations were performed using rabbit polyclonal antibodies directed against the NF-κB p65 subunit (Santa Cruz), acetylated histone H3K18 (Millipore), and pre-immune rabbit IgG (negative control antibody—Santa Cruz). Each ChIP was performed with experimental duplicates. Precipitated DNA from each sample was interrogated using quantitative PCR with technical triplicates for each sample, using standard curve methodology to determine relative DNA quantities (Table 1). Samples from either treatment condition (±LPS) were normalized by dividing their calculated qPCR input DNA by the relative concentrations of input chromatin used for immunoprecipitation. Raw data and calculations are shown for

Table 1
Quantitative PCR analysis of chromatin immunoprecipitated DNA

Sample #	Ct	Log input	Input	Norm factor	Normalized signal	Description	Average	STD dev
1	34.24172	−2.26639	0.0054	1	0.005415191	WT, IgG	0.00405	0.002544
1	35.0022	−2.46289	0.0034	1	0.003444358			
1	35.45836	−2.58076	0.0026	1	0.002625658			
2	38.49398	−3.36516	0.0004	1	0.000431364			
2	33.606804	−2.10233	0.0079	1	0.007900852			
2	34.560196	−2.34868	0.0045	1	0.00448044			
3	30.697153	−1.35048	0.0446	1	0.044618794	WT, p65	0.036451	0.01495
3	30.600996	−1.32564	0.0472	1	0.047245938			
3	30.367378	−1.26527	0.0543	1	0.054291341			
4	32.023323	−1.69316	0.0203	1	0.020269359			
4	31.127092	−1.46158	0.0345	1	0.034548012			
4	32.248333	−1.7513	0.0177	1	0.01772957			
5	27.59919	−0.54998	0.2819	1	0.281851876	WT, Ac H3K18	0.29434	0.028184
5	27.56017	−0.5399	0.2885	1	0.288471933			
5	27.25541	−0.46115	0.3458	1	0.345821916			
6	27.658571	−0.56532	0.2721	1	0.27206776			
6	27.46168	−0.51445	0.3059	1	0.305881384			
6	27.659328	−0.56552	0.2719	1	0.271945249			
7	34.537003	−2.34269	0.0045	1.1	0.004129723	WT+LPS, IgG	0.002909	0.001782
7	35.76611	−2.66028	0.0022	1.1	0.001987577			

(continued)

Table 1 (continued)

Sample #	Ct	Log input	Input	Norm factor	Normalized signal	Description	Average	STD dev
7	37.155346	−3.01926	0.001	1.1	0.000869661			
8	34.094677	−2.22839	0.0059	1.1	0.005372996			
8	36.460762	−2.83978	0.0014	1.1	0.001314705			
8	34.684925	−2.38091	0.0042	1.1	0.003781799			
9	29.941816	−1.15531	0.0699	1.1	0.063577238	WT+LPS, p65	0.057529	0.014729
9	29.990963	−1.16801	0.0679	1.1	0.061745063			
9	29.506319	−1.04277	0.0906	1.1	0.082382051			
10	30.524164	−1.30578	0.0495	1.1	0.044959856			
10	30.459415	−1.28905	0.0514	1.1	0.046725702			
10	30.493612	−1.29789	0.0504	1.1	0.045784603			
11	27.112698	−0.42427	0.3765	1.1	0.34224422	WT+LPS, Ac H3K18	0.320077	0.035925
11	27.470964	−0.51685	0.3042	1.1	0.276542196			
11	27.183767	−0.44264	0.3609	1.1	0.328074242			
12	26.957804	−0.38425	0.4128	1.1	0.375284135			
12	27.374641	−0.49196	0.3221	1.1	0.292853853			
12	27.303795	−0.47365	0.336	1.1	0.305462082			
Relative input chromatin								
13	21.101713	1.128947	13.457	1(=14.60483/ 14.60483)		WT input chromatin	14.60483	1.132864
13	20.960613	1.165407	14.635					
13	20.840244	1.19651	15.722					
14	21.00112	1.15494	14.287	1.1(=16.09/14.60)		WT+LPS input chromatin	16.08942	1.862023
14	20.813364	1.203455	15.976					
14	20.612291	1.255412	18.006					
Calculated STD curve								
Slope = −3.87002 cycles/log decade								
Y-Intercept = 25.47076								
$R^2 = 0.974$								

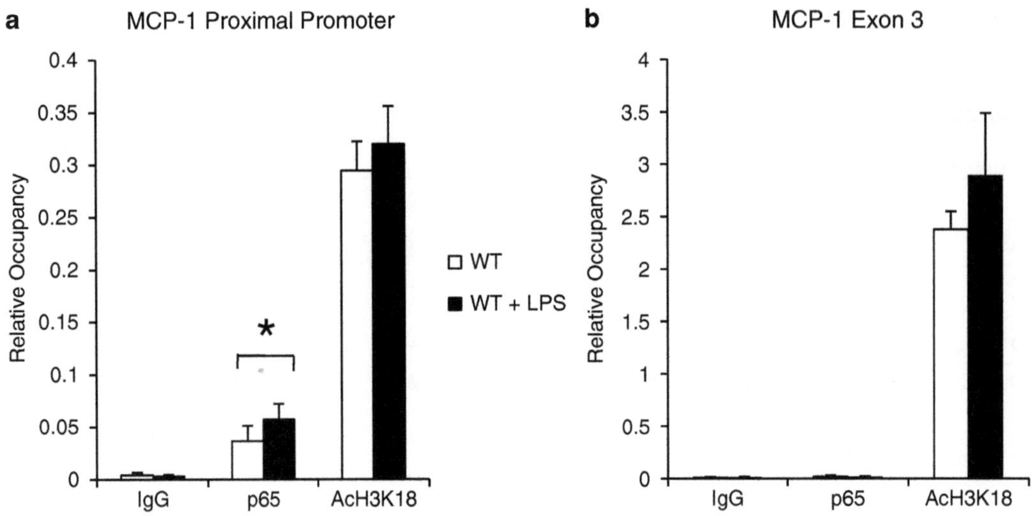

Fig. 3 Chromatin immunoprecipitations to assess occupancy of the MCP-1 gene promoter by NF-κB p65 in bone marrow differentiated macrophages before or after stimulation with lipopolysaccharide. (**a**) MCP-1 proximal promoter DNA is enriched from chromatin immunoprecipitations with antibodies to p65 or acetylated histone H3K18 relative to rabbit IgG (negative control). MCP-1 promoter DNA is significantly enriched in LPS-stimulated (WT + LPS) versus unstimulated cells (WT) ChIP'd with p65 antibody (p-value <0.05), reflecting enhanced NF-κB p65 occupancy. (**b**) No enrichment of MCP-1 Exon 3 DNA (negative control DNA site, located >1,500 base pairs away from the NF-κB binding site in the MCP-1 proximal promoter) is detected in chromatin immunoprecipitations with p65 antibody, although enrichment is observed with antibody to acetylated histone H3K18 (positive control antibody)

qPCR amplification of the mouse MCP-1 proximal promoter. Log input values for each experimental sample are calculated using the standard curve generated equation, (Ct − 25.47076)/−3.87002. The inverse log of this value is the DNA input for a given sample. A 2-tailed Student's t-test confirms statistically significant differences in the amounts of MCP-1 promoter DNA precipitated by p65 antibodies in untreated versus LPS-stimulated cells.

As shown in this example, there is enrichment of MCP-1 proximal promoter DNA in samples precipitated with antibodies directed against p65 or acetylated histone H3K18 (positive control ChIP antibody) compared to samples precipitated with rabbit preimmune antibodies (negative control ChIP) (Fig. 3a). Moreover, there is significantly increased enrichment of MCP-1 promoter DNA by immunoprecipitation with p65 antibodies following LPS stimulation (Fig. 3a). Together, these findings indicate that p65 binds to the MCP-1 proximal promoter, and its occupancy is enhanced by LPS stimulation. Similar analysis of the third Exon of MCP-1, using primers amplifying a region >1,500 bases upstream from the proximal promoter NF-κB binding site, shows no enrichment of DNA in p65 ChIP'd samples (Fig. 3b), providing evidence that p65 binding is specific to the MCP-1 proximal promoter.

4 Notes

1. We use the geWorkbench (https://cabig.nci.nih.gov/tools/geWorkbench) program for prediction of potential transcription factor binding sites. A number of alternative DNA binding motif analysis tools are available online.
2. Criteria for choosing the appropriate cell type include expression of the DNA-binding protein being ChIP'd, relevance of the cell type to the proposed studies, and availability of antibody to recognize the protein of interest in the cell.
3. Longer cross-linking times or higher percentages of formaldehyde may be used and may be required for some proteins, particularly indirectly bound cofactors. Longer cross-linking times will, however, negatively impact the efficiency of chromatin shearing and can potentially mask protein epitopes, resulting in diminished antibody binding and immunoprecipitation.
4. Alternatively, dual cross-linking, utilizing a protein–protein cross-linker followed by formaldehyde (which cross-links protein to protein as well as protein to DNA) can facilitate ChIP of indirectly DNA-associated proteins or transcription factors that are difficult to cross-link with formaldehyde alone. Dual cross-linking is extensively discussed elsewhere [7, 8]. Multiple NHS esters and imidoesters are commercially available, with covalent spacer arms up to eight times longer than formaldehyde. To dual cross-link cells, we recommend disuccinimidyl glutarate (DSG), which has a 7.7 Å spacer arm and has been successfully used in our hands for "difficult" ChIPs. To fix cells, aspirate off the cell culture media and replace it with 2 mM DSG in phosphate buffered saline for 30 min at room temperature. Remove the DSG/PBS solution from the cells by aspirating (preceded by a centrifugation step for cells grown in suspension), then add 1 % formaldehyde in PBS and incubate for 10 min at room temperature as in Subheading 3.2.1, **step 1**. Quench the dual cross-linked cells with glycine (Subheading 3.2.1, **step 2**) and perform ChIP as otherwise described. Dual cross-linked cells are more difficult to shear than formaldehyde-only fixed cells, so shearing will need to be optimized (*see* Subheading 3.1.1) if this optional fixation method is used.
5. Input DNA isolation is described in Subheading 3.2.4. It is convenient to de-crosslink and purify the input chromatin alongside of the sample chromatin, and thus the input DNA purification (Subheading 3.2.4) should be completed while the chromatin–antibody complexes are incubated with protein A- or G-agarose beads (Subheading 3.2.3, **step 12**).

6. This step helps to reduce background by eliminating aggregated, nonspecific DNA prior to the addition of agarose beads.
7. We recommend using 200 μl gel loading tips for aspirations, which must be changed between each sample to avoid cross-contamination.

References

1. Nelson J, Denisenko O, Bomsztyk K (2006) Protocol for the fast chromatin immunoprecipitation (ChIP) method. Nat Protoc 1:179–185
2. Valasek M, Repa J (2005) The power of real-time PCR. Adv Physiol Educ 29:151–159
3. Bustin S, Nolan T (2004) Pitfalls of quantitative real-time reverse-transcription polymerase chain reaction. J Biomol Tech 15:155–166
4. Nolan T, Hands R, Bustin S (2006) Quantification of mRNA using real-time RT-PCR. Nat Protoc 1:1559–1582
5. Haring M, Offermann S, Danker T, Horst I, Peterhansel C, Stam M (2007) Chromatin immunoprecipitation: optimization, quantitative analysis and data normalization. Plant Methods 3:11
6. Struhl K (2007) Interpreting chromatin immunoprecipitation experiments. In: Zuk D (ed) Evaluating techniques in biochemical research. Cell Press, Cambridge, MA
7. Nowak DE, Tian B, Brasier AR (2005) Two-step cross-linking method for identification of NF-kappaB gene network by chromatin immunoprecipitation. Biotechniques 39:715–725
8. Zeng PY, Vakoc CR, Chen ZC, Blobel GA, Berger SL (2006) In vivo dual cross-linking for identification of indirect DNA-associated proteins by chromatin immunoprecipitation. Biotechniques 41:694, 696, 698

Chapter 16

Measurement of Lecithin–Cholesterol Acyltransferase Activity with the Use of a Peptide-Proteoliposome Substrate

Boris L. Vaisman and Alan T. Remaley

Abstract

Lecithin–cholesterol acyltransferase (LCAT) is the major enzyme responsible for the esterification of free cholesterol on plasma lipoproteins, which is a key step in the reverse cholesterol transport pathway. The measurement of plasma LCAT activity not only is important in the diagnosis of patients with genetic or acquired LCAT deficiency but is also valuable in calculating cardiovascular risk, as well as in research studies of lipoprotein metabolism. In this chapter, we describe a convenient LCAT assay based on the use of an apoA-I mimetic peptide. The proteoliposome substrate used in this assay for LCAT is easily made with the peptide and can be stored by deep freezing without significant loss of activity.

Key words LCAT activity, Proteoliposomes storage, Apolipoprotein A-I mimetic peptides

1 Introduction

Lecithin–cholesterol acyltransferase (LCAT) is a plasma protein primarily produced by liver. LCAT catalyzes the hydrolysis of phospholipids, such as phosphatidylcholine (lecithin), and transfers the liberated fatty acids to unesterified cholesterol to form cholesteryl esters (CE) on lipoproteins, particularly HDL [1]. The esterification of cholesterol by LCAT is believed to promote reverse cholesterol transport, the pathway by which excess cholesterol from peripheral cells is delivered to the liver for excretion [2]. Cholesterol esterification promotes the net efflux of cholesterol from cells, since once a cholesterol molecule is esterified it remains trapped on lipoproteins until it can be removed by the liver. Cholesteryl esters on HDL can be directly delivered to the liver, where they can be selectively imported by the SR-BI receptor. In species that contain CETP, such as rabbit, monkey, and man, but not rodents, cholesteryl esters can also be transferred from HDL to apoB-containing lipoproteins, such as LDL, in exchange for triglycerides. Cholesteryl esters on LDL can then also be delivered to the liver by the LDL-receptor uptake pathway. Approximately half the cholesterol

Lita A. Freeman (ed.), *Lipoproteins and Cardiovascular Disease: Methods and Protocols*, Methods in Molecular Biology, vol. 1027, DOI 10.1007/978-1-60327-369-5_16, © Springer Science+Business Media, LLC 2013

excreted into bile is first converted to bile acids, whereas the remainder is excreted as free cholesterol. Any cholesterol and bile acids that are not reabsorbed into the enterohepatic circulation is excreted into the stool, leading to a net loss of cholesterol from the body [2]. In humans, LCAT deficiency causes low levels of plasma HDL and, in severe cases, can cause end-stage renal disease [1, 3]. Based on animal models [4, 5], LCAT appears to protect against the development of atherosclerosis, although patients with LCAT deficiency do not appear to have a marked increase in incidence of cardiovascular disease [3].

It is now well established that the plasma level of HDL is inversely correlated with the risk of cardiovascular disease [6, 7]. In addition, because of the inability to fully reduce cardiovascular events by lowering LDL with statins, new drug strategies are actively being investigated for raising HDL. Besides its role in the reverse cholesterol transport pathway, HDL also possesses a broad spectrum of other anti-atherogenic properties that protects the endothelium and the cardiovascular system [8]. Based on the multiplicity of roles of HDL and the discordance between the elevation of HDL and cardiovascular protection for some drugs, such as the CETP inhibitor torcetrapib [9], there is now greater appreciation for the need of assays that not only measure the level of HDL but also its quality or function.

One important functional property of HDL is its ability to serve as a carrier and substrate for LCAT. The measurement of plasma LCAT activity may be relevant not only to the diagnosis of patients with genetic or acquired LCAT deficiency [3] but also for calculating cardiovascular risk [10]. LCAT assays are also frequently performed in many research studies on lipoprotein metabolism and may be also relevant in the development of new drugs that alter HDL. Thus, there is a critical need for a sensitive, precise, stable, and reliable method for measurement of LCAT activity.

One of the earliest and still most commonly used methods for measuring LCAT activity is based on the use of an artificial substrate for LCAT, namely ^{14}C-cholesterol-loaded proteoliposomes made with phosphatidylcholine and apoA-I [11, 12], the main protein component of HDL. This method, however, has several limitations. The prepared proteoliposomes are not stable; they gradually degrade and are not suitable for use after 1 week at +4 °C or 2–3 weeks at −20 °C [13]. The value of LCAT activity obtained with these proteoliposomes also will decrease with time due to the degradation of the substrate. Thus, the results are often adjusted by the inclusion of a standard plasma sample with each run. In addition, the efficiency of the proteoliposome substrate critically depends on the quality of apoA-I, which is a powerful LCAT activator [1]. There are several commercial sources of pure apoA-I, but it is relatively expensive to purchase and difficult to purify in the quantities that are needed for the LCAT assay. Finally, the quality of apoA-I as an LCAT activator can vary significantly between different preparations even from the same company.

In this chapter, we describe a LCAT assay that utilizes an apoA-I mimetic peptide, based on ETC-642. ETC-642 was a peptide–lipid complex developed by Pfizer as a possible therapeutic HDL mimetic agent [14]. The peptide in ETC-642, which we will refer to as P-642, is similar to the consensus sequence of the amphipathic helices found on apoA-I but contains several amino acid substitutions that were found to activate LCAT [14, 15]. Unlike most other apoA-I mimetic peptides [1, 15], it is nearly as good as or even better than apoA-I in activating LCAT. It is also relatively economical to produce and is easy to reconstitute with phospholipid to produce proteoliposomes, which can be stored frozen for a prolonged period of time, without significant loss of LCAT activation.

2 Materials

2.1 Reagents

1. [4-^{14}C] Cholesterol PerkinElmer, cat. no. NEC018050UC, 50 µCi in 1.25 ml of ethanol, 45–60 mCi/mmol, store at -20 °C.
2. Cholesterol—Sigma, cat. no. C8667, store at -20 °C.
3. l-α-Lecithin, Egg Yolk—EMD Millipore, cat. no. 524617, 100 mg/ml in $CHCl_3$–MeOH (3:2, v/v), store at -20 °C.
4. P-642 peptide synthesized with the following sequence of l-amino acids: PVLDLFRELLNELLEALKQKLK. The peptide dissolved in water or a dilute neutral buffer with a concentration between 0.5 and 2 mg/ml should be stored at -20 °C.
5. Sodium cholate hydrate—Sigma C6445. Store at 4 °C.
6. Bovine serum albumin, essentially fatty acid free—Sigma, cat. no. A7511.
7. 0.5 M EDTA, pH 8.0—Quality Biological, Inc., cat. no. 351-027-101.
8. 1 M Tris–HCl, pH 7.4—KD Medical, cat. no. RGF-3340.
9. 2-Mercaptoethanol—Sigma, cat. no. M3148; (14.3 M).
10. Dialysis tubes—Spectrum Laboratories, Inc., Spectra/Por membrane MWCO 6–8,000, flat width 10 mm, diameter 6.4 mm, cat. no. 132645.
11. Spectra/Por closures, 35 mm—Spectrum Laboratories, Inc, cat. no. 132736.
12. Flexible plates for TLC, PE SIL G—Whatman, cat. no. 4410221.

13. Costar microcentrifuge tubes, 2.0 ml—Corning Inc., cat. no. 3213.
14. Cholesteryl oleate—Sigma, cat. no. C9253, store at –20 °C.
15. CytoScint, MP Biomedicals, cat. no. 0188245301.
16. (Optional) Mini light box (PGC Scientific, cat. no. 57957).

2.2 Reagent Setup

The following reagents are needed for preparing proteoliposomes and the reagent mix.

1. *LCAT assay buffer*: (20 l) 10 mM Tris–HCl; 140 mM NaCl; 1 mM EDTA, pH 7.4. This buffer will be used for preparing reagents and for dialysis of proteoliposomes.

   In a 20 l carboy, add 200 ml of 1 M Tris–HCl buffer, pH 7.4; 560 ml of 5 M NaCl; 40 ml of 0.5 M EDTA; and deionized water to a final volume of 20 l. Mix the contents of the carboy and keep this ready-to-use buffer in the cold room.
2. *2 % BSA in LCAT assay buffer*. Add 2.0 g BSA to 100 ml LCAT assay buffer. Store frozen in 10 ml aliquots.
3. *2 % BSA with 10 mM 2-mercaptoethanol in LCAT assay buffer*. Prepare this solution (5–10 ml) just before use. Add 3.55 µl of 2-mercaptoethanol to 5.0 ml of 2 % BSA in LCAT assay buffer.
4. *LCAT assay buffer with 35 mM 2-mercaptoethanol*. Prepare this solution (10 ml) just before use. Add 24.83 µl of 2-mercaptoethanol per 10 ml of LCAT assay buffer.
5. *Cholesterol solution in chloroform*. Prepare 20 ml solution of unlabeled cholesterol with a concentration of 5 mg cholesterol/ml chloroform. Keep in a glass vial tightly closed and protected from light at –20 °C.
6. *Solution of sodium cholate in LCAT assay buffer, 725 mM*. Prepare this solution just before use. Place 1.25 ml of LCAT assay buffer in a 15-ml tube and then add 390 mg of sodium cholate. Completely dissolve by vortexing. Keep this ready-to-use solution on ice.
7. *Cholesterol and cholesteryl oleate markers* in chloroform with concentration of 0.1 mg/ml for TLC. Store the solution at –20 °C and protect from light.
8. *TLC running solvent*. Mix petroleum ether, diethyl ether, and acetic acid in proportion 230:60:3. Keep the solvent (300–500 ml) in a tightly closed dark bottle.

2.3 Preparing Proteoliposome Substrate

The following protocol describes the preparation of 4 ml of proteoliposomes, which is sufficient for 160 reactions.

1. In a 20 ml glass scintillation vial, place 250 μl of [4-^{14}C] cholesterol (10 μCi), 41 μl of cholesterol solution in chloroform (5 mg/ml), and 87 μl of lecithin solution (100 mg/ml in $CHCl_3$–MeOH). Cover the vial with Parafilm, punch several small holes in the cover and then dry for 10–15 min in a chemical fume hood under a flow of nitrogen, provided by inserting into the vial a Pasteur pipette connected to the source of the gas under mild pressure.
2. Add LCAT assay buffer to the vial and 1 mg of P-642 for a total volume of 3.66 ml. Next, add 0.34 ml of a freshly prepared 725 mM solution of sodium cholate. Close the vial and vortex on a VortexGenie2 at maximum speed for 1 min. If foaming occurs, briefly centrifuge the vial (500 × *g* for 3 min).
3. Cut approximately 25 cm of Spectra/Por dialysis tubing, soak it in LCAT assay buffer and close one end by using two Spectra/Por closures. Transfer the 4 ml suspension from **step 2** into the Spectra/Por dialysis tube, close the second end of the tube and dialyze in the cold room against 4.0 l of prechilled LCAT assay buffer. Change buffer for a total of four times. The total dialysis time should be approximately 15–18 h. Final concentration of proteoliposome components: P-642 peptide—0.25 mg/ml; lecithin—2,866 nmol/ml; labeled cholesterol—2.5 μCi/ml; unesterified cholesterol 180 nmol/ml or 6.96 mg/dl (for specific activity of labeled cholesterol 53 mCi/mmol). The actual concentration of cholesterol in the proteoliposomes should be verified by direct measurements, using plasma cholesterol diagnostic reagents (Wako Chemicals USA Inc., Cholesterol E kit, cat. no. 439-17501). Cholesterol measurements should be performed on proteoliposomes after dialysis but before adding 2-mercaptoethanol (*see* **Note 1**).

2.4 Preparing the LCAT Reagent Mix

The LCAT reagent mix contains proteoliposomes, after dialysis (Subheading 2.3, **item 3**), plus BSA and 2-mercaptoethanol in LCAT assay buffer. It can be used immediately or stored for a prolonged period of time (>1 year) in a liquid nitrogen tank.

1. Transfer the proteoliposomes after dialysis to a 15-ml test tube. Add an equal volume (4 ml) of 2 % BSA with 10 mM 2-mercaptoethanol in LCAT assay buffer, mix, and incubate in a shaking water bath for 20 min at 37 °C.
2. Add 1.6 ml of 35 mM 2-mercaptoethanol in LCAT assay buffer and mix. This ready-to-use LCAT assay reagent contains 2-mercaptoethanol at a final concentration of 10 mM. It should be placed in 1-ml aliquots in Sarstedt 1.5-ml tubes with screw caps or in cryogen tubes of similar volume and stored in a liquid nitrogen tank (*see* **Note 2**).

3 Methods

3.1 LCAT Reaction

1. Reactions should be run in at least two parallel tubes. For measurement of LCAT activity in plasma, it is recommended to use 1 μl of plasma. The precision of the measurements will be higher if 5 μl of fivefold diluted plasma (diluted in physiological saline) is used for the assay. Keep diluted plasma on ice.
2. Take from the freezer the number of tubes with reagent mix that will give enough reagent mix for the assay. Thaw the tubes, mix, then for each assay place 60 μl of reagent mix in a Costar 2-ml microcentrifuge tube and keep the tubes on ice.
3. Add to the tubes containing reagent mix 5 μl of fivefold diluted plasma and mix; keep the tubes on ice until the entire set is ready. Place the tubes in a floating rack and incubate in a shaking water bath at 37 °C for 30 min.
4. Stop the reaction by placing the tubes on wet ice and adding 1 ml of 100 % cold ethanol. Thoroughly mix the content of the tubes and keep them at least 2 h or longer (overnight) at −20 °C, or put tubes on dry ice for about 30–60 min.

3.2 Extracting the LCAT Reaction Products

1. Precipitate the denatured proteins by centrifuging the tubes at 13,000 × *g* for 10 min at 4 °C. Pour out the supernatant into new tubes. Small losses of supernatant when removing the precipitate will not affect the measured LCAT activity, because the calculations are based on ratio of the counts of labeled cholesterol and CE, rather than their absolute values.
2. Evaporate the supernatant to complete dryness in a SpeedVac set to warming—medium; Radiant Cover—off. Approximate drying time is 1 h 40 min.
3. Add to the tubes 30 μl of cholesterol and CE markers in chloroform, mix, and briefly spin the tubes. If necessary, keep the tubes at −20 °C.

3.3 TLC Separation of CE from Cholesterol

1. Mark with a soft pencil on the flexible TLC plate the position of the start line, located at 2.5 cm from the edge of the flexible plate. Eight samples can be placed at a 2.5 cm distance from each other on one 20 × 20 cm plate. The first and last samples will be located 1 cm from the edge. It is convenient to use a portable light box during sample loading. To avoid having to measure each distance for every TLC plate, the distances can be marked directly on the portable light box with a permanent marker.
2. In the chemical fume hood, load the samples on the plate by applying small portions (7–9 μl). Wait until the chloroform of the loaded portion is dry and then repeat the operation until

the entire volume of the sample is loaded onto the plate. It is possible to load three samples at once by using a manual 10-μl 8-channel pipette in 3-tips mode (with tips in 1, 4, and 7 positions). The plate will be ready for TLC when the chloroform from the loaded samples has evaporated.

3. The TLC plate should be developed in a glass tank with the walls inside lined by Whatman 3MM paper. Before placing the plate with samples into the tank, the tank should first be equilibrated with the solvent. Add into the tank approximately 100 ml of running solvent and tightly close the tank. The solvent will rise up the Whatman paper and will start to evaporate, creating the necessary saturated atmosphere. 20 min after closing the tank, the tank will be ready for TLC. Place the plate into the tank and close the lid. The start line should be approximately 2 cm higher than the level of running solvent in the tank. Let the solvent move to about 3–4 cm near the top of the plate, which will take approximately 25 min.
4. Remove the plate, briefly air-dry it in the hood and place for 5–10 min in another closed glass tank containing crystals of iodine on the bottom. Yellow/brown spots of cholesterol will be located near the start position, whereas CE spots will be close to the front of the running solvent. Note that standards were added at Subheading 3.2, **step 3**, so each running sample has unlabeled cholesterol and cholesteryl ester standards for easy visualization of the spots.
5. Remove the plate after iodine staining and outline spots by pencil. Keep pencil line at least 5 mm away from lipid spots. Allow iodine to evaporate in the fume hood and then write the identification numbers near each spot on the surface of the plate free from any radioactivity, such as the region of the plate lower than the start line or the region higher than the front reached by the running solvent. The cholesterol spots are the substrate and typically have significantly higher counts than the CE product of the LCAT reaction; therefore, in order to decrease any risk of cross-contamination it is better to first cut out the CE spots with scissors before removing the cholesterol spots. Alternatively, two separate sets of scissors should be used for the cholesterol spots and the CE spots. Place the pieces of the TLC plate containing the spots into a scintillation vial labeled with their respective numbers. Add 10 ml of scintillation fluid to each vial, vortex, and count the radioactivity.

3.4 Calculations

For calculation of the specific LCAT activity, the following units of measurements should be used: quantity of cholesterol—in nmol; volume of plasma—in ml; time—in hours. Specific LCAT activity in nmol/ml/h $= (C \times R)/(V \times T)$, where C is amount of original unesterified cholesterol in reagent mix before the start of the reaction.

With a cholesterol concentration in proteoliposomes equal to 180 nmol/ml the total amount of this substrate in 60 μl of reagent mix *C* will be equal to 4.5 nmol (180×0.025).
R is the fraction of esterified cholesterol calculated as the ratio between the counts of esterified cholesterol to the sum of the counts of both unesterified (UC) and esterified cholesterol.

$$R = \mathrm{CE}_{\mathrm{CPM}} / (\mathrm{UC}_{\mathrm{CPM}} + \mathrm{CE}_{\mathrm{CPM}})$$

V is volume of plasma used (0.001 ml).
T is time (0.5 h).
(*See* **Note 3**).

4 Notes

1. The presence of 2-mercaptoethanol in the LCAT reagent mix prevents the analysis of its cholesterol concentration by the usual enzymatic diagnostic reagents, which will react with 2-mercaptoethanol. Cholesterol measurements should be performed on proteoliposomes after dialysis but before adding 2-mercaptoethanol.
2. After each cycle of freeze-thawing, there will be a partial decrease in the efficiency of LCAT activation with the reagent. It is still possible to use any leftover reagent after refreezing; however, in this case, a sample of standard plasma should be used for normalization of the obtained results. Fresh normal human EDTA-plasma kept frozen at −70 °C in 30–50 μl portions in 0.5-ml tubes with screw-on caps can be used as a normalization standard. It can be stored for several years without losing any LCAT activity.
3. The intra-assay coefficient of variation of the assay is approximately 5 %. As it was pointed out by Gillett and Owen, all existing methods of measurement of plasma LCAT activity are inherently sensitive to the amount of lipoproteins, especially HDL, in the tested plasma [Gillett MP, Owen JS (1992) Cholesterol esterifying enzyme—lecithin: cholesterol acyltransferase (LCAT) and acylcoenzyme A:cholesterol acyltransferase (ACAT). In: Converse CA, Skinner ER (eds) Lipoprotein analysis. A practical approach. IRL, Oxford, pp 187–201]. HDL cholesterol from plasma will compete with labeled substrate and decrease the measured LCAT activity. If we compare two samples of plasma with the same amount of LCAT by mass, but with different level of HDL cholesterol, the measured LCAT activity can be lower for plasma with higher HDL. In order to escape this mistake, the volume of plasma taken for analysis should be as low as sensitivity of the method permits. In calculations of the LCAT activity, concentration of free

Table 1
Effect of different apoA-I mimetic peptides on measured LCAT activity in normal human plasma

Sample ID	Peptide amino acid sequence	Specific LCAT act. (nmol/ml/h)
h-ApoA-I		117.3 ± 7.2
P-642	PVLDLFRELLNELLEALKQKLK	146.8 ± 3.9
19A	PDWLEAFYDLVKKLLEKFK	20.2 ± 0.4
37PA	DWLKAFYDKVAEKLKEAF-P-DWLKAFYDKVAEKLKEAF	16.5 ± 0.2
D-P-642	P-642 made with d-amino acids	11.5 ± 0.3
Consensus	PVLDEFREKLNEELEALKQKLK	5.9 ± 0.2

cholesterol in incubation mixture should be calculated as a sum of cholesterol in prepared labeled substrate plus cholesterol taken with analyzed plasma. Specific LCAT activity measured by using different volumes of plasma should be the same. The extent of cholesterol esterification should be kept below 10–15% to minimize product inhibition and maintain first-order kinetics.

4. In Table 1, we have compared human apoA-I with five different amphipathic peptides reconstituted with phosphatidylcholine, including P-642, for their ability to activate LCAT after reconstitution with phosphatidylcholine, as described above. All tested proteoliposomes had the same concentration of peptide or apoA-I (250 μg/ml in proteoliposomes or 104 μg/ml in the reaction tube). For proteoliposomes which included apoA-I, the molar ratio apoA-I–lecithin–cholesterol (labeled plus unlabeled) in the assay was 0.8:250:15.7. The P-642 peptide reconstituted with phospholipid was 1.25-fold more efficient in LCAT activation than proteoliposomes made with human apoA-I. In addition, P-642 was more effective than other commonly used amphipathic peptides, including the P-642 peptide made with d-amino acids (D-P-642) [Dassuex J-L et al. (1999) US Patent 6,004,925].

References

1. Jonas A (2000) Lecithin cholesterol acyltransferase. Biochim Biophys Acta 1529:245–256
2. Fielding CJ, Fielding PE (1995) Molecular physiology of reverse cholesterol transport. J Lipid Res 36:211–228
3. Calabresi L, Moleri E, Franceschini G (2006) LCAT deficiency: molecular genetics, lipid/lipoprotein phenotype and atherosclerosis. Future Lipidol 1:241–245
4. Hoeg JM, Santamarina-Fojo S, Berard AM, Cornhill JF, Herderick EE, Feldman S, Haudenschild CC, Vaisman BL, Hoyt RF, Demosky SJ Jr, Kauffman RD, Hazel CM, Marcovina S, Brewer HB Jr (1996)

Overexpression of lecithin cholesterol acyltransferase in transgenic rabbits prevents diet-induced atherosclerosis. Proc Natl Acad Sci U S A 93:11448–11453

5. Sakai N, Vaisman BL, Koch CA, Hoyt RF Jr, Meyn SM, Talley GD, Paiz JA, Brewer HB Jr, Santamarina-Fojo S (1997) Targeted disruption of the mouse lecithin:cholesterol acyltransferase (LCAT) gene. Generation of a new animal model for human LCAT deficiency. J Biol Chem 272:7506–7510
6. National Cholesterol Education Program (NCEP) Expert Panel on Detection, Evaluation, and Treatment of High Blood Cholesterol in Adults (Adult Treatment Panel III) (2002) Third Report of the National Cholesterol Education Program (NCEP) Expert Panel on Detection, Evaluation, and Treatment of High Blood Cholesterol in Adults (Adult Treatment Panel III) final report. Circulation 106:3143–3421
7. Gotto AM Jr, Brinton EA (2004) Assessing low levels of high-density lipoprotein cholesterol as a risk factor in coronary heart disease: a working group report and update. J Am Coll Cardiol 43:717–724
8. Feig JE, Shamir R, Fisher EA (2008) Atheroprotective effects of HDL: beyond reverse cholesterol transport. Curr Drug Targets 9:196–203
9. Bots ML, Visseren FL, Evans GW, Riley WA, Revkin JH, Tegeler CH, Shear CL, Duggan WT, Vicari RM, Grobbee DE, Kastelein JJ, RADIANCE 2 Investigators (2007) Torcetrapib and carotid intima-media thickness in mixed dyslipidaemia (RADIANCE 2 study): a randomised, double-blind trial. Lancet 370:153–160
10. Frohlich J, Dobiasova M (2003) Fractional esterification rate of cholesterol and ratio of triglycerides to HDL-cholesterol are powerful predictors of positive findings on coronary angiography. Clin Chem 49:1873–1880
11. Chen C-H, Albers JJ (1982) Characterization of proteoliposomes containing apoprotein A-I: a new substrate for the measurement of lecithin: cholesterol acyltransferase activity. J Lipid Res 23:680–691
12. Albers JJ, Chen CH, Lacko AG (1986) Isolation, characterization, and assay of lecithin-cholesterol acyltransferase. Methods Enzymol 129:763–783
13. Gillett MP, Owen JS (1992) Cholesterol esterifying enzyme—lecithin: cholesterol acyltransferase (LCAT) and acylcoenzyme A:cholesterol acyltransferase (ACAT). In: Converse CA, Skinner ER (eds) Lipoprotein analysis. A practical approach. IRL, Oxford, pp 187–201
14. Dassuex J-L, Sekul R, Bittner K, Cornut I, Metz G, Dufourcq J (1999) Apolipoprotein A-I agonists and their use to treat dyslipidemic disorders. US Patent 6,004,925
15. Sethi AA, Amar M, Shamburek RD, Remaley AT (2007) Apolipoprotein AI mimetic peptides: possible new agents for the treatment of atherosclerosis. Curr Opin Investig Drugs 8:201–212

Chapter 17

Native–Native 2D Gel Electrophoresis for HDL Subpopulation Analysis

Lita A. Freeman

Abstract

High-density lipoproteins (HDLs) are a heterogeneous mixture of lipoprotein particles with densities ranging from 1.063 to 1.021 g/ml. The various lipoprotein particles present in HDL have not yet been completely characterized, largely due to methodological difficulties. Asztalos and Schaefer have developed a powerful native–native 2D gel method to analyze HDL subpopulations with high resolution. In this technique, native HDL particles present in plasma are separated electrophoretically by charge in the first dimension and then by size in the second dimension. The native particles in the gel are then transferred onto a membrane using traditional Western blotting techniques. After probing the membrane with, for example, apoA-I antibodies, subpopulations of HDL particles containing apoA-I can be visualized. Traditionally, midi-sized native gels are poured manually in the laboratory and running, blotting, and probing the gels is a tricky and laborious procedure that involves the use of ^{125}I-labeled antibodies. Here we present a streamlined native–native 2D gel electrophoresis and blotting method using minigels. Traditional antibody incubation and chemiluminescent methods can be used for detection and use of ^{125}I is not required. This update of the Asztalos and Schaefer native–native 2D gel protocol renders the procedure more accessible to the nonspecialist.

Key words HDL subpopulations, Apolipoprotein A-I, α-Particles, β-Particles, HDL particles, 2D gel electrophoresis, Lipoproteins

1 Introduction

High-density lipoproteins (HDLs) are a heterogeneous mixture of lipoprotein particles with densities of 1.063–1.021 g/ml [1]. HDLs have been extensively studied, yet the compositions and functions of the various particles in the HDL density fraction are not yet completely understood. HDLs can be subfractionated by density gradient centrifugation into larger, less dense particles, termed the HDL_2 subfraction, and smaller, more dense particles, termed the HDL_3 subfraction [1]. However, both of these subfractions are complex and still contain multiple types of HDL particles. HDLs have also been fractionated according to the charge of the native particle, with pre-α HDL particles migrating fastest towards

Lita A. Freeman (ed.), *Lipoproteins and Cardiovascular Disease: Methods and Protocols*, Methods in Molecular Biology, vol. 1027, DOI 10.1007/978-1-60327-369-5_17, © Springer Science+Business Media, LLC 2013

the anode (faster than albumin), α-particles migrating with intermediate mobility (similar to albumin), and pre-β particles migrating most slowly towards the anode (slower than albumin) [2]. The major apolipoprotein present in the HDLs is apolipoprotein A-I (apoA-I), but additional proteins are also present, and indeed HDLs have also been subfractionated by apolipoprotein content. The specific particle composition of each of these HDL subfractions was unclear for some time.

A major breakthrough in understanding the composition of the HDLs was the native–native 2D gel electrophoresis procedure, which separates native HDL particles by charge in the first dimension and then by size in the second dimension [2]. After transferring the HDL particles in the gel to a membrane, the protein composition of HDL subpopulations separated by this procedure can be investigated by Western analysis. Using this technique, Asztalos and Roheim fractionated HDL (specifically, apoA-I-containing particles in plasma) into at least 12 separate subpopulations containing apoA-I [2] (Fig. 1). Asztalos, Schaefer, and colleagues further defined the content and physiological role of these particles by analyzing apoA-I-containing particles in plasma of normal individuals vs. patients with mutations in genes known to be involved in HDL metabolism (i.e., familial HDL deficiency states) [3–7]. These elegant studies have done much to elucidate precursor–product relationships between HDL subpopulations facilitated by several HDL remodeling enzymes [5, 8] and have provided critical insight into the composition of HDL subpopulations. These authors have also demonstrated that redistribution towards smaller α-migrating apoA-I containing particles and an increase in β1 particle levels is associated with greater risk for CAD [8–13].

Many characteristics of HDL particles and their redistribution in health and disease states remain to be determined. The distribution of apoA-I, apoA-II, apoA-IV, apoC-III, and apoE has been assigned to specific particles [5, 6] and several additional apolipoproteins have been assigned to specific HDL density fractions [14] but the specific particle distribution of many other proteins in the HDL density fraction in healthy control subjects and CAD patients [15] has not yet been determined. In addition, a number of disorders result in altered HDL levels and determining which HDL subpopulations are altered in the disease state could provide insight into functionality.

The traditional method of preparing and running 2D gels, especially the second-dimension nondenaturing gradient gel, is difficult and laborious and requires the use of ^{125}I-labeled antibodies to probe Western blots. Here we present a streamlined method

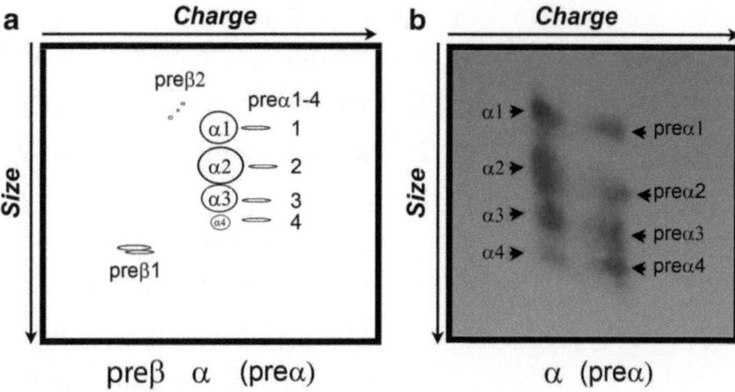

Fig. 1 Separation of native HDL particles by native–native 2D gel electrophoresis. Native HDL particles in plasma are separated by charge in the first dimension by electrophoresis on a 0.7 % agarose gel in Tris–Tricine buffer. The lane is excised and native HDL particles are then separated by size in the second dimension by electrophoresis on a 4–25 % gradient gel in TBE. Native HDL particles are transferred from the gel to a PVDF membrane, which is then probed with antibodies to apoA-I. (**a**) Schematic of native apoA-I-containing HDL particles after native–native 2D gel electrophoresis of plasma. (**b**) A Western blot of 1 μl of human plasma separated by native–native 2D gel electrophoresis, transferred to PVDF and probed with anti-apoA-I antibodies. α and preα particles are shown. preβ particles become visible after very long exposures (not shown) or after using larger volumes of plasma

using an agarose minigel in the first dimension and prepoured 3–25 % gradient minigels in the second dimension, followed by Western blotting and chemiluminescent detection of apoA-I.

2 Materials

2.1 First Dimension

1. 1.0 L of first dimension running buffer ("Tris–Tricine buffer"):

   3.0 g Tris base.

   4.5 g Tricine (Sigma/Aldrich Chemicals, St. Louis, MO).

   0.7 g calcium lactate (Spectrum Chemical Mfg. Corp, New Brunswick, NJ).

   0.5 g sodium azide.

   H_2O to 1.0 L.

   This buffer can be stored at room temperature for several months.

2. 0.7 g agarose, low endosmosis (e.g., SeaKem LE Agarose, Lonza Rockland, Rockland, ME).
3. Microwave.

4. Empty 1.5 mm disposable plastic Novex gel cassette (Life Technologies, Carlsbad, CA Catalog # NC2015).
5. 10-well 1.5 mm comb (Life Technologies, Catalog # NC3515).
6. Minigel apparatus (e.g., Novex Xcell II, Invitrogen).
7. Sample buffer:

   8 ml Tris–Tricine Buffer.

   2 ml 100 % glycerol.

   0.033 g Bromophenol Blue.

   Filter-sterilize and store in frozen aliquots at −20 °C.
8. Samples (plasma or purified HDL).
9. Flat-tipped long-tipped pipettips (PGC Scientifics, Frederick, MD).

2.2 Second Dimension

1. 3–25 % acrylamide gradient TBE minigel, 1.5 mm thick, catalog # 3 25 N 1.5 NMC1P (Jule Inc., Milford, CT). Store long-term at 4 °C. *Do not freeze.* Approximately 1 h before use, place gels on benchtop to equilibrate gels to room temperature.
2. 1.5 × 5 × 2.5 mm "stick" used to form a MW marker well in the second dimension gel, prepared by cutting off a tooth from a Hoefer 20-well 1.5 mm comb.
3. 10× TBE: 0.89 M Tris–Borate; 0.02 M EDTA (disodium salt), pH 8.3 (Accugene/Lonza, Rockland, ME).
4. 1× TBE: Dilute 10× TBE tenfold (e.g., 100 ml 10× TBE per liter). Prepare 1.0 L per two minigels; excess can be stored indefinitely at room temperature.
5. Amersham™ HMW Calibration Kit For Native Electrophoresis, Catalog 17-0445-01 (GE Healthcare, Piscataway, NJ). The markers are supplied as a lyophilized protein mixture in ten individual vials, 250 μg/vial (*see* **Note 1**). Store lyophilized vials at 2–8 °C.
6. Native HMW marker reconstitution buffer: 1× TBE containing 0.05 % (w/v) bromophenol blue. Filter-sterilize. Store at room temperature.
7. Reconstituted Native HMW markers: Remove one vial from the Amersham™ HMW Calibration Kit For Native Electrophoresis from the freezer, allow to warm to room temperature, and then add 100 μl of Native HMW marker reconstitution buffer. Rotate vial to dissolve all the lyophilized material on the tube. Once dissolved, keep on ice until used or aliquot and store at −20 °C long-term.

2.3 Transfer

1. Simply Blue[R] Coomassie Blue stain (Life Technologies).
2. 20× Tris–Glycine–SDS Western Transfer Buffer (also known as 5× Laemmli Buffer).

   Per liter:

   30.2 g Tris base.

   144.0 g glycine.

   Dissolve in 800 ml distilled water.

   pH to 8.3 with concentrated HCl (use pH paper if a pH meter with a Tris electrode is not available).

   Add 20.0 g SDS.

   Stir until dissolved.

   Sterile-filter.
3. Transfer buffer.

   Per liter:

   200 ml methanol.

   50 ml 20× Tris–Glycine–SDS Western Transfer Buffer.

   Distilled water to a final volume of 1.0 L.
4. Whatman 3MM paper (GE Healthcare).
5. Immobilon-P (PVDF) Transfer Membranes, 0.45 μm pore size (Millipore Corporation, Bedford, MA).
6. Western transfer tank—Hoefer TE22 recommended.

2.4 Blocking, Antibody Incubation, and Detection

1. Bovine serum albumin, Fraction V (BSA) (MP Biomedicals, Solon, OH). Store at 4 °C.
2. 10× PBS.
3. Tween-20 (Sigma-Aldrich, St. Louis MO).
4. PBS-T. For each liter of PBS-T, add 100 ml 10× PBS and 750 μl Tween-20 to 850 ml distilled water in a graduated cylinder and adjust the volume to 1000 ml. Mix by inverting.
5. 3 % BSA in PBS-T. Slowly add 15 g powdered BSA to 500 ml PBS-T with vigorous stirring on a magnetic stirrer and continue to stir until dissolved. Filter-sterilize. Store at 4 °C. Reserve a portion at room temperature if it is to be used that day to make blocking buffer.
6. Nonfat dried milk.
7. Blocking buffer: 3 % BSA, 5 % milk in PBS-T. Add 5.0 g nonfat dried milk per 100 ml of room-temperature, filter-sterilized 3 % BSA in PBS-T. Stir vigorously for at least 10 min to completely dissolve milk particles.
8. Antibody dilution buffer: 5 % milk, 0.3 % BSA in PBS-T. For 100 ml, dissolve 5.0 g nonfat dried milk in 100 ml PBS-T. Stir

vigorously on a magnetic stirrer for at least 10 min, then add 10 ml filter-sterilized 3 % BSA in PBS-T.

9. Primary antibody. Biodesign/Meridian Life Science (Saco, ME) and Academy Biomedical Co. (Houston, TX) are good sources of anti-apolipoprotein antibodies. See also Chapter 28, Table 1 primary antibodies preconjugated to HRP are recommended for 2D gel electrophoresis, when feasible (*see* **Note 2**). Pretest antibody specificity before using for 2D gel analysis of experimental samples (*see* **Note 3**).
10. HRP-conjugated secondary antibody (not necessary if using an HRP-conjugated or otherwise labeled primary antibody; *see* **Note 2**). To minimize unwanted cross-reactivity use secondary antibodies pre-adsorbed against human serum for human samples and secondary antibodies pre-adsorbed against mouse serum for mouse samples. Secondary antibodies must always be pretested for cross-reactivity to serum proteins (*see* **Note 3**). ^{125}I-labeled secondary antibodies can be used in place of HRP-conjugated secondary antibodies [2] (*see* **Note 4**).
11. Whatman 3 MM filter paper.
12. Western Lightning Plus-ECL Enhanced Chemiluminescence Substrate for Western Blotting (Perkin-Elmer, Waltham, MA).
13. Autoradiography cassette.
14. Autoradiography film (e.g., Kodak BioMax MR film, Carestream Health Molecular Imaging) or appropriate imaging device.

3 Methods

3.1 First Dimension

1. Make 1.0 L of first dimension running buffer ("Tris–Tricine buffer").
2. In 250 ml flask add 0.7 g agarose to 100 ml Tris–Tricine buffer (*see* **Note 5**).
3. Microwave on high setting with occasional swirling until completely dissolved, usually 2–4 min. *Caution*: Solution is hot! Handle with heat-resistant gloves (*see* **Notes 6** and 7).
4. Let cool to approximately 50–65 °C (very warm to the touch but can hold the flask) with occasional swirling (*see* **Note 8**).
5. Using a 10-ml pipette, pipette 10 ml melted agarose into an empty 1.5 mm Novex gel cassette (Invitrogen Catalog # NC2015) with a 10-well 1.5 mm comb (Invitrogen Catalog # NC3515). Insert comb immediately, avoiding bubble formation under the teeth (*see* **Note 9**). Do not discard the agarose as it will be used again in Subheading 3.2, **steps 2** and **5**, below.
6. Allow the gel to solidify completely at room temperature (approximately 45 min).

7. Remove comb carefully and clear any remaining agarose out of wells using a long-tipped flat pipettip. Be careful not to poke the bottom of the gel well with the pipettip.
8. Label the wells on BOTH sides of the cassette, with X's marking empty wells, before inserting the gel into the minigel apparatus (*see* **Note 10**).
9. Load into Novex Xcell minigel apparatus with Tris–Tricine buffer both inside and outside (*see* **Note 11**).
10. Prepare plasma for gel:
    (a) Prepare or thaw sample buffer.
    (b) Thaw plasma quickly while keeping as cold as possible. Flick very gently to mix contents. Do not vortex or cause the sample to foam, which may cause oxidation and/or denaturation of the HDL particles. Do not allow samples to sit around; use immediately after thawing.
    (c) Spin briefly at top speed at 4 °C in a refrigerated microcentrifuge. Place on ice and use *immediately*.
    (d) Prepare sample: (*see* **Note 12**).
    - For detection of well-separated α and preα particles with anti-apoA-I antibody:
        1. 9.0 μl Tris–Tricine buffer.
        2. 4.0 μl sample buffer.
        3. 1.0 μl plasma.
        4. Load 3.0 μl/well.
    - For detection of less abundant particles:
        1. 10.0 μl plasma.
        2. 4.0 μl sample buffer.
        3. Load 6–10 μl/well.
11. Electrophoresis: 100 V until dark blue dye is not quite off the gel (usually ~50 min) (*see* **Note 13**).
12. While the gel is running, prepare 1× TBE for the next step.

3.2 Second Dimension

1. Cut open a gel pouch containing the prepoured 3–25 % acrylamide gradient TBE minigel from Jule Inc. Dry the outside of the gel with a paper towel.
2. Re-microwave the 0.7 % agarose in Tris–Tricine buffer from Subheading 3.1, **step 2** until liquid (usually 1 min is fine).
3. After the agarose gel has finished running, remove it from the electrophoresis apparatus and dry it with a paper towel
4. Using a laboratory marker and a ruler, draw lines on BOTH gel plates from the top to the bottom of each gel slice that is to be excised, on the left and right sides of each gel slice. Use the gel well and the blue dye at the bottom as guides for choosing

how to excise the gel slice. *IMPORTANT*: Make sure that BOTH the front and back plates are labeled with the identity of the sample in each gel slice, as it cannot be predicted which plate the gel will stick to after separation.

5. Separate the plates of the agarose gel, pipet 1× TBE over the gel to keep it from drying out and to provide lubrication for the next step, and cut out a gel slice following the guidelines drawn on the plate. Do not let the gel slide around—hold it gently in place so that the slice you cut out does in fact correspond to the correct gel lane. Carefully note the sample identity of the gel slice.
6. Load gel slice into the 3–25 % acrylamide gradient TBE minigel. (*See* **Note 14** for tips on loading the gel slice.) Double-check that the Jule gradient gel is labeled on the front and back with the identity of the sample in the gel slice.
7. After all of the gel slices are successfully loaded, insert a 1.5 mm "stick" as a MW marker at the 3 % (soft) end of each gel. Use a 200-μl Pipetman to pipette in cooled but still liquid agarose next to the MW marker stick. Be sure to label the location of the stick on both the front and back plates, as later the slice of the gel containing the MW marker will be removed and stained with Coomassie Blue.
8. Allow the agarose to solidify completely at room temperature (approximately 15 min). Remove the MW marker stick.
9. The Jule gradient gels are poured in a "Mini Snap-A-Gel" cassette with a bottom tab that must be removed before running the gel. Follow the manufacture's instructions for removing the tab. Hold the main part of the cassette down on the bench with one hand and then wiggle the tab up and down. Usually the tab comes off easily; if not, wiggle the tab more forcefully, using pliers for added leverage. Bashing the pliers against a stubborn tab is not recommended as it is ineffectual, may crack the main part of the cassette, and can send plastic shards flying. Wear eye protection and gloves per manufacturer's instructions, and just wiggle the tab harder. It will eventually come off.
10. Check that the act of removing the bottom tab did not alter the contact of the gel slice with the top of the gradient gel (it frequently does). If it did, push the gel slice back into place. The gel slice must be snug against the top of the gradient gel before electrophoresis starts.
11. Insert the gradient gel with its loaded gel slice into a Novex Xcell II or SureLock gel apparatus, along with a second gel or a Novex spacer. The Novex gel apparatus is assembled as usual, except that a special plastic spacer (provided by Jule with each box of gels, along with instructions for use) is placed against the back of the Novex apparatus to make a tight seal (*see* **Notes 15** and **16**).

12. Add 1× TBE buffer inside and outside (*see* **Note 17**).
13. Perform one last check to make sure the gel slice is snug against the top of the gradient—it can still be gently pushed into place at this point.
14. Pipet 7.5 μl native HMW markers from Amersham, reconstituted in native HMW marker reconstitution buffer, into the MW marker well. If you did not previously mark the location of the HMW marker on the gel plate, you can mark it now at the very top of the gel, on both the front and back plates.
15. Perform electrophoresis at 200 V for 3.0–3.5 h or until blue dye is 2.0 cm from bottom of cassette.

3.3 Transfer

1. Open the gel cassette carefully.
2. Cut off the lane with the native HMW marker and drop this gel slice into a 15-ml plastic centrifuge tube filled with Simply Blue to stain the markers. Screw the top of the 15-ml tube on tightly and leave the tube on its side until ready to use. Destaining is usually not necessary.
3. Place the gel in a small dish or plastic tray containing 1× transfer buffer.
4. Transfer proteins to a PVDF membrane using 1× transfer buffer and your preferred minigel Western blotting apparatus. For best results, use a Hoefer TE22 transfer tank containing 1× transfer buffer and transfer in a cold room (4 °C) with constant stirring. We usually transfer either for 3.5 h at 80 V; alternatively, we transfer overnight at 20 V, increase the voltage to 80 V the next morning and then transfer for an additional 3.5 h (*see* **Chapter 18**). The transfer buffer does not need to be prechilled.

3.4 Blocking, Antibody Incubation, and Detection

1. Wash each blot in PBS-T ≥5 min. Gently pull off any remaining traces of agarose sticking to the membrane with forceps.
2. Block in blocking buffer on a rocker or rotating platform for at least 1 h at room temperature or overnight at 4 °C.
3. Incubate the blot in antibody dilution buffer containing the primary antibody. Primary antibody dilutions and incubation times should be empirically determined. (*For the* "secondary antibody alone" *control*, *see* **Note 18**).
4. Wash 5×, 150–200 ml PBS-T per wash, for at least 5 min and preferably 10 or more minutes per wash. (*Note*: If you used a primary antibody already conjugated to HRP, there is no need to use a secondary antibody. Skip **steps 5** and **6** and proceed to **step 7**).
5. Incubate the blot in antibody dilution buffer containing the appropriate HRP-conjugated secondary antibody at the appropriate dilution, 45 min to 1 h at room temperature.
6. Wash 5×, 150–200 ml PBS-T per wash, as in **step 4**.

7. Remove blot with forceps onto Whatman 3 MM paper. Gently blot away excess moisture by placing a second sheet of Whatman 3 MM paper on top of the blot and patting gently for a few seconds until the blot is no longer shiny. Do not let the blot dry completely.
8. Place the blot on plastic wrap on the bench top. Add 5 ml of the Enhanced Luminol Reagent (brown bottle) and 5 ml of the Oxidizing Reagent (white bottle) from the Western Lightning kit to a 50-ml tube cap the tube and vortex vigorously to prepare the Chemiluminescence Reagent. Pour the Chemiluminescence Reagent over the blot. Remove bubbles by shaking the blot and moving it as necessary, using forceps. Gently agitate the blot for 1 min in the Chemiluminescence Reagent.
9. Remove the blot from the Chemiluminescence Reagent with forceps and blot away excess moisture as in **step** 7.
10. Place the blot on a piece of plastic wrap on the bench top, cover with a second piece of plastic wrap, cut away excess plastic wrap, leaving about an inch extra on each side, and tape to the back side of an autoradiography cassette.
11. In a darkroom, place a piece of film (Kodak MR-2 film gives the best resolution) against the blot for 10 s and develop the film. Take a series of longer and/or shorter exposures as necessary. Alternatively, use an imaging device capable of detecting low levels of light ($\lambda = 428$ nm).
12. Line up the MW markers from Subheading 3.3, **step 2** with the top and bottom of the gel and mark the sizes on the autoradiogram.

This protocol can be modified to highlight whatever feature is desired—for example, the acrylamide gradient can be altered to accentuate separation of larger or smaller particles—but this protocol as written is useful for visualizing the distribution of apoA-I-containing particles.

4 Notes

1. Each vial of the Amersham High Molecular Weight Calibration Kit for native electrophoresis contains:

		MW (Da)	Diameter (nm)
Thyroglobulin	(76 μg)	669,000	17.0
Ferritin	(50 μg)	440,000	12.2
Catalase	(36 μg)	232,000	9.7
Lactate dehydrogenase	(48 μg)	140,000	8.16
Albumin	(40 μg)	66,000	7.1

These are molecular weights and diameters of the *native* particles [16, 17]. Do not treat these markers with reducing agents or detergents. Dissolve markers only in Native HMW marker reconstitution buffer.

2. Primary antibodies preconjugated to HRP are recommended for 2D gel electrophoresis since they eliminate background from the secondary antibody (*see* **Note 3**) and save time by eliminating the secondary antibody incubation and washing steps. The disadvantages to using HRP-conjugated primary antibodies are that they are generally more expensive and more dilute than unconjugated antibodies. You can label your own antibody with HRP using, for example, the EZ-Link® Activated Peroxidase Antibody Labeling Kit (#31497) from Pierce/ThermoScientific (Rockford, IL). Labeling your primary antibody with a fluorescent dye is another possible option that to our knowledge has not yet been explored; make sure that your imaging system has the capacity to (a) excite and (b) detect your fluorophore at the proper wavelengths with good sensitivity before embarking on this as-yet untested procedure. Primary antibodies can also be labeled directly with ^{125}I. Ensure that the antibody's binding characteristics have not been altered after labeling by performing a test Western on plasma electrophoresed on an SDS-PAGE gel under reducing conditions.
3. Before performing 2D gel analysis with new antibodies, pretest primary and secondary antibody specificity on an SDS-PAGE gel. Run two identical aliquots of a test sample (e.g., healthy volunteer plasma) side-by-side on a regular SDS-PAGE gel under reducing conditions. After electrophoresis, transfer the proteins onto a membrane. Mark and label the sample lane positions on the blot before transfer as they will be treated with different antibody combinations. After transfer, block the blot as usual and cut out the two strips. Place one strip into antibody dilution buffer containing the primary antibody to be tested. Place the second strip into antibody dilution buffer without primary antibody; this will be the "secondary antibody alone" test strip. Incubate the two strips side-by-side under the test conditions for the primary antibody. Wash the two strips side-by-side; incubate in HRP-conjugated secondary antibody side-by-side, and wash and develop side-by-side. Include extra-long exposures to visualize weak cross-reacting bands. The primary antibody must recognize only the apolipoprotein of interest. Ideally, the secondary antibody should not recognize any proteins present in serum. Secondary antibodies pre-adsorbed against human serum will minimize unwanted cross-reactivity if you are testing human samples. If your samples are from mice, use secondary antibodies pre-adsorbed against mouse serum. If secondary antibody cross-reactivity is still

significant, consider using a primary antibody preconjugated to HRP or a fluorescent reporter dye (*see* **Note 2**). If this is not an option, check to see whether the proteins cross-reactive to the secondary antibody migrate far away from HDL on the native–native 2D gel: run two identical aliquots of test plasma on native–native 2D gels, side-by-side; probe one blot with primary + secondary antibody and the other with secondary antibody alone (*see* **Note 18**). Frequently the signal from the secondary antibody alone is far from the primary antibody's signal and can be ignored. Again, HRP-conjugated primary antibodies that recognize only the desired protein are, if available, the ideal choice.

4. ^{125}I-labeled secondary antibodies are considerably more troublesome to work with than HRP-labeled antibodies but when used with Phosphorimager technology will give more quantifiable results. Fluorescently labeled secondary antibodies imaged with example a GE Healthcare Storm or Typhoon or LIC or ODESSY should in theory also give quantitative results but we have not yet tested this approach.

5. 100 ml is more than enough for one gel but is handy in case of mishaps. The user is also likely to need more than one first-dimension gel for several reasons. First, loading ten samples per gel is not advisable: end lanes should be empty to prevent smiling and additionally, leaving an empty lane between samples prevents cross-contamination and simplifies cutting lanes out of gels. Second, to identify preβ, α and preα particle positions for each 2D gel, it is customary to load an adjacent lane on the same gel to be run in the first dimension only; the gel strip will be cut out and then directly blotted and probed with anti-apoA-I antibodies (*see* Fig. 5 of ref. 2). Finally, since even pretty good secondary antibodies can show some cross-reactivity to miscellaneous serum proteins after long exposures, it is a good idea to run duplicate 2D gels for each sample: one with primary plus secondary antibody, and one with secondary antibody alone. This secondary-only antibody control is particularly useful while optimizing new antibodies and may not be necessary once ideal conditions (i.e., no secondary-antibody background) are achieved.

6. Disposable plastic 250 or 500 ml culture flasks are convenient but do not screw the lid on tightly. The lid must be loose, otherwise pressure will build up and the solution will boil over violently after microwaving, with risk of scalding by the superheated agarose solution.

7. Check that there are no undissolved particles left in solution. If chunks of unmelted agarose are still visible in the solution, continue microwaving in 15- to 30-s intervals with swirling until the agarose is completely dissolved.

8. Swirl vigorously at first to mix. As the solution cools, ease up on the swirling to minimize bubble formation. Before pouring, let the flask sit still on the bench for a couple of minutes to dissipate bubbles.
9. Inserting the comb into the cassette at a slight angle prior to adding the agarose and gently pushing the comb in while repositioning or jiggling it as the agarose comes into contact with the bottom of the comb helps to avoid small bubbles from settling in at the bottom of the well.
10. After the run is over and the plates are being disassembled, the gel may stick randomly to either the front or the back plate. Thus, for both the first dimension (agarose) and the second dimension (acrylamide gradient gel), always label both the front and the back plates.
11. Do not add β-mercaptoethanol, dithiothreitol, or any reducing agents or detergents to the samples or to any solutions or gels used for electrophoresis. The HDL particles must remain in native form.
12. Pre-aliquot the sample buffer, Tris–Tricine, and water into tubes at room temperature so that the plasma samples can be added immediately and loaded on the gel ASAP. A master mix may be used if multiple samples with similar sample volumes are being prepared.
13. Do not allow the dye to run off the bottom of the agarose gel. First, the dye serves as a guide in cutting out the gel slice, and second, this helps to ensure that particles do not run off the gel.
14. Tips on loading the gel slice: Position the agarose gel slice atop the acrylamide gradient gel well, with the bottom of the gel (aqua dye) to the right. Push the bottom of the gel (aqua dye) in first until it touches the top of the gradient gel. If the agarose gel slice is too thick to slide it into the gel well, stretch the gel slice by stroking it with a gloved fingertip, adding a little 1× TBE for lubrication if necessary. Avoid the temptation to grab one end of the gel slice and pull; 0.7 % agarose gel slices are fragile and break easily. After the bottom of the agarose gel slice has contacted the acrylamide gel, gently work your way leftwards along the rest of the gel slice, pushing it in gently as you go along. To avoid introducing bubbles between the first dimension agarose gel slice and the top of the second dimension gradient gel, tilt the whole assembly clockwise as you push in the agarose gel slice so that any bubbles between the gel slice and the gradient gel will rise to the top (upwards and leftwards) before you push the gel slice down against the acrylamide gel. It is almost impossible to completely avoid bubbles, but this will minimize them.

15. Even with the plastic spacer from Jule inserted into the Novex gel apparatus, the setup can leak when two gels are being run in the same apparatus. Running just one gel per apparatus is the simplest solution, but this requires more TBE, gel apparatuses and power supplies. To guard against leaks with two Jule gels in one Novex gel apparatus, first check that there are no chips or cracks in the gel cassettes that might interfere with formation of a tight seal. If a crack or chip is present in a gel cassette, electrophorese the damaged gel in a separate gel box by itself so that if there are problems during the run only that gel will be affected. Alternatively, crack open the Gel, remove the Gel slice and place the old Gel slice into a new second-dimension Gel. If the plates are undamaged, check that the gasket is in good repair. Then, make sure that the gel cassettes and gel apparatus are completely dry and that both the front and rear cassette are in full contact with the gasket. The Mini Snap-A-Gels are not an exact fit for the Novex apparatus, and the rear cassette in particular must be very carefully placed against the rear gasket and held tightly in place while being placed inside the gel apparatus. At the same time the Jule-supplied plastic spacer must be monitored such that it does not slide out of position and remains in full contact with the Novex wedge. The Jule spacer should be placed at the back of the tension wedge for SureLock units or between the wedge and the rear cassette for Xcell units. If a slow leak becomes apparent during electrophoresis, stop electrophoresis of the gel, add additional buffer as needed, and then resume electrophoresis.
16. The Mini Snap-A-Gels can be used in the BioRad Mini Protean II or III gel apparatus if the gasket in the inner frame is reversed so that the smooth side faces outwards.
17. *IMPORTANT*: The HDL particles must be kept in their native form, so do NOT add antioxidant to the inner chamber.
18. If you are performing a side-by-side control Western with secondary antibody alone, treat this control blot exactly the same way as the "real" blot throughout the procedure, but leave out the primary antibody in this step. Obtain at least 1-min, 10-min, 30-min and overnight exposures when testing secondary antibodies; sometimes the "real" Western requires long exposures that reveals background spots for even good secondary antibodies. It is also possible to perform the "secondary antibody alone" control Western first on a blot and then use that same blot for the "primary plus secondary" Western; simply rehydrate the blot in PBS-T before reprobing with your primary and then secondary antibodies.

Acknowledgments

I am indebted to Dr. Bela Asztalos and Dr. Ernst Schaefer for kindly demonstrating to me their classic native–native 2D gel electrophoresis method.

References

1. Gotto A, Pownall H (2003) Manual of lipid disorders: reducing the risk for coronary heart disease. Lippincott, Williams and Wilkins, Philadelphia, PA
2. Asztalos BF, Sloop CH, Wong L, Roheim PS (1993) 2-dimensional electrophoresis of plasma-lipoproteins—recognition of new apo-A-I-containing subpopulations. Biochim Biophys Acta 1169:291–300
3. Asztalos BF, Brousseau ME, McNamara JR, Horvath KV, Roheim PS, Schaefer EJ (2001) Subpopulations of high density lipoproteins in homozygous and heterozygous Tangier disease. Atherosclerosis 156:217–225
4. Asztalos BF, Horvath KV, Kajinami K, Nartsupha C, Cox CE, Batista M, Schaefer EJ, Inazu A, Mabuchi H (2004) Apolipoprotein composition of HDL in cholesteryl ester transfer protein deficiency. J Lipid Res 45:448–455
5. Asztalos BF, Schaefer EJ, Horvath KV, Yamashita S, Miller M, Franceschini G, Calabresi L (2007) Role of LCAT in HDL remodeling: investigation of LCAT deficiency states. J Lipid Res 48:592–599
6. Santos RD, Schaefer EJ, Asztalos BF, Polisecki E, Wang J, Hegele RA, Martinez LRC, Miname MH, Rochitte CE, Da Luz PL, Maranhao RC (2008) Characterization of high density lipoprotein particles in familial apolipoprotein A-I deficiency. J Lipid Res 49:349–357
7. Wada M, Iso T, Asztalos BF, Takama N, Nakajima T, Seta Y, Kaneko K, Taniguchi Y, Kobayashi H, Nakajima K, Schaefer EJ, Kurabayashi M (2009) Marked high density lipoprotein deficiency due to apolipoprotein A-I Tomioka (codon 138 deletion). Atherosclerosis 207(1):157–161
8. Schaefer EJ, Asztalos BF (2007) Increasing high-density lipoprotein cholesterol, inhibition of cholesteryl ester transfer protein, and heart disease risk reduction. Am J Cardiol 100:n25–n31
9. Asztalos BF, Roheim PS, Milani RL, Lefevre M, McNamara JR, Horvath KV, Schaefer EJ (2000) Distribution of apoA-I-containing HDL subpopulations in patients with coronary heart disease. Arterioscler Thromb Vasc Biol 20:2670–2676
10. Asztalos BF, Horvath KV, McNamara JR, Roheim PS, Rubinstein JJ, Schaefer EJ (2002) Comparing the effects of five different statins on the HDL subpopulation profiles of coronary heart disease patients. Atherosclerosis 164:361–369
11. Asztalos BF, Batista M, Horvath KV, Cox CE, Dallal GE, Morse JS, Brown GB, Schaefer EJ (2003) Change in alpha(1) HDL concentration predicts progression in coronary artery stenosis. Arterioscler Thromb Vasc Biol 23:847–852
12. Asztalos BF, Cupples LA, Demissie S, Horvath KV, Cox CE, Batista MC, Schaefer EJ (2004) High-density lipoprotein subpopulation profile and coronary heart disease prevalence in male participants of the Framingham Offspring Study. Arterioscler Thromb Vasc Biol 24:2181–2187
13. Asztalos BF, Collins D, Cupples LA, Demissie S, Horvath KV, Bloomfield HE, Robins SJ, Schaefer EJ (2005) Value of high-density lipoprotein (HDL) subpopulations in predicting recurrent cardiovascular events in the Veterans Affairs HDL Intervention Trial. Arterioscler Thromb Vasc Biol 25:2185–2191
14. Davidson WS, Silva RA, Chantepie S, Lagor WR, Chapman MJ, Kontush A (2009) Proteomic analysis of defined HDL subpopulations reveals particle-specific protein clusters: relevance to antioxidative function. Arterioscler Thromb Vasc Biol 29:870–876
15. Vaisar T, Pennathur S, Green PS, Gharib SA, Hoofnagle AN, Cheung MC, Byun J, Vuletic S, Kassim S, Singh P, Chea H, Knopp RH, Brunzell J, Geary R, Chait A, Zhao XQ, Elkon K, Marcovina S, Ridker P, Oram JF, Heinecke JW (2007) Shotgun proteomics implicates protease inhibition and complement activation in the antiinflammatory properties of HDL. J Clin Invest 117:746–756
16. Gonzalez-Navarro H, Nong Z, Amar MJ, Shamburek RD, Najib-Fruchart J, Paigen BJ, Brewer HB Jr, Santamarina-Fojo S (2004) The ligand-binding function of hepatic lipase modulates the development of atherosclerosis in transgenic mice. J Biol Chem 279:45312–45321
17. Warnick GR, McNamara JR, Boggess CN, Clendenen F, Williams PT, Landolt CC (2006) Polyacrylamide gradient gel electrophoresis of lipoprotein subclasses. Clin Lab Med 26:803–846

Chapter 18

Western Blots

Lita A. Freeman

Abstract

Western analysis of apolipoproteins, lipoproteins, and proteins involved in lipoprotein metabolism can be challenging due to their size, hydrophobic nature, and, in some cases, low abundance. Here we describe a Western blotting method that has been used successfully for many proteins involved in lipoprotein metabolism, as well as intact LDL or HDL particles. Proteins or lipoprotein particles separated by gel electrophoresis are transferred to a PVDF membrane in a Hoefer TE22 transfer tank with Tris–Glycine–SDS–Methanol transfer buffer. The membrane is blocked with 3 % BSA/5 % milk to prevent nonspecific binding of antibody to the membrane and is then incubated with primary antibody that binds specifically to the protein of interest. After washing away unbound primary antibody, the membrane is then incubated with an HRP-labeled secondary antibody that binds primary antibody. After washing away unbound secondary antibody, the membrane is then incubated with a substrate for HRP, generating a chemiluminescent signal at the location of the protein of interest. The protein is visualized by exposing the membrane to an autoradiography film or an imaging device. Information on the use of several human antibodies, including apoA-I, A-II, apoB, apoC-II, apoC-III, apoD, apoL1, apoM, PON1, SAA, ABCA1, nitrotyrosine, and LCAT, is provided. This method can be used for Western blotting of virtually any protein as well as native lipoprotein particles.

Key words HDL Western, LDL Western, Apolipoprotein Western, apoB-100 Western

1 Introduction

Western analysis of apolipoproteins, lipoproteins, and proteins affecting lipoprotein metabolism poses special challenges. Apolipoprotein B-100, for example, is a very large protein, prone to degradation, and difficult to transfer. The hydrophobic nature of apoB and membrane transporters such as ABCA1 can also cause protein aggregation. Some apolipoproteins are present at very low concentrations, especially in certain disease states, and sensitive methods are required for detection. Transfer of native lipoproteins such as LDL and HDL to membranes for Western analysis is not routinely performed despite its utility in biochemical characterization of lipoprotein particles. A versatile method

Lita A. Freeman (ed.), *Lipoproteins and Cardiovascular Disease: Methods and Protocols*, Methods in Molecular Biology, vol. 1027, DOI 10.1007/978-1-60327-369-5_18, © Springer Science+Business Media, LLC 2013

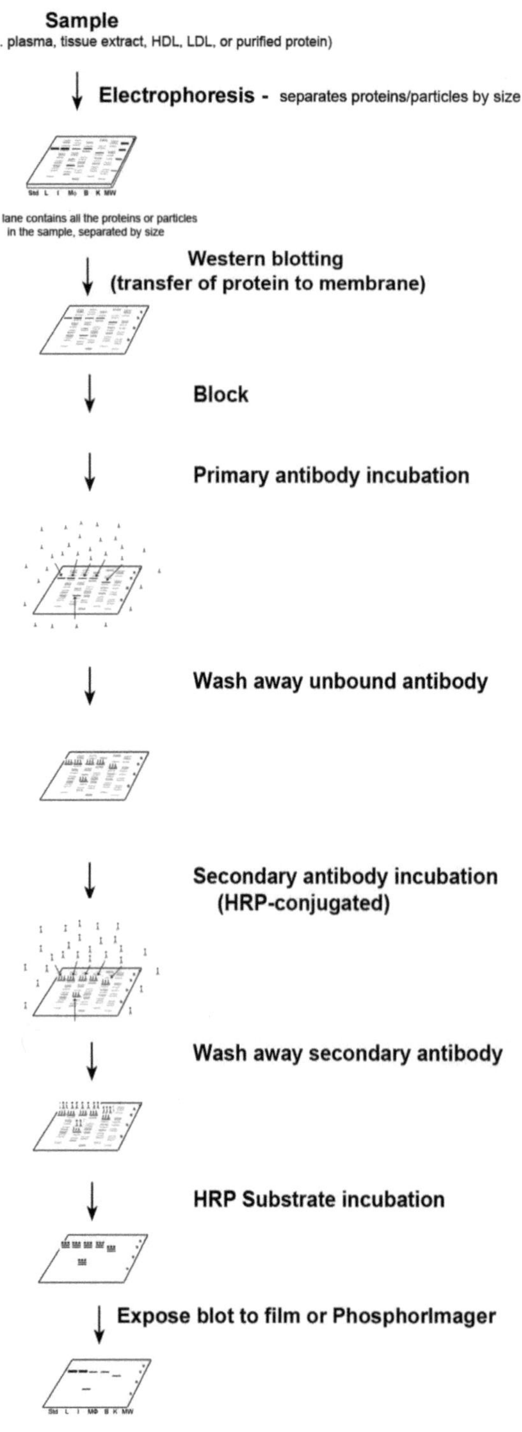

Fig. 1 Overview of Western blotting procedure

for Western analysis of lipoprotein particles and their associated proteins is critically important.

An overview of Western blotting is presented in Fig. 1. Proteins or lipoprotein particles are first separated by gel electrophoresis.

The proteins or lipoprotein particles in the gel are then transferred to a PVDF membrane and the membrane is blocked to prevent nonspecific binding of antibodies to the membrane. The blocked membrane is then incubated with primary antibody specific to the protein of interest. After washing to remove primary antibody that is not bound to the protein of interest, the membrane is then incubated with an HRP-labeled secondary antibody that binds primary antibody. Unbound secondary antibody is then removed by washing. The membrane is subsequently incubated with a substrate for HRP, generating a chemiluminescent signal at the location of the protein of interest. The protein is visualized by exposing the membrane to autoradiography film or an imaging device.

Here we present a Western blotting method that has been used successfully for many proteins involved in lipoprotein metabolism, as well as intact LDL or HDL particles.

2 Materials

2.1 Gels

2.1.1 SDS-PAGE

1. Novex NuPAGE 4–12 % Bis-Tris gels, 1.0 mm thick, 12-well (NP0322, Life Technologies, Carlsbad, CA) (*see* **Note 1**).
2. MOPS (NP0001) or MES (NP0002) SDS Running Buffer, 20× (Life Technologies). Dilute to 1× with distilled water before use.
3. Samples: Plasma, serum, FPLC fractions, cell culture supernatant, cell or tissue lysates, HDL, LDL, or VLDL. HDL or LDL isolated by ultracentrifugation should be dialyzed to remove KBr prior to electrophoresis.
4. Molecular weight markers (e.g., Novex SeeBlue Plus2 Prestained Standard, #LC5925, Life Technologies).
5. (Optional) Apolipoprotein standards (*see* **Note 2**).
6. NuPAGE LDS Sample Buffer (4×) (NP0007, Life Technologies).
7. 10× NuPage Sample Reducing Agent (NP0004, Life Technologies). Neat β-mercaptoethanol can be used as an alternative to 10× NuPage Sample Reducing Agent.
8. 1× PBS or molecular biology grade water.
9. (Recommended) Rainin L20 Pipet-Lite pipet (Rainin Instruments, Oakland, CA).
10. (Recommended) Rainin L20 Pipet-Lite low-retention pipettips (RT-L10SLR, Rainin Instruments).
11. (Optional) SimplyBlue Safe Stain (Novex) (LC6065, Life Technologies).

2.1.2 1D Native Gels

Tris–Glycine Gradient Minigels

1. Novex 4–12 % Tris–Glycine gels, 1.0 mm × 12 well (EC60352, Life Technologies).
2. Novex Tris–Glycine Native Running Buffer (10×) (LC2672, Life Technologies). Dilute to 1× with distilled water before use.
3. Novex Native Tris–Glycine Sample Buffer (2×) (LC2673, Life Technologies).

TBE Gradient Minigels

1. 10-well Tris–borate–EDTA (TBE) gels, 3–25 % acrylamide (Catalog 3 25N 1.5 NMC10P) or 3–28 % acrylamide (Catalog 3 28N 1.5 NMC10P), thickness 1.5 mM (both from Jule Inc., Milford, CT).
2. 10× TBE. Dilute to 1× TBE before use.
3. 2× TBE sample buffer:

   8 ml 2× TBE.

   2 ml 100 % glycerol.

   0.033 g Bromophenol blue.

   Mix well by vortexing.

   Filter-sterilize and store in frozen aliquots at −20 °C.
4. Amersham™ HMW Calibration Kit For Native Electrophoresis, Catalog 17-0445-01 (GE Healthcare, Piscataway, NJ). Reconstitute in 1× TBE plus 0.05 % bromophenol blue as described in Chapter 17.

Sudan Black B Staining of 1D Native Gels

1. Sudan Black B (Sigma-Aldrich, St. Louis, MO).
2. Acetone (Sigma-Aldrich) CAUTION—Flammable! Use appropriate precautions.
3. Glacial acetic acid CAUTION—Corrosive! May cause burns. Use in fume hood and wear protective clothing and eye protection.
4. Sudan Black Staining Solution (from Product Information Sheet for Sudan Black B, Sigma).
   1. Add 500 mg Sudan Black B to 20 ml of acetone with stirring.
   2. Add the Sudan plus acetone to 15 ml of glacial acetic acid and then add the mixture to 85 ml of water. CAUTION: Glacial acetic acid is corrosive and can cause burns! Wear protective clothing and eye protection.
   3. Stir the mixture for 30 min and centrifuge or filter to remove the precipitate.
5. Sudan Black Destaining Solution

   In a glass beaker or bottle combine the following:
   1. 150 ml of glacial acetic acid. CAUTION! Glacial acetic acid is corrosive and can cause burns! Wear protective clothing and eye protection.

2. 200 ml of acetone. CAUTION! Flammable! Use appropriate precautions.
3. 650 ml of water.

   Stir on a magnetic stirrer to mix. Store in a glass bottle at room temperature.

2.2 Transfer

1. 5× Laemmli buffer (prepare at least 1 day in advance).

   Per liter:

   30.2 g Tris base.

   144.0 g glycine.

   Dissolve in 800 ml distilled water.

   pH to 8.3 with concentrated HCl (use pH paper if a pH meter with a Tris electrode is not available).

   Add 20.0 g SDS.

   Stir until dissolved.

   Adjust volume to 1.0 L with distilled water.

   Sterile-filter (*see* **Note 3**).
2. Transfer buffer

   In a 2 L graduated cylinder, mix:

   400 ml methanol.

   Deionized water to ~1,800 ml.

   100 ml 5× Laemmli buffer.

   Adjust final volume to 2.0 L with deionized water.

   Mix well.
3. Membrane marking pen (Whatman # 10499001) (GE Healthcare).
4. Immobilon-P (PVDF) Transfer Membranes, 0.45 μm pore size (Millipore Corporation, Bedford, MA).
5. Western transfer tank, gel cassettes, and sponges—Hoefer TE22 recommended (*see* **Note 4**).
6. (Optional) Novex Sponge Pad for blotting (#EI9052, Life Technologies).

2.3 Blocking, Antibody Incubation, and Detection

1. Bovine serum albumin, Fraction V (BSA) (MP Biomedicals, Solon, OH or equivalent). Store at 4 °C.
2. 10× PBS.
3. Tween-20 (Sigma-Aldrich, St. Louis MO).
4. PBS-T. For each liter of PBS-T, add 100 ml 10× PBS and 750–1000 μl Tween-20 to 900 ml distilled water in a graduated cylinder. Cover with Parafilm and mix by inverting.

5. 3 % BSA in PBS-T. Slowly add 15 g powdered BSA to 500 ml PBS-T with vigorous stirring on a magnetic stirrer and continue to stir until dissolved. Filter-sterilize. Store at 4 °C. Reserve a portion at room temperature if it is to be used that day to make blocking buffer.
6. Nonfat dried milk.
7. Blocking buffer: 3 % BSA, 5 % milk in PBS-T, 50–100 ml per blot. For 100 ml blocking buffer, add 5.0 g nonfat dried milk per 100 ml of room-temperature, filter-sterilized 3 % BSA in PBS-T. Stir vigorously for at least 10 min to completely dissolve milk particles. Store at room temperature for up to 4 h or at 4 °C overnight.
8. Antibody dilution buffer: 5 % milk, 0.3 % BSA in PBS-T, 40 ml per blot. For 100 ml, dissolve 5.0 g nonfat dried milk in 100 ml PBS-T. Stir vigorously on a magnetic stirrer for at least 10 min, then add 10 ml filter-sterilized 3 % BSA in PBS, and stir for at least 5 min longer.
9. Primary antibody (*see* Table 1).
10. Secondary antibody.
11. Western Lightning Plus-ECL (Perkin-Elmer, Waltham, MA). Store at 4 °C.
12. Autoradiography film cassette.
13. Autoradiography film.

3 Methods

3.1 Gels

3.1.1 SDS-PAGE

SDS-PAGE minigels are used to separate denatured and reduced proteins by size.

Novex NuPAGE 4–12 % acrylamide Bis-Tris gels (Life Technologies) with either MOPS or MES buffer containing SDS (Life Technologies) (*see* **Note 1**) are run according to manufacturer's instructions, using reducing agent in the sample buffer. MOPS buffer is best for large proteins (e.g., apoB-100 and apoB-48). MES buffer is best for resolution of smaller apolipoproteins (apoC's and apoA-II). Midsized proteins can be separated with either MES or MOPS buffer; we usually start with MES and switch to MOPS if separation is unsatisfactory. Always include molecular weight markers and positive and negative controls on gels.

To avoid protein degradation during sample preparation, work fast and cold and ensure that protein inhibitors are present from the moment sample preparation begins. Aliquot and freeze samples immediately. Avoid multiple freeze–thaw cycles.

The amount of sample loaded per well in SDS-PAGE minigels depends on sample type and the abundance of the protein being

Table 1
Antibodies used for Western analysis of lipoproteins and proteins involved in lipoprotein metabolism

Anti-human antibody	Supplier	Catalog #	Dilution	MW	Notes
ABCA1	Novus	NB 400-105	1/1,000	220 kD	Denature samples 30 min at room temp, not 95 °C. Also reacts with mouse ABCA1. Novus also supplies the same antibody conjugated to GFP
ApoA-I HRP	Meridian	K45252P	10–100 μl → 40 ml	28 kD	
ApoA-II HRP	Meridian	K74001P	60 μl → 40 ml	8.5 kD	
ApoB HRP	Meridian	K34005G	10 μl → 40 ml	513 kD (apoB100); 241 kD (apoB-48)	
ApoC-II	Meridian	K59600R	2.5 μl → 40 ml	8.9 kD	
ApoC-III HRP	Meridian	K74140P	30 μl → 40 ml	MW ~ 8.8 kD; three sialated forms	
ApoD	SC	sc34762	40 μl → 40 ml	23 kD	Glycosylated
ApoD	Covance (Signet)	SIG-38930		23 kD	Glycosylated
ApoE HRP	Meridian	K34002G		34 kD	Multiple nonspecific bands may be present. Include purified apoE protein (Meridian) on gel as a MW standard
ApoL1	Sigma	HPA018885	10 μl → 40 ml	40 kD	2°—Abcam #7083 donkey anti-rabbit, 5μl → 30 or 40 ml

(continued)

Table 1
(continued)

Anti-human antibody	Supplier	Catalog #	Dilution	MW	Notes
ApoM	R&D Systems	AF 4550	15 μl → 40 ml	25 kD	2°—R&D #HA F016, sheep IgG-HRP
LCAT (rabbit MAb)	Epitomics	3437	20 μl → 40 ml	63–69 kD	Glycosylated
Anti-nitrotyrosine clone 1A6	Millipore	# 05-233	1/500	Many bands. Best used on purified proteins or complexes	Follow Bakillah, Clin Chem Lab Med (2009) 47:60–69; transfer in TE22 as described in this chapter
PON1	Sigma	PO123	40 μl → 40 ml	43–48 kD	2°—Abcam #7083 Donkey anti-rabbit, 5μl → 30 or 40 ml. Standard = Zeptometrix PON Standards
SAA	SC	sc20651 SAA H-84	25 μl → 50 ml	13, 17 kD	
SR-BI	Novus	NB400-101	15 μl → 40 ml	83 kD	Glycosylated

For antibodies supplied in lyophilized form, reconstitute according to manufacturer's instructions and add an equal volume of sterile glycerol for long-term storage at −20 °C or −70 °C. Dilutions in the above table refer to final fold-dilution into Antibody Dilution Buffer for antibodies supplied in liquid form by the manufacturer or supplied lyophilized and reconstituted in water or 1× PBS and diluted 1:1 with 100 % glycerol. The recommended dilutions are merely guidelines for SDS-PAGE as described in this chapter and may require optimization for specific applications or when new lots of antibodies are prepared by the manufacturer. 2D gels as described in Chapter 17 may require 3–5× more antibody. Meridian: Meridian Life Sciences. Novus: Novus Biologicals. SC: Santa Cruz Biotechnology. For antibodies not listed in this table, www.genecards.org is an excellent resource studied. A good starting point for 1.0 mm thick Novex gels, which hold a final volume of 20–25 μl/well, is 1–2 μl plasma or serum/well, 16.25 μl FPLC fractions/well, 16.25 μl cell culture supernatant/well, and 25–50 μg protein/well of cell or tissue lysates. HDL and LDL isolated by ultracentrifugation must be dialyzed to remove KBr prior to use; VLDL can be used with or without dialysis. 10–25 μg protein/well is usually sufficient.

The final volume of sample, sample buffer, and reducing agent should be 25 μl per well. Thus, per well, 6.25 μl 4× NuPAGE LDS Sample Buffer (Life Technologies), 2.5 μl 10× NuPage Sample Reducing Agent (Life Technologies), 1× PBS or molecular biology grade water, and sample are combined to give a final volume of 25 μl. The sample itself should be thawed just before use, kept on ice, and added after all other reagents have been combined to minimize protein degradation. Samples are then heated at 95 °C (see below for exceptions) 10 min to denature and reduce proteins. The tubes are then cooled to room temperature, spun briefly at room temperature to collect the tube contents, and 20 μl sample are then loaded per well. We recommend Rainin L20 Pipet-Lite pipettes with low-retention pipettips for loading protein gels, as the tips fit nicely inside the wells and samples are delivered very cleanly, with no retention of sample in the tip and no formation of bubbles while expelling samples into wells. Loading is therefore very reproducible. For very low-abundance proteins, 1.5 mm gels, which hold 30–40 μl/well, can be used to increase the sample load. Novex has recently introduced the Bolt gel system, which holds an even larger volume. For the Novex NuPAGE 4–12 % acrylamide Bis-Tris gels described here, gels are run for 200 V for 35 min (MES buffer) or 50 min (MOPS buffer) and then transferred to membranes for Western analysis (Subheading 3.2).

Some proteins involved in lipoprotein transport are prone to aggregation during sample preparation, due to interactions between hydrophobic regions. For ABCA1 and ABCG1, aggregation can be prevented by incubating samples (in sample buffer plus reducing agent) at room temperature for 30 min (rather than at 95 °C for 10 min) prior to loading the gel. Aggregation of apoB is not so easily solved. We have observed that a major determinant of apoB aggregation is its concentration when sample buffer and reducing agent are added. If the apoB concentration is so low that it can barely be seen in the Western (e.g., C56Bl/6 mice on a regular chow diet), aggregation is minimal. Under conditions where apoB is plentiful and gives a very strong signal on the gel (e.g., LDLR-KO mice on a Western diet), a greater proportion is aggregated. If apoB aggregation is severe, dilute the plasma before adding sample buffer and reducing buffer and incubate samples at room temperature for 30 min, rather than at 95 °C for 10 min, prior to loading the gel.

In addition to Western blotting, SDS-PAGE gels can be stained with protein stains. SimplyBlue Safe Stain by Novex, used according to manufacturer's instructions, is a convenient, rapid, sensitive, and safe protein stain for SDS-PAGE gels.

Additional methods for separation of apolipoproteins by polyacrylamide gel electrophoresis can be found in reference [1]. This chapter by Ordovas provides instruction on pouring gels by hand and also describes separation of apolipoproteins by IEF and denaturing 2D gel electrophoresis.

3.1.2 1D Native Gels

1D native gels (nondenaturing gradient gels) are used to separate intact lipoproteins by size [2]. The gel, sample buffer, and running buffer must not contain detergent (for example, no SDS or LDS should be present) or reducing agent (for example, no DTT or β-mercaptoethanol should be present). In addition to Western blotting, 1D native gels can be stained with a lipid stain, such as Sudan Black (Sigma), to monitor the distribution of lipoproteins.

Tris–Glycine Gradient Minigels

Electrophoresis through nondenaturing Tris–Glycine gradient minigels separates lipoprotein particles by size. LDL particles band near the top of the gel and HDL separates into distinguishable subpopulations.

1. Use Novex 12-well 4–12 % Tris–Glycine Minigels, 1.0 mm thick, and Novex Tris–Glycine Native Running Buffer (diluted to 1×) from Life Technologies according to manufacturer's instructions.
2. Dilute plasma, serum, or purified lipoproteins (dialyzed to remove KBr) 1:1 in Novex Tris–Glycine Native Sample Buffer (2×) and immediately load onto gels. Do not heat samples before loading.
3. Run gels at 35 V for 17 h before Western blotting (Subheading 3.2) or staining for lipids (Subheading "Sudan Black B Staining of 1D Native Gels").

TBE Gradient Minigels

For good separation of native HDL particles by size, use nondenaturing TBE gradient minigels custom made by Jule, Inc. (Milford, CT). Best results are achieved with 10-well Tris–Borate–EDTA (TBE) gels, 3–25 % acrylamide (Catalog 3 25N 1.5 NMC10P), or 3–28 % acrylamide (Catalog 3 28N 1.5 NMC10P), thickness 1.5 mM. The 3–28 % minigels give slightly improved size resolution of preβ HDL particles, compared to the 3–25 % minigels, but are more inclined to crack during handling. Running buffer for these gels is 1× TBE.

1. Follow manufacturer's instructions for inserting gel into electrophoresis tank, using 1× TBE as the running buffer.

2. Dilute plasma, serum, or HDL (dialyzed to remove KBr) 1:1 with 2× TBE sample buffer and immediately load onto gels. Do not heat samples before loading.
3. For the molecular weight standard, load 7.0 μl of HMW Calibration Kit For Native Electrophoresis (Amersham/GE Healthcare) reconstituted in 1× TBE plus 0.05 % bromophenol blue. This standard should be loaded on the first lane of the gel. After electrophoresis, the lane is cut off and the protein MW markers are stained with SimplyBlue SafeStain (Life Technologies). The markers are aligned with the blot after the procedure is complete. A standard serum or standard HDL sample may be run as well.
4. Run gels at 200 V for 3–4 h before Western blotting (Subheading 3.2) or staining for lipids (Subheading "Sudan Black B Staining of 1D Native Gels").

Sudan Black B Staining of 1D Native Gels

1. After electrophoresis, incubate gels in 100–200 ml molecular biology grade water and soak for at least 5 min.
2. Incubate gels overnight in Sudan Black Staining Solution (100–200 ml) with gentle shaking, in fume hood. CAUTION: Wear protective clothing, as this solution contains acetic acid and acetone.
3. Destain in three changes of Sudan Black Destaining Solution, also with gentle shaking.
4. After destaining, the gel can be photographed or imaged using transillumination by white light. Gels stained with Sudan Black cannot be used for Western analysis.

3.2 Transfer

1. Prepare transfer buffer.
2. Fill a TE22 tank with transfer buffer up to the "minimum" line (slightly more than 1 L).

   NOTE: For multiple transfers more transfer buffer is needed:

   2 L for 1 gel.

   3–4 L for 2 gels.

   5 L for 3 gels.

   6 L for 4 gels.
3. Assemble dishes (*see* **Note 5**):

   1 dish for methanol—label and fill ~1/4–1/3 full with methanol.

   1 dish for water—label and fill ~1/3 full with deionized water.

   2 dishes for each gel that will be Western blotted—fill with transfer buffer.

Label each pair:

"Gel 1" and "Blot 1"

"Gel 2," "Blot 2," etc.

1 dish for transfer buffer—fill with transfer buffer.

4. Cut PVDF membrane (Millipore Immobilon-P) to size, a little bigger than the gel on all sides, one piece per gel. USE GLOVES WHEN HANDLING PVDF MEMBRANE. Do not touch membrane with bare hands!
5. Cut Whatman 3 MM paper (four pieces per gel) to the same size as the PVDF membrane.
6. Prepare membranes for transfer:
   1. Using forceps, add one piece of precut PVDF membrane to the dish containing methanol.
   2. Incubate 30 s with agitation.
   3. Turn the membrane over, using forceps.
   4. Incubate 30 s again, with agitation.
   5. Using forceps, transfer the membrane to the dish containing water.
   6. Incubate 30 s per side with agitation, as above.
   7. Use the "wetted" side (water does not "bead" on the surface) as the side that will be placed in contact with the gel. Use a membrane marker to label this side with the date and identifying information, while the membrane is still in the water dish.
   8. Then place the membrane in a PRELABELED dish containing transfer buffer, labeled side up.

Repeat for each membrane that will be used for blotting.

7. Prepare gels (SDS-PAGE or 1D native gels) for transfer:
   1. Crack open the gel cassette with the gel opener.
   2. Separate the gel plates carefully, watching to see which plate the gel sticks to and flipping the cassette if necessary to keep the gel stuck to the bottom plate. Cut off the knobby part at the bottom of the gel, using the gel opener. If using molecular weight markers that are not prestained, cut off the appropriate lane and place it in water or staining solution.
   3. Carefully pick the gel up and place it in a PRELABELED dish containing transfer buffer.

Repeat for each gel that will be used for blotting.

8. Assemble blotting stacks: (one per gel) (Fig. 2)
   1. Open a cassette and place on lab bench, black side on the bottom.

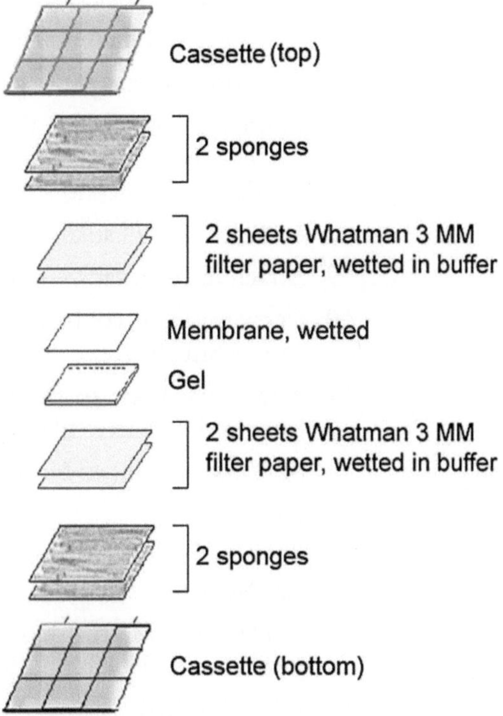

Fig. 2 Assembly of blotting stacks

2. Place sponges provided with the Hoefer TE22 apparatus on upper and lower halves of opened cassette.
3. On bottom sponge, place, in order, with debubbling between each step:
    1. Two pieces of Whatman 3 MM paper, prewetted in transfer buffer.
    2. Gel.
    3. PVDF membrane, from Subheading 3.2, **step 6.8**, above. The labeled side should contact the gel so that after transfer, labeled side up equals protein side up.
    4. Mark positions of molecular weight markers on back of membrane with membrane pen. Also mark any lanes that will be cut off later.
    5. Two pieces of Whatman 3 MM paper, prewetted in transfer buffer.
    6. Top sponge, provided with gel cassette.
    7. (Optional I not shown) White Novex blotting sponge (Novex Catalog #EI9052), aligned with the gel as best as possible. Hoefer does not recommend use of this additional sponge, as its use can warp the cassette. However, adding

this extra sponge does promote better contact between gel and membrane and hence, better transfer.

4. Close the cassette (*see* **Note 6**).
5. Insert cassette into the TE22 tank, black side in the back. (For best results, use only one blotting stack per tank, ideally the second slot back.) (*See* **Note 7**).
6. Place a small stirring bar into the tank.

9. Attach leads to tank—red in front, black in back (*see* **Note 8**).
10. Carry tanks to cold room and place each one on a magnetic stirrer. Stir slowly.
11. Attach the leads to the power supply, with positive (red) in front and negative (black) in back, and turn on power supply.

    80 V for ~3½ h (*see* **Note 9**).

    Or

    20 V for overnight transfer, then increase to 80 V for ~2–3 h (*see* **Note 9**).
12. While the gels are transferring, prepare PBS-T, 3 % BSA in PBS-T, and blocking buffer.
13. When finished:
    1. Shut off the power supply.
    2. Disconnect the leads.
    3. Carry the tanks back to the lab.
    4. Remove the leads.
    5. Take cassettes out of transfer apparatus.
    6. Remove and discard the top two pieces of Whatman paper.
    7. Place each membrane, protein side up, in a small dish containing ~100 ml PBS-T. If pieces of gel are clinging to the membrane, remove them with forceps without scratching the membrane. Wash the membrane(s) in PBS-T on a rocking or rotating platform for at least 5 min.
    8. (Optional but recommended) Stain the gel with a protein gel (e.g., Simply Blue (Life Technologies)) to ensure that most of the protein has been transferred out of the gel. Under the conditions mentioned above, no protein smaller than ~50–60 kD should be visible after transfer and higher MW proteins should be very faint after staining with Simply Blue.

3.3 Blocking, Antibody Incubation, and Detection

Membranes must be placed with protein side up throughout the remainder of the procedure.

1. After washing each blot in PBS-T, block membranes in blocking buffer (100 ml/blot) on a rocker or rotating platform for at least 1 h at room temperature or overnight at 4 °C.

2. 5 min before **step 3**, add primary antibody to Antibody Dilution Buffer. Table 1 lists antibodies we use routinely.
3. Remove blocking buffer and immediately incubate the blots, with gentle shaking, in antibody dilution buffer containing the primary antibody. Table 1 lists antibodies that we routinely use and the dilution that gives a satisfactory signal after an overnight incubation at 4 °C. Dilutions and incubation times for antibodies not included in Table 1 should be empirically determined (*see* **Notes 10** and **11**).
4. Wash 5×, 150–200 ml PBS-T per wash, for at least 5 min and preferably 10 or more minutes per wash (*see* **Note 12**).
5. Incubate the blots in antibody dilution buffer containing the appropriate HRP-conjugated secondary antibody at the appropriate dilution, 1 h at room temperature.
6. Wash 5×, 150–200 ml PBS-T per wash, as in **step 4**.
7. Remove each blot with forceps and place onto Whatman 3 MM paper, protein side up. Gently blot away excess moisture with a second sheet of Whatman 3 MM paper until the blot is no longer shiny. Do not let the blot dry completely.
8. Place blots on plastic wrap on the bench top. Per membrane, add 5 ml of the Enhanced Luminol Reagent (brown bottle) and 5 ml of the Oxidizing Reagent (white bottle) from the Western Lightning kit to a 50-ml tube and vortex vigorously to prepare the Chemiluminescence Reagent. Pour the Chemiluminescence Reagent over the blot(s). Remove bubbles by moving blots, using forceps. Gently agitate blots for 1 min in the Chemiluminescence Reagent.
9. Remove blots from the Chemiluminescence Reagent using forceps and blot away excess moisture as in **step 6**.
10. Place the blot on a piece of plastic wrap on the bench top, cover with a second piece of plastic wrap, cut away excess plastic wrap, leaving about an inch extra on each side, and tape to the back side of an autoradiography cassette.
11. In a darkroom, place a piece of film (Kodak MR-2 film gives the best resolution) against the blot for 10 s and develop the film. Take a series of longer and/or shorter exposures as necessary. Alternatively, use an imaging device capable of detecting low levels of light ($\lambda = 428$ nm).
12. If MW markers, were run in a separate lane and cut off and stained, line them up with the top and bottom of the gel and mark the sizes on the autoradiogram.

4 Notes

1. 1.0 mm thick SDS-PAGE minigels from manufacturers other than Novex have not been tried using this protocol but will likely give acceptable results. The 4–12 % acrylamide gradient and use of antioxidant in the inner chamber buffer during electrophoresis using the Novex system result in very sharp bands. The resolution is far better than that obtained with non-gradient Tris–Glycine–SDS gels. Novex has also developed a new system, BOLT™, that allows more sample to be loaded per well and holds promise for detection of very low-abundance proteins.
2. Biodesign/Meridian sells several purified apolipoproteins that can be used to positively identify bands on gels and Westerns. Genecards (www.genecards.org) conveniently lists commercial sources of purified or recombinant proteins to use as standards.
3. Adjust pH before adding SDS. After adding SDS the pH begins to drift and it becomes difficult to determine the true pH of the solution.
4. The methods in this chapter were optimized using the Hoefer TE22 transfer tank. The Hoefer TE22 holds over 1 L of transfer buffer and we have consistently noted efficient and reproducible transfer of large and small proteins and lipoproteins using this system. 1 l transfer tanks from other manufacturers have not been tested but we would expect satisfactory results after optimization. Smaller tank transfer systems holding ~200 ml transfer buffer can give results comparable to the TE22 for small, abundant proteins such as apoA-I but may not give satisfactory transfer of large proteins or lipoproteins. Semidry gel transfer systems can require considerable optimization due to "blowthrough" of smaller proteins and inefficient transfer of larger proteins and particles, and inconsistent transfer efficiency and higher background for some antibodies seems to be a common complaint. Semidry transfer is rapid and can be useful in well-optimized routine transfer of certain midsized proteins. In our experience, however, the Hoefer TE22 is the most versatile and dependable transfer system and gives highly reproducible results.
5. Plastic tops from 1-ml pipettip boxes work well.
6. If the cassette will not snap shut, trim the sponges and scoot the blotting stack up a bit so that sponges don't interfere with the bottom clasps of the cassette. If extra sponges were added to the blotting stack, remove one or two. Beware of bending in the bottom clasps inwards as they can snap.
7. Up to four gels per tank may be used but transfer efficiency and evenness across the gel will suffer.

8. In transfer buffer, proteins are negatively charged due to SDS binding and will migrate towards the positively charged electrode.
9. The optimal transfer time differs based on the size of the protein. For smaller proteins—apoCII, apoC-III, and apoA-II—2 to 3 h is acceptable. Any longer and "blowthrough" can occur. At the opposite end of the spectrum, apoB requires 3.5–4 h for complete transfer. Monitor the temperature in the tank (with the voltage turned off!) and transfer longer at a lower voltage for apoB if the temperature rises much above 55 °C. Warmer temperatures do enhance transfer but will warp the tank if overdone.
10. For new antibodies, several blots or strips should be prepared and incubated side-by-side with a series of dilutions of the primary antibody. In addition a "secondary antibody alone" control, consisting of an identical blot treated exactly the same as the others but minus the primary antibody, is absolutely essential. Much time can be wasted studying a band that is actually cross-reactivity from the secondary antibody rather than the actual desired protein, and too many people learn this the hard way!
11. For plasma, serum, or other samples containing immunoglobulins, if your protein of interest is the same size as heavy or light Ig chains, background from secondary antibody binding to immunoglobulin can confound the results. If the secondary antibody alone has a signal in this region, either use primary antibody conjugated to HRP or use an Ig-blocking solution to block signal from cross-reactive immunoglobulins.
12. If you used a primary antibody already conjugated to HRP, there is no need to use a secondary antibody. Skip **steps 5** and **6** and proceed to **step 7**.

References

1. Ordovas JM (1998) Separation of apolipoproteins by polyacrylamide gel electrophoresis. In: Ordovas JM (ed) Lipoprotein protocols. Humana, New York, NY, pp 113–130
2. Rainwater DL (1998) Electrophoretic separation of LDL and HDL subclasses. In: Ordovas JM (ed) Lipoprotein protocols. Humana, New York, NY, pp 137–151

Appendix

Working with RNA

RNases are ubiquitously present in the laboratory environment, introduced from humans, animals, bacteria, mold, or even from RNases used as laboratory reagents (in DNA minipreps and maxipreps, for example). RNases are not inactivated by boiling or even autoclaving as they can refold after heat-denaturation and regain activity. Special care must be taken to inactivate endogenous RNases present in the tissue or cell during RNA isolation, and constant vigilance is required to prevent sample contamination with exogenous RNases present in the environment.

Always:

Wear gloves and change frequently, particularly after contact with a non-RNase-free surface (e.g., skin and any untreated biological material or surfaces in the lab).

Wear a mask and a clean lab coat.

Use only clean, sterile, disposable plasticware (pipettes, pipet tips, centrifuge tubes, etc.).

Use aerosol-resistant tips and scrupulously avoid cross-contamination between samples if the sample will later be used for PCR.

Use only RNase-free reagents and solutions. Purchasing certified RNase-free reagents and solutions is certainly the best way to go. Prepare any homemade reagents in an RNase-free fashion, using sterile disposable plastic tubes or flasks, certified RNase-free water, and batches of chemicals dedicated to RNA work that have been touched only with RNase-free spatulas, etc. A solution for elimination of RNase contamination from pH meters, "Electro*Zap*," is commercially available from Ambion.

Dedicate a work area and a set of key equipment, such as Pipetmen, exclusively for RNA work, if possible. If this is not possible, try to work in an area where RNases are not used. Conversely,

Lita A. Freeman (ed.), *Lipoproteins and Cardiovascular Disease: Methods and Protocols*, Methods in Molecular Biology, vol. 1027, DOI 10.1007/978-1-60327-369-5, © Springer Science+Business Media, LLC 2013 when performing other procedures that require RNases (e.g., DNA isolation), take care not to contaminate RNase-free equipment or work areas with RNases.

Treat all surfaces, including the workbench, equipment, Pipetmen, etc., with RNaseZAP and rinse copiously with RNase-free water before use. Glassware can also be covered with aluminum foil and baked at 240 °C for 4 h overnight or filled with 0.1 % DEPC (diethylpyrocarbonate) in water, covered, left overnight, and autoclaved for 15 min. *CAUTION*: DEPC is a possible carcinogen! Handle in a fume hood and wear gloves and protective clothing.

Tissue harvesting and homogenization procedures must guard RNA from intracellular RNases that are released upon cell death and disruption. The key to preventing degradation during harvesting and homogenization is *immediate* processing of harvested tissues. Harvested tissues and cells must be either immediately homogenized in a chaotrope that inactivates RNases upon contact or immediately plunged into a stabilizing storage solution, such as RNA*later* (Ambion) or Allprotect (Qiagen), at the very instant it is harvested. RNA*later* and Allprotect are highly effective in stabilizing most tissues and protecting against RNA degradation. Note that for adipose tissue and brain, however, Allprotect should be used, as RNA*later* does not penetrate very lipid-rich tissues sufficiently to fully stabilize RNA. RNAs from tissues such as lung and intestine, as well as adipose tissue, are especially prone to degradation, and particular care must be taken with these tissues. Tissue pieces more than 5 × 5 mm in size will not be totally infused with RNA*later*, as the passive diffusion of the liquid will not proceed that far. Therefore, prepare several small pieces of tissue, immerse in RNA*later*, and snap-freeze samples as well. Once the tissue is stabilized in RNAlater or Allprotect, it can be stored for a long term at −80 °C or even −20 °C until homogenization and RNA isolation.

During homogenization, cells and tissues must be disrupted in the presence of a strong chaotrope (usually guanidine thiocyanate) that denatures and inactivates all proteins, including RNases, the instant they come into contact with the solution. Many homogenization and RNA isolation procedures have been developed; Chapters 2 and 3 of this volume include tried-and-true RNA isolation procedures for cells and tissues relevant to lipoprotein research. Careful adherence to these procedures will yield high-quality RNA suitable for virtually any application.

Purified RNA must be handled with great care to avoid degradation. It must be kept frozen in aliquots at −80 °C until just before use, and when thawed kept on ice for the minimal amount of time before use. Multiple freeze–thaws and alkaline conditions must be avoided. Utmost care must be taken to prevent contamination of purified RNA with environmental RNases. Work clean, cold, and fast, and RNA degradation will be minimal.

Index

Lita A. Freeman (ed.), *Lipoproteins and Cardiovascular Disease: Methods and Protocols*, Methods in Molecular Biology, vol. 1027, DOI 10.1007/978-1-60327-369-5, © Springer Science+Business Media, LLC 2013

MIX
Papier aus verantwortungsvollen Quellen
Paper from responsible sources
FSC® C105338

If you have any concerns about our products,
you can contact us on
ProductSafety@springernature.com

In case Publisher is established outside the EU,
the EU authorized representative is:
Springer Nature Customer Service Center GmbH
Europaplatz 3, 69115 Heidelberg, Germany

Printed by Libri Plureos GmbH
in Hamburg, Germany